Earth and Environmental Sciences Library

Earth and Environmental Sciences Library (EESL) is a multidisciplinary book series focusing on innovative approaches and solid reviews to strengthen the role of the Earth and Environmental Sciences communities, while also providing sound guidance for stakeholders, decision-makers, policymakers, international organizations, and NGOs.

Topics of interest include oceanography, the marine environment, atmospheric sciences, hydrology and soil sciences, geophysics and geology, agriculture, environmental pollution, remote sensing, climate change, water resources, and natural resources management. In pursuit of these topics, the Earth Sciences and Environmental Sciences communities are invited to share their knowledge and expertise in the form of edited books, monographs, and conference proceedings.

Nitish Kumar
Editor

Mercury Toxicity Mitigation: Sustainable Nexus Approach

Springer

Editor
Nitish Kumar
Department of Biotechnology
Central University of South Bihar
Gaya, Bihar, India

ISSN 2730-6674 ISSN 2730-6682 (electronic)
Earth and Environmental Sciences Library
ISBN 978-3-031-48816-0 ISBN 978-3-031-48817-7 (eBook)
https://doi.org/10.1007/978-3-031-48817-7

This Springer imprint is published by the registered company Springer Nature Switzerland AG
The registered company address is: Gewerbestrasse 11, 6330 Cham, Switzerland

Preface

Mercury (Hg) exists naturally and as a man-made contaminant. Mercury exists in various forms: elemental (or metallic) and inorganic (to which people may be exposed through their occupation); and organic (Methylmercury (MeHg), to which people may be exposed through their diet). It is released into the environment from volcanic activity, weathering of rocks, and as a result of human activity. Human activity is the main cause of mercury releases, particularly coal-fired power stations, residential coal burning for heating and cooking, industrial processes, waste incinerators and as a result of mining for mercury, gold, and other metals. The release of processed mercury can lead to a progressive increase in the amount of atmospheric mercury, which enters the atmospheric–soil–water distribution cycles where it can remain in circulation for years. Mercury poisoning is the result of exposure to mercury or mercury compounds resulting in various toxic effects depending on its chemical form and route of exposure. The major route of human exposure to methylmercury is largely through eating contaminated fish, seafood, and wildlife which have been exposed to mercury through ingestion of contaminated lower organisms. methylmercury toxicity is associated with nervous system damage in adults and impaired neurological development in infants and children. Ingested mercury may undergo bioaccumulation leading to progressive increases in body burdens.

The book sheds light on this global environmental issue and proposes solutions to contamination through multidisciplinary approaches. This book contains three parts. The first part describes the different sources and distribution of mercury in the environment. The second part explains the health risks linked to mercury toxicity. The third part addresses sustainable mitigation strategies and solutions using recent technology. This book provides an overview of the bioremediation phytoremediation, nano-adsorbents, and plant growth-promoting bacteria in remediation of areas polluted by mercury. These restoration processes have the advantage of being environmentally friendly and cost-effective solutions that exploit plants to immobilize and extract contaminants from soil and water, and fungi and bacteria to degrade them. This book is a valuable resource to students, academics, researchers, and environmental professionals doing fieldwork on mercury contamination throughout the world.

Gaya, Bihar, India Nitish Kumar

Acknowledgments

First and foremost, I would like to praise and thank God, the Almighty, who has granted countless blessings, knowledge, and opportunities to accomplish the project work. Thanks to all the authors of the various chapters for their contributions. It had been a bit of a long process from the initial outlines to developing the full chapters and then revising them in the light of reviewer's comments. We sincerely acknowledge the author's willingness to go through this process. I also acknowledge the work and knowledge of the members of our review panels, many of which had to be done at short notice. Thanks to all the people at Springer Nature especially Dr. Qiao Shu and Mrs. Saranyaa Vasuki Balasubramanian with whom I corresponded for their advice and facilitation in the production of this book. I am grateful to my family members, Mrs. Kiran (Wife), Miss Kartika Sharma and Laavanya Sharma (Daughters), and parents, for their incredible and selfless support all the time.

Gaya, Bihar, India Nitish Kumar

Contents

About the Editor

Dr. Nitish Kumar is a Senior Assistant Professor at the Department of Biotechnology, Central University of South Bihar, Gaya, Bihar, India. He completed his doctoral research at the Council of Scientific and Industrial Research–Central Salt & Marine Chemicals Research Institute, Bhavnagar, Gujarat, India. He has more than 12 years of research and teaching experience in the field of plant and microbial biotechnology, and bioremediation. He has published more than 70 research articles in leading international and national journals, more than 20 chapters, and 10 books with Springer and Taylor & Francis. He was a recipient of the Young Scientist Award from the Science and Engineering Research Board (SERB) in 2014. He has received many awards/fellowships/projects from various organizations, for example, the CSIR, DBT, ICAR and SERB-DST, BRNS-BARC, among others. He is an active reviewer for journals, including Biotechnology Reports, Aquatic Botany, Industrial Crops and Products, PLoS One, Plant Biochemistry and Biotechnology, and 3Biotech. He also serves as an Associate Editor of the journal Gene (Elsevier)

Source and Distribution of Mercury in the Environment

Source and Distribution of Mercury in Environment—A Review

Shihab Uddin, Sumona Khanom, and Md. Rafiqul Islam

Abstract Mercury (Hg) is an unescapable pollutant across the globe due to its unique cycling process. Understanding the causes and distribution of Hg pollution in the environment is essential for managing it efficiently and reducing its adverse consequences. Although Hg naturally occurs in the crust of the Earth, human activity has greatly increased the amount of Hg released into the environment. While some Hg is naturally released into the atmosphere as a result of volcanic activity and other geological processes, human activities such as industrial processes, small-scale gold mining, the production of cement and non-ferrous metals, and coal combustion have significantly increased the amount of Hg in the atmosphere globally. Mercury contamination is a genuine worldwide environmental hazard since it may travel great distances and eventually deposit in water, soil, and biota. This chapter provides a thorough review of the sources and distribution of Hg in the environment. It looks at how much anthropogenic and natural sources of Hg contribute to the overall environmental burden. Additionally, this chapter looks at the variables that affect how Hg is transported and changed in various environmental compartments. Regular monitoring of its concentrations is necessary to pinpoint highly contaminated locations and comprehend how Hg spreads throughout the environment. The development of monitoring techniques and modeling strategies used to forecast Hg transport within the environment will also be covered in this chapter. Overall, the goal of this chapter is to better understand the origins, modes of transportation, and distribution of Hg in the environment. It seeks to establish a basis for policymakers, researchers, and environmental practitioners to build informed solutions to address this enduring environmental issue and defend ecosystems and human health for future generations.

Keywords Mercury · Natural sources · Anthropogenic sources · Long-range transport · Dry and wet deposition · Volatilization

---

S. Uddin (✉) · Md. R. Islam
Department of Soil Science, Faculty of Agriculture, Bangladesh Agricultural University, Mymensingh 2202, Bangladesh
e-mail: shihab.ss@bau.edu.bd

S. Khanom
Department of Plant Pathology, Faculty of Agriculture, Bangladesh Agricultural University, Mymensingh 2202, Bangladesh

N. Kumar (ed.), *Mercury Toxicity Mitigation: Sustainable Nexus Approach*, Earth and Environmental Sciences Library, https://doi.org/10.1007/978-3-031-48817-7_1

1 Introduction

One of the 10 compounds that pose a considerable risk to human health, mercury (Hg), continues to be a major environmental concern on a global scale [126]. Widespread anthropogenic Hg emissions and use over the past few decades has resulted in Hg pollution of numerous ecosystems on land and in water [101, 110]. Mercury is one of the heavy metals that poses substantial risk to both human health and the environment due to its capacity to bioaccumulate and persist in ecosystems. Mercury can be released into the environment through both natural and human activities. When released into the environment, Hg can transform into methylmercury (Me–Hg), a highly toxic form known as a potential neurotoxin that accumulates in different foods. People who consume foods contaminated with Me–Hg are at high risk of Hg poisoning, which can cause neurological and developmental issues, particularly in fetuses and young children. Understanding the sources of Hg helps authorities monitor and regulate emissions to minimize human exposure. Governments and international organizations have established regulations and guidelines to limit Hg emissions and discharges from various sources, like industrial processes, coal, gas, or oil-fired power plants, waste incineration, and artisanal gold mining. Understanding the sources of Hg is critical for enforcing these regulations and ensuring compliance, as well as for implementing effective pollution control technologies and practices.

The atmospheric pollutant Hg is well-known for its ability to travel great distances through the sky. Due to the patterns of air circulation, it may be emitted in one area and later deposited in another, far from its original location. This phenomenon is known as global atmospheric Hg transport. As a result, even areas that are far from industrial activities geographically may nonetheless experience Hg pollution. It is easier to pinpoint potential sources of Hg and evaluate its effects on a larger scale if you are aware of its global distribution. It is easier to direct efforts to manage and limit Hg inputs to these habitats when one is aware of the distribution of Hg in the environment. To undertake risk evaluations, scientists and policymakers need to be aware of the origins and spread of Hg. These analyses examine the possible risks that Hg pollution poses to the environment and human health. By measuring exposure levels and weighing the hazards associated with different Hg sources, authorities can develop effective management plans to lower exposure and diminish the harmful consequences of Hg. Climate change is one of the most concerning issues of our time. Climate change may have an impact on the cycle and distribution of Hg in the environment. For instance, variations in temperature and precipitation patterns might impact the speed of Hg methylation and demethylation in both aquatic and soil systems. Changes in weather patterns can also affect the geographic distribution of Hg emissions. Understanding the relationships between Hg contamination and climate change aids in forecasting potential shifts in Hg distribution and developing adaptable solutions to the problem. Overall, understanding the sources and distribution patterns of Hg in the environment is crucial for protecting human health, safeguarding ecosystems, complying with regulations, assessing risks, and managing Hg pollution effectively. This in-depth knowledge will aid policymakers, scientists, and the public in making informed decisions to reduce Hg pollution and its associated impacts. Insights into Hg sources and distribution in the environment will be addressed in this chapter.

2 Characteristics and Available Forms of Mercury

Mercury, originally named after the Greek words "hydr-" (water) and "argyros" (silver), is a fascinating element with an atomic number of 80 and a standard atomic weight of 200.59. Often referred to as liquid silver or quicksilver, it has unique properties that set it apart from other elements. Boasting a boiling point of 356.7 °C and a density of 13.534 g cm^{-3} at 25 °C, Hg captivates with its high level of flammability. At ambient temperature, Hg stands alone as the only metal that exists in liquid form, appearing heavy, dazzling, and silvery-white with a subtle bluish tinge. Moreover, it exhibits a high surface tension, making it thick and somewhat stable. In its solid state, Hg takes the form of a white substance, while as a gas, it becomes a colorless vapor. While Hg serves as a fair electrical conductor, it falls short as a heat conductor. This peculiarity in its characteristics stems from its unusual electrical configuration. The element boasts seven stable isotopes, each with different abundance percentages in accordance with its atomic weight. These isotopes include ^{196}Hg (0.16%), ^{198}Hg (10.04%), ^{199}Hg (16.94%), ^{200}Hg (23.14%), ^{201}Hg (13.17%), ^{202}Hg (29.74%), and ^{204}Hg (6.82%). However, despite its intriguing properties, Hg poses significant health risks to humans. Designated as an immunotoxin and neurotoxin, it has been listed by the World Health Organization as one of the ten compounds with the most significant impact on human health [126].

With its ubiquity in air, water, and soil, mercury demands careful consideration due to its hazardous nature, prompting researchers to explore its various forms and their implications for human and environmental health. Mercury is not confined to its liquid form; it can be found in various states and compounds in the environment. Beyond the pure, elemental version of silver (Ag), mercury also exists in organic and inorganic forms, such as elemental or metallic mercury (Hg^0), inorganic mercury (Hg^+, Hg^{2+}), alkylated compounds, and organic mercury. Among the different Hg species, common compounds include mercuric sulfide (HgS), mercuric chloride ($HgCl_2$), mercuric oxide (HgO), Me–Hg, dimethylmercury ($(CH_3)_2Hg$), ethylmercury (C_2H_5Hg), and phenylmercury (C_6H_5Hg). Hg^0 is a liquid that readily evaporates at room temperature. Mercury is more dangerous when it is vaporized than when it is liquid. When a Hg container ruptures, Hg rushes out, and inhaling this powerful vapor of Hg can be fatal. Organic Hg compounds such as Me–Hg and ethylmercury (Et–Hg) are more dangerous than inorganic Hg molecules. According to increasing Hg toxicity, the hierarchy of distinct forms of Hg is as follows: Hg0 $<$ Hg^{2+}, Hg^+ $<$ Me–Hg [6].

3 Natural Sources of Mercury

Mercury emissions from natural sources are those that result from entirely natural processes lacking any human interference. Atmospheric Hg levels due to natural sources are known as background Hg levels. Mercury is naturally obtained by geothermal processes, volcanic eruptions, naturally occurring ocean surface volatilization, and the weathering of Hg-containing minerals. When compared to the estimated total yearly worldwide Hg emissions of 5000–8000 tons, natural emissions of Hg into the atmosphere are relatively small, about 10% (Fig. 1a).

3.1 Mercury in the Earth's Crust and Geological Processes

Geothermal processes are one of the most significant natural sources of background Hg in the environment, contributing about 9% of total natural Hg emissions (Fig. 1b) [121]. As a result of the long-term high-temperature dissolution of underlying minerals, geothermal fluids have been found to be abundant in trace metals such as Ag, gold (Au), arsenic (As), antimony, and Hg. Temperature, pH, redox state, quantities of prevalent complexing agents, and the kinds of underlying minerals present in the environment all affect how easily and where Hg dissolves and moves through geothermal fluids. Geochemical simulations demonstrate that Hg^0 is plentiful at temperatures higher than 200 °C whenever reducing geothermal fluids (HS^- is dominant over $SO_4{}^{2-}$) are present [120]. The formation of Hg^0 is favored by rising temperature and pH, as well as falling ionic strength, pO_2, and total sulfur. These geothermal fluids combine with acidic or oxidizing water, favoring the production of Hg^{2+}, which causes cinnabar to reprecipitate. Therefore, the thermal expulsion of Hg from the heated, Hg-enriched substrate in contact with the decreasing geothermal fluid, subsequently followed by the volatilization of Hg^0 to the environment driven by heat flow, results in Hg emissions from these systems [43]. The hot springs

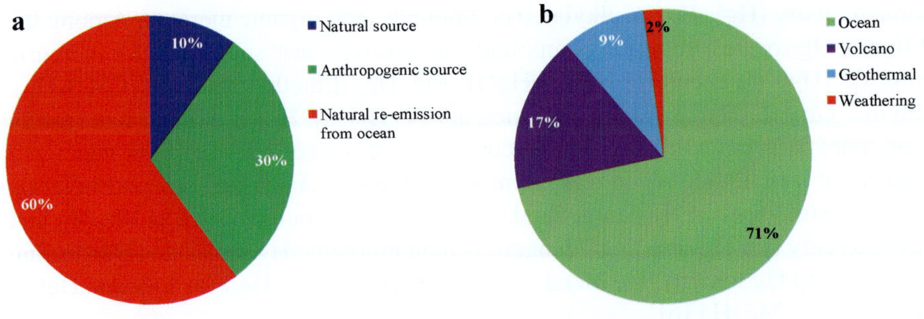

Fig. 1 Relative contribution of Hg emissions to the atmosphere. **a** All possible sources [118] and **b** Natural sources [121]

and fumaroles that are quiescently degassing exhibit regional and temporal diversity in their Hg emissions, which is influenced by the age and type of the geothermal system. According to current statistics, geothermal activity contributes 60 tons of Hg to the atmosphere each year [121]. If geothermal energy is aggressively sought as an alternative energy source, the amount of Hg released into the atmosphere through geothermal activities may be of environmental concern on a local or regional level, even though it is not significant on a global scale.

In addition, Hg is naturally present in the earth's crust in trace amounts, typically occurring as minerals such as livingstonite ($HgSb_4S_8$), corderoite ($Hg_3S_2Cl_2$), and cinnabar (HgS). Mercury concentrations in the earth's crust ranged from 0.021 to 0.056 mg kg^{-1} [125]. The most prevalent Hg ore, cinnabar, is typically distributed throughout the lithosphere and is often associated with volcanic and alkaline hot springs, as well as sedimentary rock formations [36]. Even today, the primary metal mined to make Hg^0 is cinnabar, which was once widely utilized as a pigment (vermillion). Since the Hg content is not widely dispersed or mobile in the environment, these inherent forms do not pose a serious concern to the ecosystem. Despite this, metals other than ferrous and fossil fuels, particularly coal, also contain Hg as an impurity. Mercury is released from rocks and minerals into the environment because of weathering and erosion, two natural processes. Hg-containing minerals begin to degrade over geological timeframes when exposed to constituents like water, oxygen, and acids. The release of Hg^0 into the soil and water during weathering processes like chemical reactions and disintegration. Then the released Hg compounds can subsequently go through other processes, such as microbial conversion to more dangerous forms of Hg like Me–Hg.

It is possible that sediment transport and wind-blown dust will cause low-temperature direct emissions from the weathering of mineralized Hg deposits into aquatic and terrestrial habitats. Cinnabar is insoluble in water and very resistant to low temperature oxidation and weathering processes. If a sizable amount of the Hg present in the mineral is in elemental form, volatilization may result in the low-temperature release of Hg from mineral deposits directly into the environment. The amount of Hg^0 present in the mineral deposit as well as meteorological factors like temperature and wind speed would then affect the pace of volatilization. Although Hg^0 has been discovered in highly concentrated Hg mineral deposits, such as those at the New Idria Hg mine in California, these regions have also been the sites of mining activity, and the Hg emissions are primarily attributed to these human-caused activities. There is scant proof that minerals or soils outside of these localized, enriched mineral occurrences contain Hg^0 in proportions that are environmentally important. According to estimates, mineralized areas naturally emit between 10 and 20 tons of Hg into the atmosphere each year, or about 2% of all Hg emissions to the atmosphere from all natural sources (Fig. 1b) [121]. However, how much of these releases are solely the result of natural processes is still a matter of debate.

3.2 Volcanic Emissions and the Natural Degassing of Mercury

Volcanic activity represents a substantial natural source of Hg emissions into the atmosphere, contributing about 17% of all natural emissions (Fig. 1b). During volcanic eruptions, large amounts of Hg vapor and particles are released. Similar to geothermal systems, active volcanoes produce Hg in a similar manner. Under reducing conditions, Hg^0 is volatilized from the molten magma and released into the atmosphere together with other hot gases. In accordance with the quantity, location, and eruptive stage of active volcanoes, the amount of Hg emissions from volcanoes varies through space as well as time. The atmospheric load from a single significant volcanic eruption lasts roughly two years, and a strong eruption can cause Hg levels to be 4–6 times higher than they were before the event. About 50–70 terrestrial volcanoes are presently in some state of activity. The exact amount of Hg released during eruptions can vary based on the type of volcano, the eruption style, and the presence of Hg-rich minerals in the volcanic rocks. One to around 700 tons per year are currently estimated to be the total annual global Hg emissions from volcanic activity [36]. Due in part to the broad variance in the types and locations of the eruptions analyzed, there is a wide range of estimations. About 75 tons of Hg are released a year from continually erupting and degassing volcanoes, while 10–100 tons are released annually from intermittent, smaller eruptions [36]. Large, explosive eruptions are only responsible for 15% of the total emissions of volcanic Hg. However, if there were to be multiple significant eruptions each century, 1 ton of Hg might be released into the atmosphere, quickly outpacing the entire atmospheric load. According to estimates, volcanic activity contributes roughly 112 tons of Hg to the atmosphere annually [121]. Mercury vapor is transported to the atmosphere by volcanic gases like carbon dioxide, water vapor, sulfur dioxide, and hydrogen sulfide. These gases can also interact with Hg-containing minerals in volcanic rocks, leading to the release of Hg^0 during volcanic degassing.

3.3 Mercury Release From Oceans

The emissions of volatile Hg species from the ocean floor are arguably the biggest source of natural Hg emissions to the atmosphere (Fig. 2). Submarine hydrothermal vents, volcanic eruptions on the ocean surface, and the atmospheric deposition of naturally occurring Hg are some of the natural sources of Hg in the ocean [104]. In volcanically active regions, frequently on mid-ocean ridges where there are tectonic plate gaps, hydrothermal vents develop. Under the intense pressure of the ocean floor, the hydrothermal fluids in these vents can reach temperatures of up to 400 °C. High temperatures and lowering conditions can leach Hg^0 from the Hg-enriched substrate, just like in terrestrial geothermal systems. It has been reported that Hg concentrations in hydrothermal fluids are 1,000 times higher than those in ambient saltwater [122].

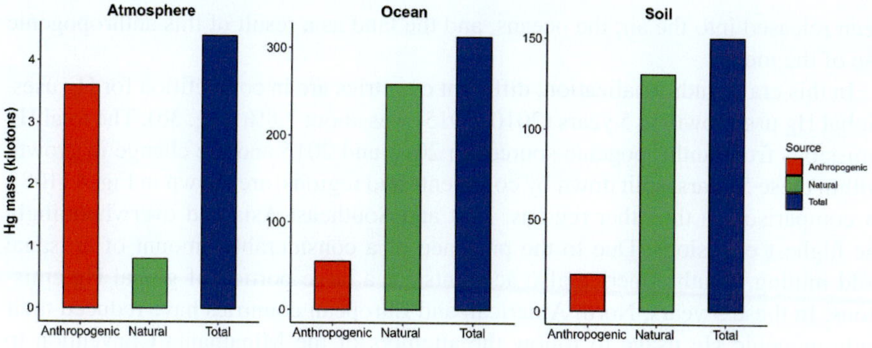

Fig. 2 Updated estimates of total, anthropogenic and natural Hg masses in global air, soils and oceans [118]

Hg^{2+} is created as these geothermal fluids combine with cold, oxidized seawater, precipitating cinnabar back to the ocean floor and raising Hg concentrations near the hydrothermal vents. The eventual burial of these particulate matter particles in deep marine sediments could restore Hg to the mineral reservoir and serve as a long-term sink for environmental Hg. The latest projections place the annual amount of Hg-containing sediments buried now between 180 and 260 tons [3]. It is also possible to convert this particulate, Hg^{2+}, back into Hg^0 and let it re-enter the water column. It has been demonstrated that the thermophilic bacteria that surround such hydrothermal vents are capable of converting Hg^{2+} into Hg^0, purifying the immediate area while releasing the flammable elemental Hg into the seafloor, where it can be transported to surface seas. It is predicted that the surface ocean releases 456 tons of naturally produced Hg into the atmosphere per year [83]. 71% of the Hg emissions from natural sources into the atmosphere can be attributed to this (Fig. 1b).

4 Anthropogenic Sources of Mercury

Anthropogenic Hg sources are those where there is direct human involvement. Anthropogenic sources include artisanal gold production, small and large-scale gold production, cement production, combustion of coal, oil, and gas in households, industry, or power plants, chlor-alkali production, waste incineration, non-ferrous metal production, etc. The comparison of emissions in 2000 and 2015 among different sources is shown in Fig. 3. On account of human activity, total atmospheric Hg levels have increased by around 4.5 times compared to natural levels in the last few decades [49, 73, 118]. Streets et al. [107] estimated that over the past 500 years, human activities have resulted in the release of 1–3 million tons of Hg into the environment. Due to the special characteristics of Hg, it is incredibly adaptable and has been used for a variety of things throughout the course of millennia. Large amounts of Hg have been released into the air, the oceans, and the land as a result of this anthropogenic use of the metal.

In this era of industrialization, different countries are in competition for Hg uses. Gobal Hg use growth in 5 years (2010–2015) was about 1.8% (Fig. 3b). The total Hg emissions from anthropogenic sources in 2010 and 2015 and the change in growth within these 5 years, split down by continents and regions, are shown in Fig. 4 [108]. In comparison to the other regions, East and Southeast Asia had overwhelmingly the highest emissions. Due to the presence of a considerable amount of artisanal gold mining, South America also accounts for a large portion of global Hg emissions. In these 5 years, North American and European countries have reduced their anthropogenic Hg usage to follow the attempts of the Minamanta Convention to reduce global Hg load. However, countries in Southeast Asia are still emitting a considerably higher amount of Hg. The sector-wise Hg emissions and their relative contribution in 2015 are presented in Table 1.

4.1 Artisanal Small-Scale Gold Mining and Large-Scale Gold Production

It was recognized in the past that the process utilized to generate pure forms of gold may pollute the environment with Hg. This procedure, which involves leaching and amalgamating the metals from ores, is reasonably quick and inexpensive. The use of Hg in this method of getting gold entails dissolving the metals in amalgams, which are subsequently recovered through Hg evaporation [55]. It is a sector of the economy that uses the most metallic Hg, or roughly 1,000 tons per year. Around the world, there are 10–15 million small businesses that mine and process gold across 70 different nations. According to [112], they generate more than 350 tons of this precious metal each year. As a result of this source, Hg emissions from China, Brazil, Indonesia, Columbia, Bolivia, Venezuela, and the Philippines are significantly higher than those from other nations. According to [111], this accounts for more than 30% of the Hg used in all industrial operations. Around 50 A.D., the Romans most likely utilized the method for the first time. Spaniards were the first to employ the amalgamation technique in modern times, initially in Mexico (starting in 1554) and then in Peru and Bolivia, mostly to create gold [72]. The amalgamation process used to produce gold had previously been by far the biggest source of Hg emissions. Although its percentage has dropped since 1950, it is still substantial [105]. A "gold rush" that began in the early nineteenth century and ended in the early twentieth century saw large-scale gold mining operations set up in North America and Australia. However, a fresh "gold rush" started in emerging nations after gold prices on the global market rose from USD 58 for an ounce in 1972 to USD 447 for an ounce in 1987. In this instance, modest and very tiny production facilities mostly use the amalgamation method to generate gold [93]. In the twentieth century, a less expensive approach utilizing cyanides eventually replaced amalgamation technology in industrial gold

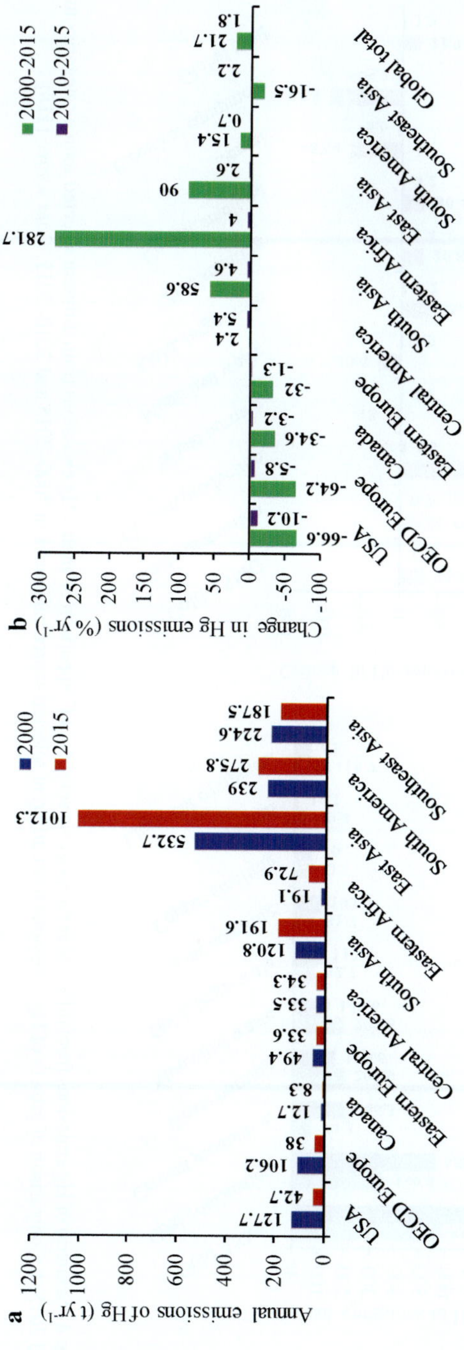

Fig. 3 Scenario of Hg emissions in different regions of the world. **a** Comparison of annual Hg emissions in selected regions of the world in 2000 and 2015; **b** Comparison of change in Hg emissions in selected regions of the world in 2000–2015 and 2010–2015. *Data source* [118]

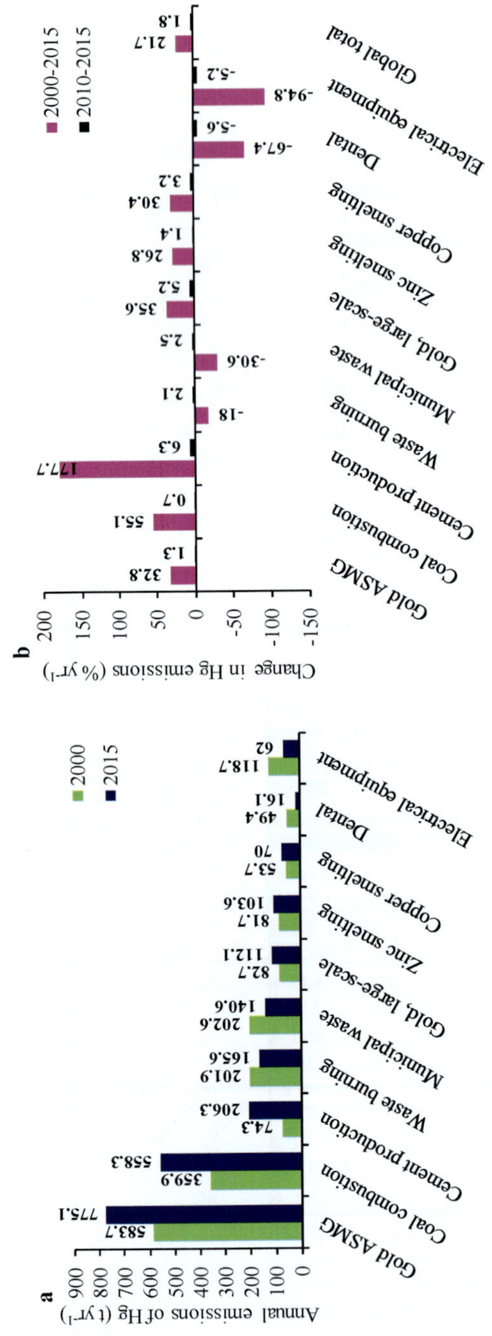

Fig. 4 Scenario of Hg emissions from different anthropogenic sources. **a** Comparison of annual Hg emissions from different selected source categories in 2000 and 2015; **b** Comparison of growth of Hg emissions from different selected source categories in 2000–2015 and 2010–2015. *Data source* [118]

Table 1 Mean Hg emissions from different anthropogenic sources and their relative contribution in 2015

Source	Mean Hg emissions (t)	Relative contribution (%)
Artisanal and small-scale gold mining	838.0	37.7
Large scale gold production	84.5	3.8
Cement production	232.0	10.5
Stationary combustion of coal, oil, and gas		
Residential/Transport/Domestic	58.67	2.64
Industry	127.52	5.74
Power plant	294.8	13.23
Biomass burning	51.9	2.33
Non-ferrous metal production (primary Al, Cu, Pb, Zn)	228	10.3
Waste and waste incineration	162	7.27
Vinyl-chloride monomer	58.2	2.6
Chlor-alkali production	13.5	0.68
Pig iron and steel production (primary)	29.8	1.34
Cremation emissions	3.77	0.17
Mercury production	13.8	0.62
Oil refining	14.4	0.65
Secondary steel production	10.1	0.46
Total	2220.0	100.0

Data source [118]

production. With this new method, there was no need to employ Hg during the manufacturing process. According to estimates, this source released between 650 and 1000 tons of Hg into the atmosphere annually in 2015, with an average of 838 tons (Table 1). About 60–80% of the emissions from them are thought to be released directly into the atmosphere, with the balance ending up in soil, rivers, lakes, and landfills.

4.2 Cement Production

One prominent anthropogenic source of Hg pollution in the environment is mentioned as being the industrial cement-producing industry. Burning powdered and mixed raw materials (coal, sand, shale, limestone, and gypsum) is the production method. The process's high temperature, which can exceed 1000 °C, is advantageous for Hg emissions. According to [118], the total global Hg emissions from cement manufacturing in 2015 were predicted to be 233.1 tons on average, with a range of 117.0–782.0 tons, or around 10.5% of all anthropogenic emissions. With an average share of 72%,

Asia outside of Russia dominated the market, and Europe outside of Russia came in second with a share of 9.9% [79]. In terms of states, China accounts for the highest portion of the global emissions from this business, or 14.8%, as well as 5.7% of all emissions in this nation [109]. The manufacturing of cement in China is also increasing rapidly at the same time. Recent estimations indicate that this source's growth rate was 177% from 2000 to 2010 and 6.3% from 2010 to 2015, respectively (Fig. 4). Gaseous elementary Hg accounts for 80%, gaseous organic Hg for 15%, and total particulate Hg for 5% of the Hg emissions from cement manufacturing [80]. According to [78], the emission factor is 1 g of Hg per ton of cement manufactured.

4.3 *Non-ferrous Metal Production*

Non-ferrous metals, including aluminum, copper, zinc, and lead, all have admixtures of other metals that were recovered during manufacture. Another admixture that frequently appears in ores but is unfavorable is Hg. Depending on the type of ore, it contains a wide range of variations. According to [68], this variation is connected to the many ore mineralization generations that have occurred in these deposits. Asian zinc ores have a concentration of 86.6 t/g [106], whereas [77] reported 20.0 g/t. As much as 0.6–5.8% more Hg can be found in gold ores from the RSA [34]. The amount of Hg in the ore, the principal technological process, any potential use for scrap, and the kind of equipment used to reduce emissions all affect Hg emissions [78]. Recent data [118] show that the production of non-ferrous metals (aluminum, copper, zinc, and lead) led to 228 tons of global Hg emissions (Table 1). Currently, 80 and 20%, respectively, of the emissions from the primary source and processing of Hg are made up of gaseous Hg^0 and gaseous organic Hg [80].

4.4 *Waste and Waste Incineration*

Updated estimates show that in 2015, waste creation and incineration contributed roughly 162 tons of yearly global emissions, or 7.27% of all anthropogenic sources (Table 1). The amount of Hg emitted into the air by incineration processes depends on the waste composition, the methods employed, and the machinery utilized to reduce pollutants in waste gases. The burning procedure is often carried out at high temperatures of about 1000 °C, which renders nearly all the Hg into gaseous form. Items with high quantities of Hg that are disposed of in landfills include thermometers, batteries, different kinds of lamps, Hg relays, office trash from dentists and doctors, measuring equipment, etc. Sewage sludge from wastewater treatment plants can be a significant source of mercury. According to Pirrone et al. [84], municipal waste and sewage sludge have high emission factors of 1 and 5, respectively.

4.5 Combustion of Coal, Oil, and Gas

The amount of mercury (Hg) in coal varies depending on the kind and origin of coal and is mostly found in inorganic compounds such as HgS, chlorides, and sulfates. The remaining Hg is found in organic molecules. In the Nowa Ruda Mine, hard coal contained the highest concentrations of Hg in Poland, ranging from 0.81 to 9.67 mg kg^{-1} [15]. According to [81], the Hg content of lignite from Polish deposits is four times higher than that of hard coal. In contrast, Australian hard coal reserves have extremely low Hg content, averaging 0.04 mg kg^{-1} (0.01–0.13 mg kg^{-1}), compared to other countries' hard coal deposits. According to [35], 99% of the Hg in coal combustion waste gases is present as Hg^0. When compared to coal, crude oil has significantly less Hg. The content of a deposit can change between deposits by several orders of magnitude, depending on where it was found. The Hg concentration, according to most authors, is between 1 and 5 ng kg^{-1}. Crude oil also contains mostly Hg from Hg^0 species. The ratios between the various Hg species, according to [13], are as follows: 33% gaseous organic Hg, 76% gaseous Hg^0, and 1% Me–Hg. Direct Hg emissions into the air, water, solid waste, and goods themselves are all part of the problem of Hg in the atmosphere during crude oil refining. According to [127], in the USA, 55% of Hg is still present in products, whereas 20% is released into the air, 25% is released into the waters, and 10% is released as solid waste. There is some Hg in natural gas, too. In contrast, organic Hg compounds make up a sizable portion of condensates and liquids made from petroleum and can even be the major species of Hg [134]. When coal, oil, or gas containing Hg to some extent are combusted in industry or power plants, significantly higher amounts of Hg are released into the environment. In 2015, stationary combustion of coal, oil, and gas in homesteads, industry, and power plants was responsible for 480.99 tons, which was 21.61% of the total anthropogenic Hg emissions (Table 1). However, the growth of this source reduced significantly in 2010–2015 compared to 2000–2010 (Fig. 4).

4.6 Other Anthropogenic Sources

These sources include anthropogenic sources other than the above-described sources. Mercury can also be introduced into the environment from biomass burning, vinyl-chloride monomer production, chlor-alkali manufacturing, pig iron and steel manufacturing, cremation processes, Hg production, secondary steel production, use of Hg in electric and electronic devices, lighting equipment, batteries, laboratories, cosmetics, thermometers and sphygmomanometers, dental amalgams, etc. These sources can also significantly contribute to the atmospheric Hg load.

5 Global Distribution of Mercury

5.1 Global Mercury Cycling

The latest global Hg budget is presented in Table 2. Most of the Hg (>90%) in the atmosphere is present as Hg^0, which predominates in the gaseous phase and allows for long-distance Hg movement on a worldwide basis [19]. According to current scientific knowledge, Hg persists in the natural environment for up to 3,000 years [99], moving continually through water, air, and land in accordance with the so-called global Hg cycle (Fig. 5). After getting released from the geological reservoirs via either natural or man-made processes, Hg is spread across the entire global environment. Once spread, it undergoes an intricate series of chemical and physical modifications as it moves through the environment, land, and surface waters. These changes have a significant impact on chemical attributes, ecological consequences, and biological toxic effects. Although the atmosphere is the primary means of Hg transport on a global scale, the effects of Hg on the environment and human health have no direct connection with atmospheric loads [66]. Rather, it enters both aquatic and terrestrial environments through a variety of mechanisms after being deposited in the atmosphere. It can undergo transformations and continue to accumulate in the aquatic food web after deposition; alternatively, it may be resuspended and returned to the atmosphere for additional conveyance and redeposition. The cycle of deposition and resuspension is crucial because it enables Hg to be transferred far from the source, even though bioaccumulation in aquatic ecosystems constitutes the main route of exposure for people and wildlife [36]. The chemical characteristics and reactivities of the various Hg chemical species in each of the environmental compartments also differ substantially, for reasons that aren't all fully understood. These chemical variations have an impact on the residence times and biological toxicities of the various Hg species. So, the complex processes that control how mercury moves through the atmosphere and how it settles, changes chemically, is re-emitted, and finally builds up in living things as Me–Hg also control how it affects ecosystems and how dangerous it is for people to be exposed to it [115].

5.2 Mercury Distribution in Different Ecosystems

5.2.1 Mercury in Air

A crucial pathway for the biogeochemical cycling of Hg is the atmosphere. According to studies by [86, 103, 25], background air Hg levels typically range from 1.1 to 1.3 ng m^{-3} in the Southern Hemisphere and between 1.3 and 1.7 ng m^{-3} in the Northern Hemisphere. The background level of total gaseous Hg is typically 1.5 to 2.0 ng m^{-3} worldwide [44]. Mercury concentrations in the atmosphere typically vary from 0.01 to 0.06 ng m^{-3} in remote locations, but between 0.09 and 38 ng m^{-3}

Table 2 The latest global Hg budget

Source/Sink	Amount (t yr^{-1})
Emission	
1. Natural (geogenic)	500
2. Re-emission/global transport	
Biomass burning	600
Soil and vegetation	1000
3. Anthropogenic	2500 (2000–3000)
4. Ocean, net evasion (gaseous Hg^0)	3400 (2900–4000)
Total	7000
Deposition	
(i) To land/freshwater	3600
(ii) To ocean	3880
Total	7480

Data source [118]

in metropolitan or polluted areas [52]. Hg is mostly found in oxidized form in the environment; its three oxidation states are Hg^0, Hg^+, and Hg^{2+}. Gaseous, oxidized mercury makes up around 98% of the total mass of Hg in the atmosphere. Over 95% of the mass of Hg is in the form of Hg^0, which predominates. In contrast, very small amounts of the first oxidation state, Hg^+, and extremely small amounts of the second oxidation state, $Hg2^+$, are found [97]. Hg^0, among the Hg compounds, has a significantly greater vapor pressure than the rest, although Hg^{2+} continues to be in second place. According to [89], the vapor pressure for Hg at room temperature (25 °C) is 0.2 pa for Hg^0 and 0.1 pa for Hg^{2+}.

How long Hg will stay in the atmosphere depends on several factors. In addition to the weather, factors including the quantity and kind of air pollutants are considered. According to estimates from [66], the residence durations for Hg^0 range between 6 and 18 months, but they are only a few hours or days for total particulate Hg and gaseous oxidized Hg. These findings imply that Hg^0 can travel over long distances due to its delayed atmospheric removal. Mercury is currently released from natural sources, is currently released from human-made activities, and is re-emitted from previously deposited natural and human-made sources [118]. According to evidence from lake sediments, peat cores, and ice cores given in several studies, atmospheric Hg deposition has increased globally since preindustrial times (about 1850) by around a factor of three, with larger increases in areas receiving enhanced deposition from regional sources. It is exceedingly challenging and complex to measure the amount of Me–Hg in the environment. The emission level from various sources is frequently estimated using a variety of models, including the CTM–Hg, GEOS-Chem, and GRAHM [44]. According to [118], yearly emissions of Hg into the air from all sources are estimated to be between 5000 and 7000 tons (Table 2). On a worldwide scale, total emissions have an impact on net atmospheric Hg deposition; locally, atmospheric processes and the speciation of regional and local emissions have an

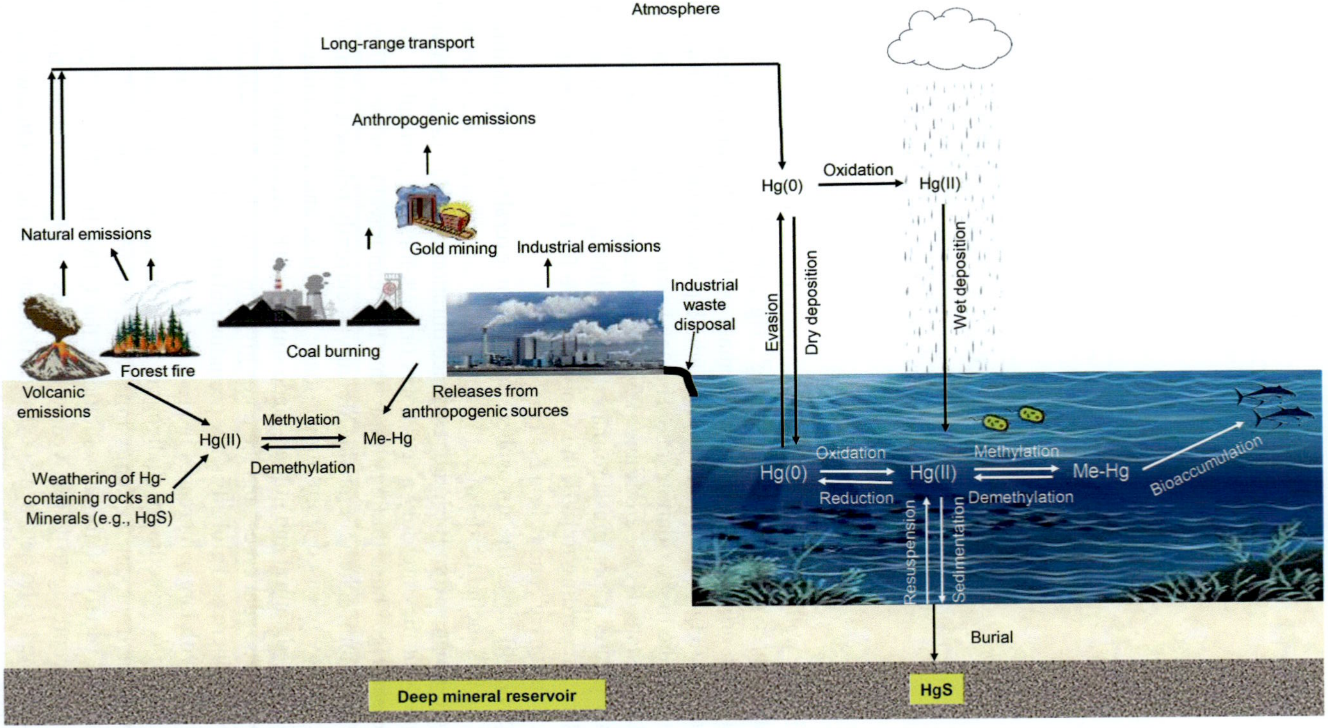

Fig. 5 Global Hg cycle showing the transformations and transport of Hg in different ecosystems

impact on deposition. For the creation of precise predictive models that guide global Hg policy, an understanding of these processes is consequently essential. Models of atmospheric Hg must accurately depict emission levels, transport, redox chemistry, and deposition.

5.2.2 Mercury in Soil and Sediment

Soil is one of the key settings that play a significant role in ecological services and food supply. Mercury is a naturally occurring soil contaminant, and the concentration may vary from region to region depending on the availability of Hg sources and the intensity of Hg use for different purposes. According to [52], the global mean concentration of Hg in soils is predicted to be 1.1 mg kg^{-1}, ranging from 0.01 to 1.5 mg kg^{-1}. However, in recent years, significantly higher results have been reported in soil near the Hg sources. Previous reports suggested that these Hg contaminations of soil are mostly caused by using sewage water for irrigation, mining of Hg, and using too many Hg-containing fertilizers and fungicides in farming fields [61, 135]. Researchers from around the world measured the levels of Hg in various soil ecosystems close to industrial Hg sources. Some of the country-based data are summarized in Table 3. According to estimates by [7, 26, 90], the median Hg concentrations in topsoil are 0.038, 0.02, and 0.01 3 g kg^{-1}, respectively, across Europe, the United States, and Australia. [92] reported 0.04–1.99 mg kg^{-1} Hg in soil near a gold mining activity in Brazil. China is the Hg hub of Asia and is responsible for significantly higher Hg emissions into the atmosphere. Several authors reported Hg concentrations in soil near anthropogenic Hg sources like chlor-alkali, thermometer factories, the steel industry, artisanal gold mining, etc. in different regions of China. The Hg concentration in soil varied between 0.05 and 3601.0 mg kg^{-1} across different regions of China (Table 3). In China, about 1.6% of the cropland soils are contaminated with Hg [70]. Soils contaminated with Hg are becoming a major environmental pollution issue across the globe because of the rapid economic growth and ongoing urbanization of the countries. [74] discovered that soil Hg concentrations significantly varied among land covers, with wooded upland having the highest concentrations, followed by planted or cultivated upland, herbaceous upland or shrubland, and then bare soils. There were about 2.5 times as many concentrations in forests as there were in desert areas. Assessments of the intensity and heterogeneity of soil Hg contamination have been the main areas of focus up until the present time. The biological availability of Hg and the variables affecting its transfer and modification in soil-crop systems have been the subject of infrequent research.

The ultimate Hg sink is the ocean, and sediment burial of Hg is a significant sink. But according to reports, only 30% of the mercury flow that arrives at the deep ocean is preserved in sediments, while an estimated 96% of the mercury that is deposited there is lost through evasion from the surface [64]. Mercury can exist in sediments in a variety of states, with Me–Hg concentrations only accounting for 0.1–1.0% of the overall Hg content. Although researchers are attempting to develop data based on modeling methodologies, which are essentially overestimations compared to the

Table 3 Various instances of topsoil contamination with Hg in different countries near important sources of emissions

Country	Location	Source of pollution	Time	Total Hg (mg kg^{-1})	Reference
Brazil	Paracatu	Gold mining	2010–11	0.04–1.99	[92]
China	An Ning	Chlor-alkali		0.09–1.30	[102]
	Chongqing	Thermometer factory		0.06–0.88	[124]
	Guizhou	Acetic acid	2016	1.09–3.71	[58]
	Huludao	Chlor-alkali Zn-smelting	2006–08	0.05–14.6	[136]
	Kunming	Chlor-alkali polyvinyl chloride		0.15–4.79	[138]
	Lianyuan	Coal mine, steel industry	2015	1.20–3601.0	[60]
	Tongguan	Artisanal gold mining		0.69–23.7	[131]
	Wanshan	Hg-mining, artisanal gold	2012	0.5–187.0	[132]
	Wuchuan	Hg-mining	2003	0.33–320.0	Qui et al. (2006)
	Xiaoqinling	Gold mining		0.04–61.2	[129]
	Zhuzhou	Zn/Pb smelter	2012	0.62–2.61	[130]
Czech Republic	Bohemian	Non-ferrous metal	2017	6.49	[71]
France	Vosges Mountains	Chlor-alkali	2002	0.16–3.99	[48]
	Grenoble	Chlor-alkali		1.3–10.0	[40]
Germany		Chlor-alkali		0.5–4.2	[12]
	Rhine-Westph	Floodplain soil	2017	31.2	[9]
Italy	Etna	Volcano		0.1–0.4	[16]
	Mt. Amiata	Hg mining, volcano-geothermal		2.4–68.0	[21]
	Vallalta	Old Hg-mining		6.0–21.0	[123]
Kazakhstan	Pavlodar	Chlor-alkali		0.93–22.3	[116]
Maxico		Cinnabar mine	2012	2.4–4164.0	[63]
Netherlands		Chlor-alkali	2004	4.3–1150.0	[11]
Portugal	Caveira	Sulfide mine		1.1–76.5	[91]
Poland	Warsaw	Thermometer factory	2005	122–393.0	[17]
Pakistan	Karachi/ Lahore	Highway		61.5–144.0	[53]
Slovakia	Rudnany	Hg-mining		9.1–54.3	[8]
Slovenia	Podljubelj	Hg-mining	2003–04	0.35–244.0	[113]

(continued)

Table 3 (continued)

Country	Location	Source of pollution	Time	Total Hg (mg kg^{-1})	Reference
Spain	Almaden	Hg-mining		6.0–8889.0	[47]
	Caunedo	Old Hg-mining		0.09–50.0	[14]
	Flix	Chlor-alkali		0.04–12.9	[28]
Switzerland	Canton of Valais	Industrial region	2017	0.2–390.0	[76]
Turkey	Halıköy	Hg mining	2004	0.10–33.0	[37]
USA	Alaska	Hg-mining		0.05–5326	[5]
	Texas	Hg-mining		3.8–11.0	[42]
Tanzania	Geita	Gold mining		0.003–0.1220	[94]

actual value, there is insufficient observed data on sediment Hg concentration [133]. In a rising area and two significant river mouths, [133] examined the accumulation of Hg in seafloor sediments. They reported median Hg concentrations of 90.7, 93.4, and 35.8 $\mu g\ kg^{-1}$ in the Peru Margin, Congo Basin, and Amazon Fan, respectively.

5.2.3 Mercury in Water

Water bodies are another significant medium in the global Hg cycle. Mercury enters the water body from natural and anthropogenic sources like the weathering of minerals, volcanic eruptions, dry and wet atmospheric deposition, runoff from contaminated sites, direct discharge of industrial waste, etc. Inorganic forms of Hg, such as Hg^{2+} and Me–Hg, can be directly deposited into reservoirs by wet and dry depositions from the atmosphere [62]. However, Hg^{2+} and Me–Hg are introduced into water reservoirs by surface runoff in addition to leaching from topsoil layers into groundwater and later surface waters. Depending on the oxidative-reductive environment, Hg can be found in the aquatic environment in a variety of forms. In favorable oxidative environments, the forms $HgCl_4{}^{2-}$ and $HgOH^-$ predominate, whereas in favorable reductive environments, the sulfur-related forms HgS^{2-} and CH_3HgS^- dominate. The alkyl forms of Hg, MeHgCl, and EtHgCl are most frequently encountered in intermediate circumstances [51]. Mercury can often be found in soluble forms, including $[HgOH]^+$, $[HgCl]^+$, $[HgCl_2]$, $[HgCl_3]$, $[HgCl]^{2-}$, and $[HgS_2]^{2-}$. Mercury solid phases dissolve more readily in the presence of higher quantities of Cl ions, which yield highly enduring complexes with Hg, such as $HgCl_3{}^-$, $HgCl_2{}^-$, $HgCl_4{}^{2-}$, or $HgBrCl^-$ [41]. Additionally, various organic and inorganic molecules, as well as soluble non-ionic organic substances, may contain Hg. Mercury is subjected to several chemical and biological reactions in the aquatic environment, which affect its speciation and movement across the solid and liquid states [33]. In the aquatic environment (water, sediments, aquatic animals, and flora), the bulk of Hg is present in both inorganic and organic forms of divalent Hg and Hg^0, a form of Hg that dissolves in water [117]. The spatial distribution of different forms of Hg in the various components of the aquatic environment is principally influenced by the processes of Hg adsorption and desorption. These processes regulate the transport, transformation, and uptake of Hg by living beings, as well as the toxicity of Hg in water.

The concentration of Hg in different water bodies, including steam water, the water of ponds, the water of lakes, water from rivers, ocean water, etc., depends on the source of the water, the existence of Hg contamination hotspots nearby, the distance from the Hg-contaminated location, etc. Several thousand tons of Hg from air deposits enter aquatic habitats each year. The primary routes for Hg transmission into marine environments include surface waters with Hg content that are either moving or still, such as estuaries, rivers, and streams. The concentration of Hg in different ocean waters was reported in several research papers. The average total Hg concentration among all the specific oceans is 1.5 pmol (pM), while there are differences between them [56]. While the Antarctic Ocean had a lower concentration of 0.8 pM [110], the Mediterranean Sea had a higher concentration of 2.5 pM [22], and the northern Atlantic had a higher concentration of 2.0 pM [67]. [39] measured the Hg levels in the water column in locations both upstream and downstream of gold mining activities in the Peruvian Amazon. In comparison to upstream river sites, the overall Hg concentrations in river water were roughly ten times higher downstream from mining operations (downstream river: 10.1 ± 2.8 ng Hg L^{-1} and upstream river: 0.9 ± 0.5 ng Hg L^{-1}).

6 Regional Hotspots of Mercury

Certain regions around the world are known as "hotspots" for Hg contamination. These hotspots are areas where Hg concentrations are particularly high, often due to local industrial activities, mining, or other sources of Hg pollution. Additionally, areas with significant volcanic activity may experience higher levels of Hg emissions naturally. Hotspots can lead to widespread contamination of ecosystems, impacting both aquatic and terrestrial environments and posing health risks to local communities. Some examples of regional hotspots include:

6.1 Minamata Bay, Japan

The most deadly Hg poisoning outbreak occurred in the 1950s on the Japanese island of Kyushu [29]. There had been a lot of unchecked chemical waste discharge into the water from a neighboring chemical plant that produced industrial chemicals, synthetic resins, plasticizers, and chemical fertilizers. As a result of the use of Hg as a catalyst in the synthesis of acetaldehyde, the local fish and shellfish population, which was predominantly consumed by the people of Kyushu, became contaminated with Hg waste. The first strange occurrences were observed in the local animal populations: fish rotated continuously and floated belly-up; shellfish opened and decomposed; birds fell from the sky; and a large number of cats died unnaturally. Along with sensory, auditory, and visual problems, ataxia, dysarthria, and other symptoms, many residents also reported them [45]. Due to a nerve injury, dysarthria is characterized by word mispronunciation. The Minamata Sickness Epidemic was so named. By 1959, Hg had been identified as the culprit, thanks in part to tests on deceased cats and the brains of Minamata illness patients who had passed away. Mercury levels in human organs were likewise found to be extremely high, with ranges of 2.6–24.8, 22.0–70.5, and 21.2–140.0 ppm, respectively, in the brain, liver, and kidneys. Human autopsies reveal that the central nervous system has suffered major harm, especially in the form of severe lesions in the cerebral cortex and granular cell layers of the cerebellar cortex [30]. Hair Hg levels in Minamata illness patients reached levels of up to 705, which is more than 50 times the 10–20 ppm limit advised for pregnant women, women of reproductive age, and young children [50]. Mercury concentrations in the bay's marine life were found to range from 5.61 to 35.7 ppm [45], which is much higher than the FDA's stated action threshold for Hg in fish that is intended for human consumption of 1.0 ppm [32]. Mercury levels have been found to reach 2010 ppm in sludge near the plant's drainage channel.

6.2 *Amazon Basin*

When it comes to sustaining environmental services like biodiversity preservation, biogeochemical cycling, and climate regulation, the Amazon rainforest can be regarded as the most significant ecosystem on earth. Due to the breadth of the Amazon biome, a variety of components, including geology, soil properties, and climate, can have an impact on the geochemistry, speciation, and behavior of potentially dangerous elements like Hg. Due to inappropriate land use, there is currently a lot of worry regarding preserving and conserving the Amazon rainforest. Mercury use in artisanal gold extraction from the 1970s to the 1980s had an adverse effect on the Amazon region's ecosystem [83]. Intact forests next to artisanal gold mining may lower the risk of Hg to surrounding aquatic ecosystems and to the global atmospheric Hg pool by sequestering atmospheric Hg. If these trees are cut down to make room for further mining or farming, legacy Hg could be transported from the terrestrial to the aquatic ecosystem by forest fires, evasion, and/or runoff [54]. Hg is present in Amazon soils due to atmospheric deposition, anthropogenic causes, and unique parent materials [20]. [73] claim that over 80% of the mercury (Hg) in surface soils originates via air sources. Hg may be carried by the atmosphere in the Amazon basin as a result of biomass burning, where elevated temperatures cause Hg to volatilize, as demonstrated by [4]. According to [38], the Amazon rainforests are the main repository for atmospheric Hg contamination from artisanal gold mining. They found 71 μg Hg m^{-2} yr^{-1} in the precipitation water and 138–155 ng Hg g^{-1} in the soil at a depth of up to 45 cm. The mining districts and geological formations influenced by Andean sediments, where Hg values of 960 and 8100 g kg^{-1}, respectively, were found by [24, 75], are some hotspots for Hg contamination in the Amazon biome. However, understanding the relationships between levels of Hg and species, as well as the geological makeup of Amazon soils, is necessary to ascertain the likelihood of Hg accumulating in the Amazon ecosystem.

6.3 The Arctic

The Arctic is one of the most active Hg hotspots in the world. Although there are no natural sources or anthropogenic activities in the Arctic, it still turns into a global Hg hub. This is the outcome of the distinctive worldwide cycle of Hg. Atmospheric deposition, which originates from remote global anthropogenic and historical causes, is the principal source of Hg in the Arctic. In remote areas far from industrial sources, Hg can be transported through the atmosphere from distant regions, leading to elevated levels in Arctic ecosystems. This phenomenon is known as the "Arctic Mercury Cycle." According to some researchers, between 50 and 90% of the Hg entering water bodies is the result of direct wet deposits [57, 65], whereas other researchers believe that most of the Hg deposition comes from dry deposition [100, 128]. According to [23], worldwide anthropogenic Hg emissions were responsible for 32% of yearly Hg deposition in the Arctic, while legacy Hg emissions were responsible for 64% (6–10% from forest fire), and geogenic Hg emissions were responsible for 4% of the total deposit. On a regional basis, East Asia (32%), Commonwealth countries (12%), and Africa (12%), in that order, contributed the most relative anthropogenic contributions to Arctic deposition. According to the latest Hg assessment, only a small portion (<1%) of the global anthropogenic Hg came from the Arctic region. However, the Arctic acts as a reservoir for Hg, which, in turn, will be released into the environment. Mercury concentration varies between locations and seasons [2].

The origins, transport routes, and duration of Hg in environmental media both within and beyond the Arctic are related to the Hg buildup in various Arctic regions (land, air, and waterways). Based on the most recent data available, several authors outlined the Arctic Ocean Hg balance as follows: inflow flux (55 ± 7 t yr^{-1}), outflow flux (73 ± 8 t yr^{-1}), pan-Arctic riverine input (44 ± 4 t yr^{-1}), and oceanic sediment burial (28 ± 13 t yr^{-1}) [82, 114, 139]. Different modeling approaches are now employed to estimate the distribution of Hg in arctic regions. The Hg concentration of arctic ocean surface water (0–20 m) was determined and is much higher (1.21 ± 0.60 pM, $n = 159$) compared to the lower part [18]. Schartup et al. [96] reported that the average arctic sea-ice Hg concentration is 3.0 pM. In northern Alaska, along a transect of around 500 km, [98] evaluated the amounts of Hg in 13 northern permafrost cores (collected from a depth of 0–300 cm). They reported 17–207 μg kg^{-1} (average 64 μg kg^{-1}) total Hg concentrations in permafrost. Climate change influences the Arctic region to a great extent. The ice cap is melting at a quick rate, and if it continues, Hg will be released into the environment, which in turn will be a great threat to human existence.

Along with the above-mentioned Hg hotspots, there are some other hotspots depending on the presence of Hg sources. Some examples of country-specific Hg hotspots are listed in Table 4.

Table 4 List of some country specific Hg hotspots of the world

Country	Region	Source of Hg	Mean hair Hg conc (mg kg^{-1}, ww)	Reference dose level (mg kg^{-1})
Albania	Vlora Bay	Contaminated Sites	–	1.0*
Cameroon	Douala	Mixed-use chemical industrial sites	1.93	
Cook island	Muri	South Pacific Ocean	3.30	
Czech Republic	Spolana, Neratovice Spolchemie, Ústínad Labem	Chlor-Alkali Facilities	–	
Indonesia	Sekotong Poboya	Artisanal Small-Scale Gold Mining	3.60 5.00	
Japan	Tokyo	Pacific Ocean	2.70	
Mexico	Coatzacoalcos and Minatitlán	Mixed-use chemical industrial sites	1.75	
Russia	Volgograd	Chlor-Alkali Facilities	1.93	
Tanzania	Matundasi and Makongolosi	Artisanal Small-Scale Gold Mining	2.74	
Thailand	Tha Tum	Coal-Fired Power Plants	4.50	

*EU and WHO recommended value (EU and WHO 2008). *Data source* [31]

7 Factors Influencing Mercury Transport and Transformation

7.1 Physical and Chemical Processes Affecting Mercury Fate and Behavior

Volatilization: Mercury can be transformed from its liquid form to gasous Hg^0 and released into the atmosphere, a process known as volatilization. This process is particularly significant in areas with high temperatures and low atmospheric pressure. Volatilization has an effect on the transport and transformation of Hg by releasing Hg^0 into the atmosphere.

Atmospheric transport: Hg^0 can travel over great distances once in the atmosphere, frequently on a global scale. Long-distance transfer enables Hg to accumulate in pristine habitats, particularly those that are geographically removed from significant pollution sources. As a result, Hg pollution can occur even in places that are thought to be mostly unaffected by human activity.

Oxidation: Hg^0 can be converted to Hg^{2+} compounds, primarily mercuric chloride ($HgCl_2$), when it interacts with atmospheric oxidants like ozone (O_3). These oxidized forms can easily be deposited in aquatic ecosystems and are more water-soluble.

Wet deposition: Precipitation like rain and snow can remove Hg from the atmosphere. Wet deposition, a mechanism that is responsible for transporting Hg to surface water bodies, is very important.

Dry deposition: Mercury can also fall from the sky directly onto land and ocean surfaces without causing precipitation. Particle size distribution, surface properties, and meteorological conditions all have an impact on dry deposition.

Methylation: Specific bacteria can convert Hg^{2+} into Me–Hg, a highly toxic and bioaccumulative form of Hg, in aquatic environments. Methylation happens anaerobically, typically in marshes and sediments.

Bioaccumulation: In aquatic food systems, Me–Hg has a great propensity to bioaccumulate. As larger aquatic species eat smaller ones, the Me-Hg content of each trophic level rises, resulting in higher Hg concentrations in carnivorous fish and mammals.

7.2 Climate Change and Its Potential Impacts on Mercury Distribution

Altered precipitation patterns: Increases in the frequency and intensity of rainfall episodes or droughts may have an impact on Hg transfer. The input of Hg into aquatic ecosystems may increase as a result of increased rainfall due to higher levels of Hg runoff from contaminated areas. On the other hand, in some areas, changed precipitation patterns might result in droughts or prolonged dry periods. Less rain falls during these times, making it harder for Hg to be washed from the atmosphere or removed from the soil. As a result, Hg might accumulate in dry soils and sediments over time, possibly becoming more concentrated. Modified precipitation patterns can have an impact on the growth and spread of vegetation. In ecosystems, the cycling of Hg depends heavily on vegetation. Some plant species are known to be capable of absorbing and storing Hg from the soil. The general dynamics of Hg in an ecosystem can change when its abundance and distribution change. In some areas, excessive precipitation or altered precipitation can accelerate erosion rates. Sediments polluted with Hg can be moved from one place to another as a result of soil erosion. This may cause Hg to be redistributed and introduced to new locations. Changes in the chemistry of Hg and its atmospheric destiny, as well as its deposition rates and durations, can result from altered precipitation.

Rising temperatures: The distribution of Hg in the environment can be significantly impacted by a temperature rise in a number of ways. The effects are intricate and can change based on the particular environmental circumstances and the type of Hg involved. The volatilization of Hg from numerous sources, like soil, water, and vegetation, may quicken as temperatures rise. In particular, with Hg^0, higher temperatures can accelerate evaporation. As a result, the amount of Hg in the atmosphere may increase. Increased air circulation due to rising temperatures can result in more broad and effective long-range transmission of Hg. Mercury might get stuck in ice and snow in colder areas. The released Hg can reach neighboring ecosystems and water bodies when ice and snow melt and temperatures rise, potentially causing localized Hg contamination. Organic materials in soil and aquatic systems can decompose more quickly in warmer climates. This procedure has the potential to re-release previously bound Hg into the environment, raising Hg concentrations in the area's ecosystems. A rise in temperature can have an impact on aquatic species' behavior, metabolism, and distribution. Temperature can change microbial processes that impact Hg reactions. Rising temperatures and changes in the distribution of microbial populations may have an impact on Me–Hg synthesis in sediments and wetlands. The bioaccumulation and biomagnification of Hg in the aquatic food chain can then impact the temperature rise.

Changes in biogeochemical processes: Weathering, or the degradation of rocks and minerals, can fluctuate depending on the climate. For instance, rising temperatures and precipitation can hasten chemical weathering, causing the release of nutrients and trace metals like Hg into the environment. Increased erosion may cause the movement of soil and sediment containing Hg from one place to another, which may contaminate places downstream. The conversion of Hg between different chemical forms depends heavily on redox processes. The mobility and bioavailability of Hg can be decreased by the adsorption of soil and sediment particles. Changes in the environment, particularly pH and the presence of organic matter, can have an impact on the fate and transit of Hg in soils and sediments. Climate change can modify biogeochemical processes in aquatic ecosystems, which can lead to changes in Hg methylation rates. Due to the complexity of biogeochemical processes, it is challenging to predict the precise impacts of climate change and changes brought on by humans on the distribution of Hg and other elements.

Sea-level rise: In addition to causing land loss and the release or reintroduction of previously buried Hg into the marine ecosystem, sea level rise can worsen coastal erosion. Rising sea levels can result in saltwater intrusion into freshwater habitats, influencing Hg methylation and other processes that affect Hg speciation and bioavailability in these ecosystems. Changes in water depth, salinity, and nutrient inputs can affect the distribution and behavior of Hg in these habitats, which may have an impact on the bioaccumulation of Hg in aquatic animals. Sea level rise may result in an increase in flooding incidences in low-lying coastal areas. Flooding can cause Hg that has been buried in sediment and soil to become mobile and move to nearby water bodies. This may result in higher Hg levels in surface waters and a chance of bioaccumulation in aquatic life. Storm surges during tropical cyclones and other extreme weather events may become more intense due to rising sea levels. Storm surges have the ability to transport significant amounts of silt and mercury into coastal waters, temporarily raising Hg concentrations there.

Forest fires: In addition to serving as a significant sink of HgS, forest ecosystems also act as a potential source of Hg due to the burning of biomass (reemission) and subsequent processes throughout time, such as soil accumulation, runoff, and bioaccumulation into the food chain. Forest fires are intimately correlated with the mobilization and agglomeration of Hg in the environment. Biomass burning contributes roughly 13% of the natural Hg sources. The frequency and severity of wildfires can alter due to climate change, and these fires have the potential to release huge amounts of Hg, which were previously deposited in the soil and vegetation, into the atmosphere. Elevated levels of Hg are caused by the volatilization of Hg^0 and other Hg compounds, which can happen when there is a fire. Then these Hgs will be transported and deposited in other places. Thus, forest fires represent a serious threat to both humans and wildlife since they help spread Hg in other components of the environment.

It is significant to emphasize that there are complicated linkages between Hg transport, transformation, and climate change, and these interactions can differ based on regional and local circumstances. For the purpose of creating efficient solutions to reduce Hg pollution and the effects it has on ecosystems and human health.

8 Monitoring and Modeling Mercury Distribution

8.1 Methods for Monitoring Mercury Levels in the Air, Water, and Soil

8.1.1 Monitoring Mercury in the Air

Through a variety of human endeavors, including industrial processes, waste incineration, and coal burning, Hg is released into the environment. Monitoring air Hg levels is crucial for understanding atmospheric Hg contamination and its potential consequences for human health and the ecosystem. The Hg concentration in the air can be measured using the following methods:

a. **Direct sampling techniques**: This technique entails physically gathering air samples at predetermined sites and evaluating them in a lab. The most popular direct sampling methods use sorbent traps for high-volume samples. This technique involves drawing enormous quantities of air through sorbent traps, which bind both gaseous and particulate Hg. Mercury from the air is absorbed by the sorbent material, which is commonly activated carbon or gold-coated quartz filters. The sorbent traps are removed and delivered to the lab for analysis after a predetermined sampling interval. Using methods like atomic absorption spectrometry (AAS), atomic fluorescence spectrometry (AFS), or cold vapor atomic absorption spectrometry, the collected sorbent traps are examined in the laboratory. By measuring the amount of Hg adsorbed on the sorbent material, these methods calculate the total Hg concentration in the air sample.
b. **Continuous monitoring systems**: Real-time information on Hg quantities in the air is provided via continuous monitoring devices. To continuously detect Hg levels, they use analytical tools like Hg vapor analyzers and atomic absorption spectrometers. Portable devices with a detector designed specifically for Hg are known as Hg vapor analyzers. They draw air samples into the apparatus, where the Hg vapor concentration is continuously measured.
c. **Passive air sampling**: In this method, Hg is passively collected over an extended period of time using tools like activated carbon or diffusion samplers. The advantage of this method is that it can give a comprehensive measurement of Hg exposure over time and across a wide area. This approach makes use of diffusion samplers, which include a sorbent substance within that allows Hg vapor to be diffusely absorbed. They are inexpensive and excellent for long-term monitoring because they do not need electricity or a pump to function. After the passive sampling time, the sorbent materials are removed, and the Hg concentration is calculated using methods like AAS or AFS in the lab.

Challenges and limitations: Because of regional sources, weather patterns, and geography, Hg concentrations can vary dramatically over short distances. Consequently, it is crucial to choose a site carefully to collect representative data. The existence of other air contaminants may have an impact on Hg measurements. These interferences must be considered using accurate calibration and data correction techniques. To achieve precise and dependable measurements, instruments must be accurately calibrated and validated. Measurement inaccuracies must be kept to a minimum through routine maintenance and calibration tests.

8.1.2 Monitoring Mercury in the Water

To evaluate water quality and identify any potential threats to aquatic ecosystems and public health, it is crucial to monitor Hg levels in water. However, Hg contents can differ greatly depending on depth and location due to the complexity of water systems. When constructing monitoring programs, it is important to take this heterogeneity into account to guarantee that representative samples are obtained. Additionally, the presence of suspended particles and other materials can obstruct Hg measurements, perhaps producing unreliable results.

a. **Grab sampling**: Grab sampling is a time-tested technique for gauging Hg levels in water quality. This method involves physically taking discrete water samples from the target water body at predetermined places and depths. The samples are then taken to a lab for a thorough examination to ascertain the quantities of Hg present. Grab sampling is beneficial for determining the current state of the water quality since it offers a snapshot of the Hg contents at the moment of the sample. However, this method only gives a snapshot of Hg levels at particular times and locations, which might not account for temporal fluctuations. Additionally, taking physical samples can be expensive and time-consuming, particularly for large reservoirs.
b. **In-situ monitoring**: With in-situ monitoring, Hg concentrations are continually monitored over a prolonged period of time using specialized sensors and devices that are put in the water body. In-situ monitoring, as opposed to grab sampling, offers real-time or nearly real-time data, enabling the tracking of trends and variations in Hg levels throughout time. Assessing temporal variations and reactions to particular events or changes in the environment is one of the main benefits of this approach. Continuous data can be obtained through in-situ monitoring, but precise and dependable measurements depend on well-calibrated and maintained sensors and instruments. Additionally, it can be difficult to deploy and maintain the equipment in aquatic bodies, especially under extreme weather conditions.
c. **Remote sensing**: Mercury concentrations in large water bodies can be estimated using remote sensing techniques, such as those based on satellite-based spectrometers. These techniques operate by identifying particular spectral signatures connected to Hg in the water. Researchers can determine the distribution of Hg across large areas by examining satellite data. For tracking massive bodies of water or areas where direct sampling can be difficult, remote sensing provides a broad-scale perspective of Hg contents. While remote sensing has the benefit of providing wide-ranging coverage, it does not have the necessary spatial resolution to reliably measure Hg amounts in tiny or confined locations. Additionally, ground-based data calibration and validation—which can be challenging to obtain in distant or inaccessible regions—are essential for the accuracy of remote sensing estimations.

8.1.3 Monitoring Mercury in the Soil

Agricultural practices, industrial processes, atmospheric deposition, and other factors can all contribute to the presence of Hg in soil. Assessing possible threats to ecosystems and human exposure through the food chain requires regular monitoring of soil Hg levels. The specifics of the various techniques for measuring soil Hg are as follows:

a. **Core sampling**: To determine the distribution of Hg in the soil profile, core sampling is the process of extracting soil samples from various depths. Utilizing specialist equipment like soil corers, which enable researchers to retrieve entire vertical pieces of soil, soil samples are taken. Scientists can learn how Hg is dispersed in the soil and whether certain strata may have higher concentrations due to past contamination or particular deposition episodes by evaluating the Hg concentrations at various depths.
b. **Soil pore water analysis**: The water in the soil pores must first be removed before it can be analyzed. The method reveals the mobility and bioavailability of Hg in the soil. The quantity of Hg that is readily available for absorption by plants and soil organisms can be determined by the concentration of Hg in the pore water of the soil. Assessing the potential for Hg to contaminate surface and groundwater, which may then affect surrounding ecosystems, is also helpful.
c. **Soil gas monitoring**: The Hg vapor contents in the soil are measured via soil gas monitoring. Soil contains volatile mercury compounds, and conditions like humidity and temperature can promote the vaporization of Hg. Researchers can evaluate the likelihood of Hg being released into the atmosphere by observing soil gas. Understanding Hg cycling between the soil and the atmosphere and its function in the overall Hg cycle requires knowledge of this information.

Challenges and limitations: In a limited area, soil characteristics can vary widely, making it difficult to collect representative samples. To accurately capture this diversity, proper sampling design and techniques are needed. Inorganic Hg compounds, Me-Hg, and Hg^0 are only a few of the types of Hg that can be found in soil. For a proper assessment of potential dangers, speciation analysis is crucial because each form exhibits distinct behaviors and toxicities. Because Hg is present in the environment in trace amounts, there is a high danger of contamination during sample collection, handling, and analysis. During the monitoring procedure, extra care must be taken to prevent the introduction of external Hg sources.

8.2 Advanced Modeling Techniques to Predict Mercury Transport

8.2.1 Atmospheric Mercury Transport Modeling

Utilizing computer simulations to replicate the behavior and movement of Hg in the earth's atmosphere is known as "atmospheric Hg transport modeling". With the aid of these models, scientists are better able to comprehend the global and regional transportation of Hg as well as the changes that it goes through as it travels through the atmosphere. The following are specifics regarding various models for the movement of atmospheric Hg:

a. **Lagrangian models**: The tracking of individual air parcels or particles is done via lagrangian models, a subset of atmospheric transport models. These models track the motions of distinct air masses through time to replicate the movement of Hg. For analyzing the long-range movement of Hg and comprehending how it is moved between different regions and across continents, lagrangian models are very helpful.
b. **Eulerian models**: Grid-based models, commonly referred to as Eulerian models, separate the atmosphere into a grid of cells. To replicate the movement of Hg within each grid cell, these models employ differential equations. For large-scale simulations, Eulerian models are effective and give data on the distribution of Hg concentrations across various spatial areas. They are frequently employed in regional and worldwide investigations of the movement of atmospheric Hg.
c. **3D atmospheric models**: Three-dimensional (3D) atmospheric models take into account both vertical and horizontal fluctuations in the atmosphere. These models provide a comprehensive picture of how mercury moves through the atmosphere, capturing not only its horizontal motion but also its vertical dispersion and interactions with different layers of the atmosphere. Although more complicated than 2D models, 3D atmospheric models provide a more accurate portrayal of the processes in the atmosphere controlling Hg transport.

Challenges and limitations: Mercury changes dramatically in the atmosphere due to oxidation and reduction processes, which have an impact on its chemical makeup and behavior. For models to make accurate predictions, these transitions must be captured precisely. Understanding atmospheric dynamics, Hg interactions with aerosols, and precipitation interactions is necessary for modeling these deposition processes. Input data for atmospheric Hg models, such as information on Hg emissions from various sources, meteorological data, and details on atmospheric reactions, must be precise and up-to-date. The quantity and caliber of such data can have an impact on the model's accuracy.

8.2.2 Hydrological Mercury Fate and Transport Modeling

Understanding the flow and behavior of Hg in aquatic systems, such as rivers, lakes, and estuaries, is largely dependent on hydrological models. In order to analyze the effects of Hg on water quality and possible dangers to ecosystems and human health, scientists and policymakers use these models to simulate the fate and transit of Hg. The specifics of several hydrological Hg models are as follows:

a. **Watershed-scale models**: To estimate the movement of Hg from different sources to water bodies within a watershed, watershed-scale models integrate meteorological, hydrological, and biogeochemical processes. These models take into account things like air deposition, runoff, and interactions between Hg and various watershed components, including soils and vegetation, as well as Hg emissions from both natural and human-made sources. Watershed-scale models can help pinpoint key locations for management and mitigation activities and are crucial for understanding the sources and pathways by which Hg enters water bodies.

b. **Fate and transport models**: The main goal of fate and transport models is to forecast the mobility of Hg within water bodies based on its chemical makeup and interactions with sediments and biota. These models take into account the numerous species of Hg, including Hg^0, inorganic Hg, and Me-Hg, each with its own unique toxicity and behavior. The processes governing Hg methylation and demethylation in aquatic environments are better understood with the aid of fate and transport models. These activities are important determinants of Hg bioaccumulation in aquatic animals.
c. **Coupled watershed-atmospheric models**: To investigate the relationships between atmospheric deposition and Hg in aquatic systems, coupled watershed-atmospheric models take into consideration the interactions between the atmosphere and watersheds. These models take into account the fate and transit of Hg within watersheds after it enters the atmosphere through wet and dry deposition. It is possible to have a more complete picture of how Hg moves through the environment and affects water bodies by linking atmospheric and hydrological processes.

Challenges and limitations: Environmental circumstances and microbial activity are just two of the many variables that might affect the complicated methylation and demethylation processes of Hg. It is still difficult to adequately represent these processes in models. Data on Hg emissions, atmospheric deposition, hydrological parameters, and biogeochemical processes are needed for hydrological models. It might be difficult to gather correct and representative data, which can cause uncertainty in model projections. Water flow rates, sedimentation, and biological interactions are just a few of the many variables that have an impact on highly dynamic aquatic ecosystems.

8.2.3 Ecological Mercury Fate and Transport Modeling

The construction of models to analyze how Hg travels and accumulates in ecosystems, affecting a variety of animals and food webs, is known as "ecological Hg fate and transport modeling." These models have improved our understanding of the dynamics of Hg in our surroundings and its potential impacts on ecological systems. The specifics of the various ecological Hg fate and transport modeling elements are as follows:

a. **Bioaccumulation models**: The main goal of bioaccumulation models is to forecast how Hg will move from water and sediments to aquatic species. These models consider a number of variables, including the content of Hg in water and sediments, the rates at which various organisms absorb and eliminate Hg, and the trophic level of the organism. By showing how Hg builds up in different stages of the food chain, bioaccumulation models can help find species that may be more sensitive to Hg exposure and biomagnification.

b. **Food web models**: Models of the food chain explain how Hg moves across various trophic levels in ecosystems. These models take into account the interactions and feeding linkages between diverse species. Through the food chain, Hg can be passed from prey to predators, which can result in biomagnification, where the content of Hg rises at higher trophic levels. Understanding the potential effects on biodiversity and the general health of the ecosystem is made easier with the use of food web models, which shed light on how Hg builds up and distributes across various species in an ecosystem.
c. **Ecotoxicological models**: Ecotoxicological models are used to assess the toxic effects of Hg on various populations and species. These models take into account things like the potential for sublethal and lethal impacts on people and communities, the toxicity of various Hg forms (such as Me-Hg), and the susceptibility of different living things to Hg exposure. For risk evaluations of ecosystems exposed to high amounts of Hg, ecotoxicological models may be employed to determine safe exposure limits for various species.

Challenges and limitations: Complex food webs can be challenging to simulate due to the high number of connected species and the diversity of their feeding connections. Extensive data and reliable modeling methods are needed to capture these dynamics accurately. Different species may exhibit different behaviors and ecological preferences that influence how much Hg they are exposed to. Due to data shortages and uncertainties, it might be difficult to incorporate species-specific behavior into models. The calibration and validation of Hg fate and transport models primarily rely on field data. However, it might be difficult to gather thorough and long-term data for model validation, particularly in distant or understudied habitats. Despite these difficulties, ecological Hg destiny and transport modeling remain crucial tools for comprehending the complicated behavior of Hg in ecosystems and evaluating its possible effects on biodiversity and human health. These models aid in the oversight and restoration of ecosystems affected by Hg contamination.

8.3 Combination of Monitoring and Modeling

8.3.1 Data-Driven Models

Data-driven models belong to a class of models that base their development, calibration, and validation on observable data from monitoring programs. Unlike conventional mechanistic models, these models discover patterns and correlations immediately from the data rather than explicitly including theoretical concepts or physical equations. Here are some specifics on how data-driven models function:

Using monitoring data for calibration: Real-world information gathered from monitoring programs is vital for calibrating data-driven models. In order to reduce the discrepancy between predictions made by the model and the actual data collected, calibration entails modifying the parameters of the model and the initial conditions. The accuracy and predictive power of the model are improved by adjusting these parameters to fit the real-world data. The calibration procedure aids the model in learning from the data at hand and enhances the model's capacity to represent the system's underlying processes.

Uncertainty and sensitivity analysis: Analyses of sensitivity and uncertainty are crucial elements in determining the dependability and consistency of data-driven models. In order to assess the uncertainty relating to model predictions caused by the data's inherent unpredictability and the model's simplifications, uncertainty analysis is used. Contrarily, sensitivity analysis aids in locating the most vital input variables that profoundly affect the model's output. By being aware of these concerns, researchers can concentrate on enhancing data gathering or prioritizing the quality of specific input parameters to improve the performance of the model.

Model validation: A crucial stage in the establishment of data-driven models is model validation. It entails contrasting the predictions of the model with external data that was not incorporated into the calibration of the model. These independent data offer a measurable indicator of how effectively the model predicts unknown or improbable outcomes. When a model's predictions and validation data agree well, it can be said to be reliable for making predictions under comparable circumstances and to have strong generalization capabilities.

It is crucial to remember that data-driven models have a number of advantages and disadvantages. They are especially helpful when dealing with complicated systems for which it may be difficult to develop physical equations or when the underlying processes are not well understood. They are less useful for projecting circumstances beyond the recorded data range, though, because they only consider historical data. Furthermore, because data-driven models do not explicitly define the underlying mechanisms regulating the system, they might not offer the same level of interpretability as mechanistic models. Nevertheless, data-driven models, when correctly calibrated and validated, can be useful resources for comprehending and forecasting a variety of environmental phenomena, including those connected to water quality and Hg monitoring.

9 Conclusion

In conclusion, it is crucial to comprehend the complex network of sources and entry points that Hg uses to enter our environment. Mercury pollution is a complicated global problem with extensive effects on ecosystems and human health. It results from both natural events like volcanic emissions and the myriad contributions of human activities. This chapter emphasizes the significance of comprehending the complex distribution of Hg, its capacity to biomagnify in food chains, and its propensity to migrate over great distances. Despite the reduction in Hg emissions brought about by international agreements and regulatory measures, careful environmental stewardship and continued research are still necessary to address this pervasive and important environmental issue.

Acknowledgements The authros are expressing their immense gratitude to all the persons involved in the preparation of the final report of the "Golbal Mercury Assessment 2018" and sharing this report with all the people across the globe. The authors also thank all the researchers and scientists all over the world who are working to explore the source and distribution of Hg in the environment and are warning against reducing Hg exposure to humans and wild life.

References

1. Al-Saleh I, Nester M, Abduljabbar M, Al-Rouqi R, Eltabache C, Al-Rajudi T, Elkhatib R (2016) Alterations in biochemical markers due to mercury (Hg) exposure and its influence on infant's neurodevelopment. Int J Hyg Environ Health 219:898–914
2. AMAP (2021) AMAP Assessment 2021: mercury in the arctic. Arctic Monitoring and Assessment Programme (AMAP), Tromsø, Norway
3. Amos HM, Jacob DJ, Streets DG, Sunderland EM (2013) Legacy impacts of all-time anthropogenic emissions on the global mercury scale. Glob Biogeochem Cycl 27:1–12
4. Artaxo P, de Campos RC, Fernandes ET, Martins JV, Xiao Z, Lindqvist O, Fernández-Jiménez MT, Maenhaut W (2000) Large scale mercury and trace element measurements in the Amazon basin. At Environ 34(24):4085–4096
5. Bailey EA, Gray JE, Theodorakos PM (2002) Mercury in vegetation and soils at abandoned mercury mines in southwestern Alaska. USA Geochem Explor Environ Anal 2:275–285
6. Balali-Mood M, Naseri K, Tahergorabi Z, Khazdair MR, Sadeghi M (2021) Toxic mechanisms of five heavy metals: mercury, lead, chromium, cadmium, and arsenic. Front Pharmacol 12:643972. https://doi.org/10.3389/fphar.2021.643972
7. Ballabio C, Jiskra M, Osterwalder S, Borrelli P, Montanarella L, Panagos P (2021) A spatial assessment of mercury content in the European Union topsoil. Sci Total Environ 769:144755. https://www.sciencedirect.com/science/article/pii/S0048969720382887
8. Banásová V (1999) Vegetation on contaminated sites near an Hg mine and smelter. In: Ebinghaus RR, Turner RR, de Lacedra LD, Vasiljev O, Salomons W (eds) Mercury contaminated sites. Springer, Berlin, pp 321–335
9. Beckers F, Awad YM, Beiyuan J, Abrigata J, Mothes S, Tsang DC et al (2019) Impact of biochar on mobilization, methylation, and ethylation of mercury under dynamic redox conditions in a contaminated floodplain soil. Environ Inter 127:276–290
10. Beckers F, Rinklebe J (2017) Cycling of mercury in the environment: sources, fate, and human health implications—a review. Crit Rev Environ Sci Technol 47(9):693–794

11. Bernaus A, Gaona X, van Ree D, Valiente M (2006) Determination of mercury in polluted soils surrounding a chlor-alkali plant: direct speciation by X-ray absorption spectroscopy techniques and preliminary geochemical characterisation of the area. Anal Chim Acta 565:73–80
12. Biester H, Müller G, Schöler HF (2002) Binding and mobility of mercury in soils contaminated by emissions from chlor-alkali plants. Sci Total Environ 284:191–203
13. Bloom NS (2000) Analysis and stability of mercury speciation in petroleum hydrocarbons. Fres J Anal Chem 366(5):438–443
14. Boente C, Albuquerque MTD, Gerassis S, Rodríguez-Valdés E, Gallego JR (2019) A coupled multivariate statistics, geostatistical and machine-learning approach to address soil pollution in a prototypical Hg-mining site in a natural reserve. Chemosphere 218:767–777
15. Bojakowska I, Sokołowska G (2001) Rtęć w kopalinach wydobywanych w Polsce jako potencjalne źródło zanieczyszczenia środowiska (Mercury in the extracted substances extracted in Poland as a potential source of environmental pollution—in Polish). Biul Panstw Inst Geol 394:5–53
16. Bonanno G, Giudice RL, Pavone P (2012) Trace element biomonitoring using mosses in urban areas affected by mud volcanoes around Mt. Etna. The case of the Salinelle. Italy Monit Assess 184:5181–5188
17. Boszke L, Kowalski A, Siepak J (2004) Grain size partitioning of mercury in sediments of the middle Odra River (Germany/Poland). Water Air Soil Poll 159:125–138
18. Bowman KL, Collins RE, Agather AM, Lamborg CH, Hammerschmidt CR, Kaul D, Dupont CL, Christensen GA, Elias DA (2020) Distribution of mercury-cycling genes in the Arctic and equatorial Pacific Oceans and their relationship to mercury speciation. Limnol Oceanogr 65(S1):S310–S320
19. Carocci A, Catalano A, Lauria G, Sinicropi MS, Genchi G (2016) A review on mercury toxicity in food. Food Toxicol 315
20. Carpi A, Fostier AH, Orta OR, dos Santos JC, Gittings M (2014) Gaseous mercury emissions from soil following forest loss and land use changes: Field experiments in the United States and Brazil. Atm Environ 96:423–429
21. Chiarantini L, Rimondi V, Benvenuti M, Beutel MW, Costagliola P, Gonnelli C et al (2016) Black pine (*Pinus nigra*) barks as biomonitors of airborne mercury pollution. Sci Total Environ 569:105–113
22. Cossa D, Martin JM, Takayanagi K, Sanjuan J (1997) The distribution and cycling of mercury species in the western mediterranean. deep-sea research Part II: Top. Stud Oceanogr 44:721–740
23. Dastoor A, Wilson SJ, Travnikov O, Ryjkov A, Angot H, Christensen JH, Steenhuisen F, Muntean M (2022) Arctic atmospheric mercury: sources and changes. Sci Total Environ 839:156213. https://doi.org/10.1016/j.scitotenv.2022.156213
24. de Souza JG, Robinson M, Maezumi SY, Capriles J, Hoggarth JA, Lombardo U, Novello VF, Apaéstegui J, Whitney B, Urrego D, Alves DT (2019) Climate change and cultural resilience in late pre-Columbian Amazonia. Nat Ecol Evol 3:1007–1017
25. Dommergue A, Sprovieri F, Pirrone N, Ebinghaus R, Brooks S, Courteaud J, Ferrari CP (2010) Overview of mercury measurements in the Antarctic troposphere. Atmos Chem Phys 10:3309–3319
26. Eagles-Smith CA, Wiener JG, Eckley CS, Willacker JJ, Evers DC, Marvin-DiPasquale M, Obrist D, Fleck JA, Aiken GR, Lepak JM, Jackson AK (2016) Mercury in western North America: a synthesis of environmental contamination, fluxes, bioaccumulation, and risk to fish and wildlife. Sci Total Environ 568:1213–1226
27. Eisler R (2006) Mercury hazards to living organisms. CRC Press, Boca Raton
28. Esbrí JM, López-Berdonces MA, Fernández-Calderón S, Higueras P, Díez S (2015) Atmospheric mercury pollution around a chlor-alkali plant in Flix (NE Spain): an integrated analysis. Environ Sci Pollut Res 22:4842–4850
29. Eto K (2000) Minamata disease. Neuropathol 20:S14–S19

30. Eto K, Takeuchi T (1978) A pathological study of prolonged cases of Minamata disease. With particular reference to 83 autopsy cases. Acta Pathol Jpn 28:565–584
31. Evers DC, DiGangi J, Petrlík J, Buck DG, Šamánek J, Beeler B, Turnquist MA, Hatch SK, Regan K (2014) Global mercury hotspots: New evidence reveals mercury contamination regularly exceeds health advisory levels in humans and fish worldwide. Biodiversity Research Institute. Portland, Maine. IPEN. Göteborg, Sweden. BRI-IPEN Sci Commun (34):1–20
32. FDA (2000) Action levels for poisonous or deleterious substances in human food and animal feed. Accessible at: http://www.cfsan.fda.gov/~lrd/fdaact.html#merc.
33. Fitzgerald WF, Lamborg CH, Hammerschmidt CR (2007) Marine biogeochemical cycling of mercury. Chem Rev 107:641–662
34. Frimmel HE, Gartz VH (1997) Witwatersrand gold particle chemistry matches model of metamorphosed, hydrothermally altered placer deposits. Miner Dep 32(6):523–530
35. Furimsky E (2000) Characterization of trace element emissions from coal combustion by equilibrium calculations. Fuel Process Technol 63:29–44
36. Gaffney JS, Marley NA (2014) In-depth review of atmospheric mercury: sources, transformations, and potential sinks. Energy Emiss Control Technol 2:1–21
37. Gemici Ü, Tarcan G, Somay AM, Akar T (2009) Factors controlling the element distribution in farming soils and water around the abandoned Halıköy mercury mine (Beydağ, Turkey). Appl Geochem 24:1908–1917
38. Gerson JR, Szponar N, Zambrano AA, Bergquist B, Broadbent E, Driscoll CT, Erkenswick G, Evers DC, Fernandez LE, Hsu-Kim H, Inga G (2022) Amazon forests capture high levels of atmospheric mercury pollution from artisanal gold mining. Nat Commun 13(1):559. https://www.nature.com/articles/s41467-022-27997-3
39. Gerson JR, Topp SN, Vega CM, Gardner JR, Yang X, Fernandez LE, Bernhardt ES, Pavelsky TM (2020) Artificial lake expansion amplifies mercury pollution from gold mining. Sci Adv 6(48):eabd4953
40. Grangeon S, Guédron S, Asta J, Sarret G, Charlet L (2012) Lichen and soil as indicators of an atmospheric mercury contamination in the vicinity of a chlor-alkali plant (Grenoble, France). Ecol Indic 13:178–183
41. Grassia S, Nettib R (2000) Sea water intrusion and mercury pollution of some coastal aquifers in the province of Grosseto (southern Tuscany, Italy). J Hydrol 237:198–211
42. Gray JE, Theodorakos PM, Fey DL, Krabbenhoft DP (2015) Mercury concentrations and distribution in soil, water, mine waste leachates, and air in and around mercury mines in the Big Bend region. Texas Environ Geochem Hlth 37:35–48
43. Gustin MS, Lindberg SE, Weisberg PJ (2008) An update on the natural sources and sinks of atmospheric mercury. Appl Geochem 23:482–493
44. Gworek B, Dmuchowski W, Baczewska AH, Brągoszewska P, Bemowska-Kałabun O, Wrzosek-Jakubowska J (2017) Air contamination by mercury, emissions and transformations—a review. Water Air Soil Pollut 228:1–31
45. Harada M (1995) Minamata disease: methylmercury poisoning in Japan caused by environmental pollution. Crit Rev Toxicol 25:1–24
46. He F, Gao J, Pierce E, Strong PJ, Wang H, Liang L (2015) In situ remediation technologies for mercury-contaminated soil. Environ Sci Pollut Res Int 22:8124–8147
47. Higueras P, Oyarzun R, Biester H, Lillo J, Lorenzo S (2003) A first insight into mercury distribution and speciation in soils from the Almadén mining district. Spain J Geochem Explor 80(1):95–104
48. Hissler C, Probst JL (2006) Impact of mercury atmospheric deposition on soils and streams in a mountainous catchment (Vosges, France) polluted by chlor-alkali industrial activity: the important trapping role of the organic matter. Sci Total Environ 361:163–178
49. Hsu-Kim H, Eckley CS, Acha´ D, Feng X, Gilmour CC, Jonsson S, Mitchell CPJ (2018) Challenges and opportunities for managing aquatic mercury pollution in altered landscapes. Ambio 47:141–169
50. JECFA (2003) Joint FAO/WHO Committee on Food Additives: Sixty-first meeting, Rome. Summary and conclusions. Accessible at: www.who.int/entity/ipcs/food/jecfa/summaries/en/summary_61.pdf.

51. Kabata-Pendias A (2011) Trace elements in soils and plants, 4th edn. CRC Press/Taylor & Francis, Boca Raton
52. Kabata-Pendias A, Szteke B (2015) Trace elements in abiotic and biotic environments. Taylor & Francis, pp 468
53. Khan MN, Wasim AA, Sarwar A, Rasheed MF (2011) Assessment of heavy metal toxicants in the roadside soil along the N-5, National Highway. Pakistan Environ Monit Assess 182:587–595
54. Kumar A, Wu S, Huang Y, Liao H, Kaplan JO (2018) Mercury from wildfires: global emission inventories and sensitivity to 2000–2050 global change. Atmos Environ 173:6–15
55. Lacerda LD (1997) Global mercury emissions from gold and silver mining. Water Air Soil Pollut 97(3–4):209–221
56. Lamborg CH, Fitzgerald WF, Damman AWH, Benoit JM, Balcom PH, Engstrom DR (2002) Modern and historic atmospheric mercury fluxes in both hemispheres: global and regional mercury cycling implications. Glob Biogeochem Cycl 16:1104–1114
57. Landis MS, Keeler GJ (2002) Atmospheric mercury deposition to Lake Michigan during the Lake Michigan mass balance study. Environ Sci Technol 36(21):4518–4524
58. Li Q, Tang L, Qiu G, Liu C (2020) Total mercury and methylmercury in the soil and vegetation of a riparian zone along a mercury-impacted reservoir. Sci Total Environ 738:139794. https://doi.org/10.1016/j.scitotenv.2020.139794
59. Li R, Li R, Wu H, Ding J, Fu W, Gan L, Li Y (2017) Mercury pollution in vegetables, grains and soils from areas surrounding coal-fired power plants. Sci Rep 7(1):46545. https://doi.org/10.1038/srep46545
60. Liang J, Feng C, Zeng G, Gao X, Zhong M, Li X et al (2017) Spatial distribution and source identification of heavy metals in surface soils in a typical coal mine city, Lianyuan. China Environ Pollut 225:681–690
61. Lin Y, Vogt R, Larssen T (2012) Environmental mercury in China: a review. Environ Toxicol Chem 31:2431–2444
62. Lina CJ, Pongprueksaa P, Lindberg SE, Pehkoned SO, Byune D, Jang C (2006) Scientific uncertainties in atmospheric mercury models I: model science evaluation. Atm Environ 40:2911–2928
63. Martínez-Trinidad S, Hernández Silva G, Ramírez Islas ME, Martínez Reyes J, Solorio Munguía G, Solís Valdez S, García Martínez R (2013) Total mercury in terrestrial systems (air-soil-plant-water) at the mining region of San Joaquín, Queretaro. Mexico Geofísica Int 52(1):43–58
64. Mason R, Sheu GR (2002) Role of the ocean in the global mercury cycle. Glob Biogeochem Cycl 16:1–14
65. Mason RP, Lawson NM, Sullivan KA (1997) The concentration, speciation and sources of mercury in Chesapeake Bay precipitation. Atm Environ 31(21):3541–3550
66. Mason RP, Pirrone N, Hedgecock I, Suzuki N, Levin L (2010) Conceptual overview. In: Pirrone N, Keating T (eds) Hemispheric transport of air pollution—part B. United Nations Publication, New York, pp 1–19
67. Mason RP, Rolfhus K, Fitzgerald WF (1998) Mercury in the North Atlantic. Mar Chem 61:37–53
68. Mayer W, Sass-Gustkiewicz M (1998) Geochemical characterization of sulphide minerals from the Olkusz lead-zinc ore cluster, Upper Silesia (Poland), based on laser ablation data. Miner Polon 29:87–105
69. Meija J, Yang L, Sturgeon RE, Mester Z (2010) Certification of natural isotopic abundance inorganic mercury reference material NIMS-1 for absolute isotopic composition and atomic weight. J Anal Atom Spectrom 25:384–389
70. Meng W, Wang Z, Hu B, Wang Z, Li H, Goodman RC (2016) Heavy metals in soil and plants after long-term sewage irrigation at Tianjin China: a case study assessment. Agric Water Manag 171:153–161
71. Navrátil T, Burns DA, Nováková T, Kaňa J, Rohovec J, Roll M et al (2018) Stability of mercury concentration measurements in archived soil and peat samples. Chemosphere 208:707–711

72. Nriagu JO (1993) Legacy of mercury pollution. Nature 363:589. https://www.nature.com/articles/363589a0
73. Obrist D, Kirk JL, Zhang L, Sunderland EM, Jiskra M, Selin NE (2018) A review of global environmental mercury processes in response to human and natural perturbations: changes of emissions, climate, and land use. Ambio 47:116–140
74. Obrist D, Pearson C, Webster J, Kane T, Lin CJ, Aiken GR et al (2016) A synthesis of terrestrial mercury in the western United States: spatial distribution defined by land cover and plant productivity. Sci Total Environ 568:522–535
75. Oliveira LC, Dietz JM (2011) Predation risk and the interspecific association of two Brazilian Atlantic forest primates in Cabruca agroforest. Am J Primatol 73(9):852–860
76. Osterwalder S, Huang JH, Shetaya WH, Agnan Y, Frossard A, Frey B et al (2019) Mercury emission from industrially contaminated soils in relation to chemical, microbial, and meteorological factors. Environ Pollut 250:944–952
77. Pacyna EG, Pacyna JM (2002) Global emission of mercury from anthropogenic sources in 1995. Water Air Soil Pollut 137:149–165
78. Pacyna EG, Pacyna JM, Steenhuisene F, Wilsond S (2006) Global anthropogenic mercury emission inventory for 2000. Atm Environ 40(22):4048–4063
79. Pacyna EG, Pacyna JM, Sundsetha K, Munthec J, Kindbomc K, Wilsond S et al (2010) Global emission of mercury to the atmosphere from anthropogenic sources in 2005 and projections to 2020. Atm Environ 44(20):2487–2499
80. Pacyna JM, Munthe J, Wilson S (2008) Global emission of mercury to the atmosphere. In: AMAP/UNEP, technical background report to the global atmospheric mercury assessment, arctic monitoring and assessment programme. UNEP Chemical Branch, pp 64–72
81. Pasieczna A, Bojakowska I, Paolo A (2007) Arsenic and mercury in brown and hard coals from deposits of Poland. Goldschmidt conference abstracts. Geochimica et Cosmochimica Acta 71(159):A762
82. Petrova MV, Krisch S, Lodeiro P, Valk O, Dufour A, Rijkenberg MJA, Achterberg EP, Rabe B, Rutgers van der Loeff M, Hamelin B, Sonke JE, Garnier C, Heimbürger-Boavida LE (2020) Mercury species export from the arctic to the atlantic ocean. Mar Chem 225:103855. https://doi.org/10.1016/j.marchem.2020.103855n
83. Pinto LDC, Dórea JG, Bernardi JVE, Gomes LF (2019) Mapping the evolution of mercury (Hg) research in the Amazon (1991–2017): a scientometric analysis. Int J Environ Res Pub Health 16(7):1111. https://pubmed.ncbi.nlm.nih.gov/30925692
84. Pirrone N, Cinnirella S, Feng X et al (2010) Global mercury emissions to the atmosphere from anthropogenic and natural sources. Atmos Chem Phys 10:5951–5964
85. Pirrone N, Cinnirella S, Feng X, Finkelman RB, Friedli HR, Leaner J et al (2010) Global mercury emissions to the atmosphere from anthropogenic and natural sources. Atm Chem Phys 10:5951–5964
86. Pirrone N, Hedgecock IM, Sprovieri F (2008) New Directions: Atmospheric mercury, easy to spot and hard to pin down: Impasse? Atmos Environ 42:8549–8551
87. Qiu G, Feng X, Wang S, Shang L (2006) Environmental contamination of mercury from Hg-mining areas in Wuchuan, northeastern Guizhou. China Environ Pollut 142:549–558
88. Raj D, Maiti SK (2019) Sources, toxicity, and remediation of mercury: an essence review. Environ Monit Assess 191:566. https://doi.org/10.1007/s10661-019-7743-2
89. Rayaboshapko AG, Korolev VA (1997) Mercury in the atmosphere: estimation of model parameters. Meteorological Synthesizing Centre—East, EMEP/MSC-E, Report 7/97, Moscow
90. Reimann C, de Caritat P (2017) Establishing geochemical background variation and threshold values for 59 elements in Australian surface soil. Sci Total Environ 578:633–648
91. Reis AT, Rodrigues SM, Davidson CM, Pereira E, Duarte AC (2010) Extractability and mobility of mercury from agricultural soils surrounding industrial and mining contaminated areas. Chemosphere 81:1369–1377
92. Rezende PS, Silva NC, Moura WD, Windmöller CC (2018) Quantification and speciation of mercury in streams and rivers sediment samples from Paracatu, MG, Brazil, using a direct mercury analyzer®. Microchem J 140:199–206

93. Roulet M, Lucotte M, Farella N, Serique G, Coelho H, Sousa Passos CJ et al (1999) Effects of recent human colonization on the presence of mercury in Amazonian ecosystems. Water Air Soil Pollut 112(3):297–313
94. Sanga TR, Maseka KK, Ponraj M, Tungaraza C, Mwakalapa EB (2023) Accumulation and distribution of mercury in agricultural soils, food crops and associated health risks: a case study of Shenda gold mine-Geita Tanzania. Environ Challen 11:100697. https://doi.org/10.1016/j.envc.2023.100697
95. Satoh H (2013) Mercury. In: Weiss B (ed) Aging and Vulnerability to environmental chemicals. Royal Society of Chemistry, Cambridge
96. Schartup AT, Qureshi A, Dassuncao C, Thackray CP, Harding G, Sunderland EM (2018) A model for methylmercury uptake and trophic transfer by marine plankton. Environ Sci Technol 52(2):654–662
97. Schroeder WH, Anlauf KG, Barrie LA, Lu JY, Steffen A, Schneeberger DR et al (1998) Arctic springtime depletion of mercury. Nature 394:331–332
98. Schuster PF, Schaefer KM, Aiken GR, Antweiler RC, Dewild JF, Gryziec JD, Gusmeroli A, Hugelius G, Jafarov E, Krabbenhoft DP, Liu L, Herman-Mercer N, Mu C, Roth DA, Schaefer T, Striegl RG, Wickland KP, Zhang T (2018) Permafrost stores a globally significant amount of mercury. Geophys Res Lett 45(3):1463–1471
99. Selin NE (2009) Global biogeochemical cycling of mercury: a review. Ann Rev Environ Resour 34:43–63
100. Selin NE, Jacob DJ (2008) Seasonal and spatial patterns of mercury wet deposition in the United States: constraints on the contribution from North American anthropogenic sources. Atm Environ 42(21):5193–5204
101. Selin NE, Jacob DJ, Park RJ, Yantosca RM, Strode S, Jaeglé L et al (2007) Chemical cycling and deposition of atmospheric mercury: global constraints from observations. J Geophysic Res 112:1–14
102. Song Z, Li P, Ding L, Li Z, Zhu W, He T, Feng X (2018) Environmental mercury pollution by an abandoned chlor-alkali plant in Southwest China. J Geochem Explor 194:81–87
103. Sprovieri F, Pirrone N, Ebinghaus R, Kock H, Dommergue A (2010) A review of worldwide atmospheric mercury measurements. Atmos Chem Phys 10:8245–8265
104. Stoffers P, Hannington M, Wright I, Herzig P, de Ronde C (1999) Elemental mercury at submarine hydrothermal vents in the Bay of Plenty, Taupo volcanic zone, New Zealand. Geology 27:931–934
105. Streets DG, Devane MK, Lu Z, Bond TC, Sunderland EM, Jacob DJ (2011) All-time releases of mercury to the atmosphere from human activities. Environ Sci Technol 45(24):10485–10491
106. Streets DG, Haob J, Wuc Y, Jiangb J, Chand M, Tianb H et al (2005) Anthropogenic mercury emissions in China. Atm. Environ. 39:7789–7806
107. Streets DG, Horowitz HM, Jacob DJ, Lu Z, Levin L, Ter Schure AF, Sunderland EM (2017) Total mercury released to the environment by human activities. Environ Sci Technol 51(11):5969–5977
108. Streets DG, Horowitz HM, Lu Z, Levin L, Thackray CP, Sunderland EM (2019) Global and regional trends in mercury emissions and concentrations, 2010–2015. Atmos Environ 201:417–427
109. Streets DG, Zhang Q, Wu Y (2009) Projections of global mercury emissions in 2050. Environ Sci Technol 43(8):2983–2988
110. Sunderland EM, Mason RP (2007) Human impacts on open ocean mercury concentrations. Glob Biogeochem Cycl 21(4):GB4022
111. Swain EB, Jakus PM, Rice G, Lupi F, Maxson PA, Pacyna JM et al (2007) Socioeconomic consequences of mercury use and pollution. Ambio 36:45–61
112. Telmer KH, Veiga MM (2009) World emissions of mercury from small scale and artisanal gold mining. In: Pirrone N, Mason R (eds) Mercury fate and transport in the global atmosphere: emissions, measurements and models. Springer, New York, pp 131–172
113. Teršič T, Gosar M, Šajn R (2009) Impact of mining activities on soils and sediments at the historical mining area in Podljubelj, NW Slovenia. J Geochem Explor 100:1–10

114. Tesán Onrubia JA, Petrova MV, Puigcorbé V, Black EE, Valk O, Dufour A, Hamelin B, Buesseler KO, Masqué P, Le Moigne FAC, Sonke JE, Rutgers van der Loeff M, Heimbürger-Boavida LE (2020) Mercury export flux in the arctic ocean estimated from 234Th/238U Disequilibria. ACS Earth Space Chem. 4(5):795–801
115. Uddin S, Afroz H, Hossain M, Briffa J, Blundell R, Islam MR (2023) Heavy metals/metalloids in food crops and their ımplications for human health. In: Hossain MA, Hossain AKMZ, Bourgerie S, Fujita M, Dhankher OP, Haris P (eds) Heavy metal toxicity and tolerance in plants: a biological, omics, and genetic engineering approach, 1st edn. Wiley, UK, pp 59–86. https://doi.org/10.1002/9781119906506.ch3
116. Ullrich SM, Ilyushchenko MA, Kamberov IM, Tanton TW (2007) Mercury contamination in the vicinity of a derelict chlor-alkali plant. Part I: sediment and water contamination of Lake Balkyldak and the River Irtysh. Sci Total Environ 381:1–16
117. Ullrich SM, Tanton TW, Abdrashitova SA (2001) Mercury in the aquatic environment: a review of factors affecting methylation (Review). Crit Rev Environ Sci Technol 31(3):241–293
118. UNEP (2019) Global mercury assessment 2018. Geneva. www.unep.org
119. UNEP/WHO (2008) Guidance for identifying populations at risk from mercury exposure. In: UNEP chemicals branch and WHO department of food safety, zoonoses and foodbourne diseases. Geneva, Switzerland, p 176
120. Varekamp JC, Buseck PR (1984) The speciation of mercury in hydrothermal systems, with applications to ore deposition. Geochim Cosmochim Acta 48:177–185
121. Varekamp JC, Buseck PR (1986) Global mercury flux from volcanic and geothermal sources. Appl Geochem 1:65–73
122. Vetriani C, Chew YS, Miller SM et al (2005) Mercury adaptation among bacteria from a deep sea hydrothermal vent. Appl Environ Microbiol 71:220–226
123. Wahsha M, Maleci L, Bini C (2019) The impact of former mining activity on soils and plants in the vicinity of an old mercury mine (Vallalta, Belluno, NE Italy). Geochem-Explor Environ Anal 19:171–175
124. Wang D, Shi X, Wei S (2003) Accumulation and transformation of atmospheric mercury in soil. Sci Total Environ 304:209–214
125. Wedepohl KH (1995) The composition of the continental crust. Geochim Cosmochim Acta 59:1217–1232
126. WHO (2017) Ten chemicals of major health concern. World Health Organization
127. Wilhelm SM (2001) Estimate of mercury emissions to the atmosphere from petroleum. Environ Sci Technol 35(24):4704–4710
128. Wright LP, Zhang L, Marsik FJ (2016) Overview of mercury dry deposition, litterfall, and throughfall studies. Atm Chem Phys 16(21):13399–13416
129. Wu G, Kang H, Zhang X, Shao H, Chu L, Ruan C (2010) A critical review on the bio-removal of hazardous heavy metals from contaminated soils: issues, progress, eco-environmental concerns and opportunities. J Hazard Mater 174:1–8
130. Wu Q, Wang S, Wang L, Liu F, Lin CJ, Zhang L et al (2014) Spatial distribution and accumulation of Hg in soil surrounding a Zn/Pb smelter. Sci Total Environ 496:668–677
131. Xiao R, Wang S, Li R, Wang JJ, Zhang Z (2017) Soil heavy metal contamination and health risks associated with artisanal gold mining in Tongguan, Shaanxi. China Ecotox Environ Safe 141:17–24
132. Yin D, He T, Yin R, Zeng L (2018) Effects of soil properties on production and bioaccumulation of methylmercury in rice paddies at a mercury mining area. China J Environ Sci 68:194–205
133. Zaferani S, Biester H (2021) Mercury accumulation in marine sediments–a comparison of an upwelling area and two large river mouths. Front Mar Sci 8:732720. https://doi.org/10.3389/fmars.2021.732720/full
134. Zettlizer M, Scholer HF, Eiden R, Falter R (1997) Distribution of elemental, inorganic and organic mercury in north german gas condensates and formation brines. In: Society of petroleum engineers ınternational symposium on oilfield chemistry. Houston, Texas

135. Zhao L, Qiu GL, Anderson CWN, Meng B, Wang DY, Shang LH, Yan H, Feng X (2016) Mercury methylation in rice paddies and its possible controlling factors in the Hg mining area, Guizhou province, Southwest China. Environ Pollut 215:1–9
136. Zheng N, Liu J, Wang Q, Liang Z (2011) Mercury contamination due to zinc smelting and chlor-alkali production in NE China. Appl Geochem 26(2):188–193
137. Zheng Y, Jensen AD, Windelin C, Jensen F (2012) Review of technologies for mercury removal from flue gas from cement production processes. Prog Energ Combust 38:599–629
138. Zhu W, Li Z, Li P, Yu B, Lin CJ, Sommar J et al (2018) Re-emission of legacy mercury from soil adjacent to closed point sources of Hg emission. Environ Pollut 242:718–727
139. Zolkos S, Krabbenhoft DP, Suslova A, Tank SE, McClelland JW, Spencer RGM, Shiklomanov A, Zhulidov AV, Gurtovaya T, Zimov N, Zimov S, Mutter EA, Kutny L, Amos E, Holmes RM (2020) Mercury export from Arctic great rivers. Environ Sci Technol 54(7):4140–4148

Mercury Contamination and It's Dynamics in Soil–Plant Systems

Mehak Maqsood, Noreen Zahra, Abida Kausar, Sobia Shahzad, Aaliya Batool, and Rubina Naseer

Abstract Mercury (Hg) is a naturally occurring heavy metal with wide distribution in the environment in various forms such as elemental, organic, and inorganic. Higher concentrations of mercury radiation, fluctuates from >6000 to 8000 Mg Hg/year, are regarded as a global contaminant which accumulates in the atmosphere because of anthropogenic and natural processes. The soil, water, and air resources are heavily contaminated by these toxic substances due to their high persistence and toxicity. Hg accumulation within plants magnifies the presence of harmful substances in the food web and has the potential to be genotoxic, neurotoxic, and oncogenic. However, higher Hg concentrations are also harmful to plants as it hampers growth-related attributes and impair their physiological and metabolic activities. Consequently, it promotes the risk to food security and the flow of toxicity from targeted sources towards untargeted sources. However, soil cleaning, solidification, thermal and biological treatments are the main topics of discussion to overcome the Hg mobilization in soil.

Keywords Hg contamination · Thermal treatment · Biological treatments · Plants · Physiology · Growth

M. Maqsood · N. Zahra (✉) · A. Kausar
Department of Botany, Government College Women University Faisalabad, Faisalabad 38000, Pakistan
e-mail: noreenzahra59@gmail.com

N. Zahra · A. Batool
Department of Botany, University of Agriculture Faisalabad, Faisalabad 38000, Pakistan

S. Shahzad
Department of Botany, Islamia University of Bahawalpur, Bahawalnagar Campus, Bahawalpur 62300, Pakistan

R. Naseer
National Institute for Biotechnology and Genetic Engineering, Faisalabad 38000, Pakistan

N. Kumar (ed.), *Mercury Toxicity Mitigation: Sustainable Nexus Approach*, Earth and Environmental Sciences Library, https://doi.org/10.1007/978-3-031-48817-7_2

1 Introduction

Global contaminant mercury (Hg) is highly toxic and has injurious effects on the health of all living biota [40, 88]. Hg is an extremely volatile element having atomic number 80, has a distinctive quality of liquid and metal at ambient temperature. It is silvery-white, sparkling having slight bluish tint appearance [9]. Hg when in gaseous form, travels across great distances in the atmosphere before settling into the soil. In 2015, 220 tonnes of mercury were added to the atmosphere because of human activity, according to UNEP (United Nations Environment Programme). Several processes like Industrial waste, burning of fossil fuels, mining operations, mineral draw out, the use of insecticides and fertilizers comprising mercury, and waste discharge are the main anthropogenic causes of Hg transmission [101]. Moreover, coal-fired power stations alone contributes 65% of the world's mercury emissions [114].

Hg discharge even at low concentrations is extremely persistent, neurotoxic, and bio-accumulative in the food chain, and can cause permanent failure of central nervous systems in humans and mammals [19]. Hg is therefore one of the "top ten chemicals of concern," and Hg contamination in the earth's ecosystem has become a popular research subject. The natural occurrence of Hg emissions, movement, conversion, and installation in the environment is known as the global Hg cycle [100]. Nevertheless, since the Industrial Rebellion, human practices, particularly, have dramatically boosted the construction and usage of mercury [63] and regenerated the least active mercury from the environment to take part in the mercury cycle around the world [9], leading to altered biogeochemical procedures and ultimately global Hg allocation in the atmosphere [77].

With 250–1000 Gg, soil is the greatest mercury reservoir in the world's ecosystem, while human activities including the manufacturing of cement, gold mining, coal burning, and chloralkali processing contribute more than 86 Gg of that amount [110]. It is obvious that human activity contributes significantly to global Hg buildup and modifies the global Hg cycle pattern through influencing Hg biogeochemical processes in soil [78]. More over 3 billion people worldwide rely on the soil-rice distribution system as their primary food supply [53] so, it is regarded as the largest artificial wetland ecosystem in the world displaying characteristics of the soil–water interface [92]. Rice plants have become essential to the global Hg cycle and serve as a crucial link between the atmospheric and soil Hg flow [47]. Recent studies have demonstrated that rice produced in Hg-polluted areas may contribute significantly to global methylmercury (MeHg) contact [93], especially in Asia and Africa. The danger of human MeHg contact in Central Asia, Europe, and Africa is further raised as a result of the global rice trade [62].

Multiple phenomena, including global warming and human activities, can affect the global HgO greenery sink, deforestation, CO_2 fertilization, vegetation (biome) alterations, water deficit, storms of ice, or occurrence of fire are some examples of environmental factors. The rainforest of Amazon contributed 29% of the overall HgO terrestrial basin. As a result of continuous deforestation and global warming, the stability of the tropical forest and, consequently, its value as a substantial Hg basin, is in danger. The GEOS-Chem model estimated that 400 Mg of HgO wouldn't be stored in amazon soils yearly, but rather would be primarily deposited to oceans under an extreme negative scenario when the amazon is totally transformed into savannah.

Deciduous leaves showed 3.2 times advanced application than coniferous leaves in terms of foliar stomatal HgO application and in case of foliar HgO, the availability of nutrients (leaf nitrogen) and moisture generally increased the absorption. The GEM-MACH-Hg model's simulation of global vegetation HgO uptake (1700–2100 Mg $year^{-1}$) evaluates yearly HgO arid installation changes to the world's main biomes which ranges from temperate broadleaf/mixed forests (18 lg m^{-2} $year^{-1}$) to hot broadleaf (26 lg m^{-2} $year^{-1}$) (16 lg m^{-2} $year^{-1}$) to hot grasslands conifers that are temperate (14 lg m^{-2} $year^{-1}$) to tundra (4 lg m^{-2} $year^{-1}$) from temperate grasslands (9 lg m^{-2} $year^{-1}$) [118].

A rise in net land primary yield and a projected overall rise in vegetative HgO consumption of 140 Mg $year^{-1}$ have resulted from the continuous northward shift in world biomes since 1990 [47]. According to a current modelling research [29], an international reforestation initiative may further support Hg repossession in soils and lower Hg flows to the ocean by 98 Mg $year^{-1}$.

2 Mercury Producing Sources

2.1 Regular Sources

Typical contextual levels of mercury (Hg) in various categories of soils from around the domain range from 0.58 to 1.8 mg/kg, with the global nasty level projected to be 1.1 mg/kg. Histosols and Cambisols have been found to have higher Hg contents [48]. Hg is emitted naturally from a wide range of sources through a number of different processes. This includes Hg released during sporadic occurrences like volcanic eruptions, geothermal activity, or the constant Earth's crust's Hg-containing rocks weathering [6].

According to [69], current Hg emanations from natural sources are expected to be released between 80 and 600 t/year into the atmosphere. Depending on various elements, such as the existence of volcanic straps, the intensity of geothermal activity, rock formations, and the amount of natural wildfires, these contributions vary throughout time and space [30, 31, 83]. Mercury travels a great distance in the atmosphere before chronic to the earth's surface by damp and arid deposition. As a result, the terrestrial environment absorbs more than 90% of the emitted Hg, along with soils become the main receiver [61].

2.2 Anthropogenic Sources

According to AMAP/UNEP [6], present anthropogenic bases, which contain a number of industrial point causes, are expected to discharge about 1960 t of mercury annually. According to [69], the main industries mentioned include cement manufacture, artisanal and small-scale gold withdrawal (ASGM), coal burning, non-ferrous metal manufacture (including large-scale gold production, copper, lead, zinc, and aluminium), non-ferrous metal production, and left-over disposal that involves mercury (Hg). The fresh information on the usage about Hg in ASGM in some places is mostly to blame for the emissions related with ASGM activities being much higher than previously reported. Even though there are significant regional variations, cement manufacturing is another significant source of Hg and has improved by over 30% between 2005 and 2009 [104].

2.3 Re-Additions

In the framework of the global Hg cycle, re-emanation is described as Hg emanations that originate from earlier natural and anthropogenic sums. Hg that has already been deposited on the Earth's surface might be able to be released into the atmosphere once more under the correct conditions by a new transport mechanism. Mercury re-emissions are now estimated to be between 4000 and 6300 t/year [6, 69], and a sizeable portion of this mercury may eventually accumulate in surface soils.

3 Implications of Climate Change on the Biogeochemical Cycling of Mercury

The Intergovernmental Panel on Climate Change (IPCC) claims that human activity has contributed to global warming [46]. There are numerous weather and climatic extremes, such as heat waves, a lot of precipitation, droughts, and tropical cyclones, in every region of the world that posed serious impact on the mercury cycle. These adjustments are marine ecosystems, biota, and eventually human experiences are all impacted by Hg levels, however, the extent and direction of the effects varies. It is now challenging to assess the collective impact of fluctuations in climate. For instance, more current global-scale weather change models have not been the foundation for widely used Hg emissions projections. As a result, future estimates of anthropogenic Hg emissions and estimates of climate change cannot be directly compared [95].

However, the movement of pollution from significant emissions source locations would be impacted by climate change. For arctic amplification, or the Arctic is being warmed excessively, affects the transfer of pollution poleward by changing mid-latitude circulation patterns [17, 21]. Due to decreases in vertical transport and deep convection-based dispersion, transport changes may also result in higher levels of superficial pollution in the tropics in the absence of tougher mitigation efforts [24]. The amount of Hg emissions, circulation patterns, and type of land shield are only a few of the activities that can vary and affect Hg installation, the mechanism through which Hg arrives ecosystems.

The annual mean rainfall flow continues to grow across most of the continent and to decrease spanning the majority of the tropical and mid-latitude waters. Changes in plant type and density lead to an increase in HgO dry deposition velocity, which causes variations in land use and land shelter to have an overall rise in HgO dry deposition flux with notable geographic heterogeneity. Due to the combined properties of anticipated variations in land use, land shelter, and temperature, Hg testimony to the continental surface increased.

4 Mechanism of Hg Toxicity

It was long believed that mercury toxicity resulted in neuropathological difficulties; however, the same mechanism that produces any other heavy metal's harmful potential also produces mercury's deadly potential: intake, supply, metabolism, and elimination [88]. Because of its unusual properties, mercury (HgO) is a fluid at room temperature. Due to its low vapour pressure, it rapidly evaporates and is easily inhaled by the lungs. It travels through the lungs from the urinary tract to the nervous system.

Bacteria give rise to it, and it commonly travels up the food chain in water. It bioaccumulates in fish and other long-lived predators' tissues in aquatic habitats. Human gastrointestinal system absorbs 95% of MeHg when consumed orally. As glutathione and selenium do, it is supposed to attach the cell proteins, amino acid carriers, and sulhydryl (–SH) chains within the blood. About 20% of the MeHg soluble in water in the brain was discovered to be contained in a MeHg-glutathione combination, and 10% of the load is in the skull [19].

5 Dynamics in Soil–Plant System

5.1 Thresholds for Mercury in Soils

The designation of mercury as a hazard is not generally accepted in the European Union. The accessibility and mobility of mercury differ according to the soil's characteristics, such as its organic content of carbon, pH, texture, and environmental factors, so instructions and thresholds have been developed as serves of these soil properties, which vary from country to country [15]. According to [15], the "Predicted No Effect Concentrations (PNEC)" figure established by the "Registration, Evaluation, The Authorization and Limitation of Chemicals (REACH)" regulation is roughly 22 g kg in Germany. The most common norms and mercury threshold values were provided by the statutes of Finland and Sweden regarding soil contamination.

5.2 Plant Toxicity of Mercury (Hg) and Its Occurrence in the Food Chain

Hg damages both plants and animals and contaminates the environment even in little amounts. This metal arises into direct interaction with the plant system since it is widely recycled as seed disinfectants and herbicides [35]. Mercury (Hg) attaches to thiol (–SH) groups in cellular structures to form HgSH, which disrupts the protein structure. Mercury (Hg) slows down plant growth and development by moving essential elements, primarily cations, through changes to the protein structure. Changes in cell membrane permeability are indicative of the deleterious properties of Hg on plant absorption, which include decreased photosynthesis, decreased water acceptance, increased transpiration, and increased peroxidation of lipid. Advanced concentrations of Hg decrease the antioxidant enzyme activity of several enzymes, such as superoxide dismutase (SOD), ascorbate peroxidase (APX), and peroxidase (POD). The majority of mercury is absorbed into plants in the organic form (Hg^{2+} oxidation state), which damages grain seeds and causes abnormal seedling germination as well as radicle and coleoptile hypertrophy [41].

As mercury substitutes the magnesium atom that makes up the core of chlorophyll, it damages the light-harvesting complex and interferes with light and dark both cycles of photosynthesis. Besides inhibiting plant development, mercury also results in necrotic and chloric symptoms in plants [113]. Other research done by [71] revealed that *Vigna radiata* plants exposed to advanced Hg applications had numerous ultrastructural defects.

a. ***Effect on rice plants***

The worldwide Hg cycle has become largely dependent on rice plants, which serve as a crucial link between the soil and atmospheric Hg circulation [47]. According to recent research, particularly in Asia and Africa, rice grown in Hg-contaminated areas may have a significant impact on the amount of methylmercury (MeHg) that humans are exposed to globally [93]. The risk of spreading human MeHg exposure between different countries is also increased as a result of the global rice trade. These regions include Central Asia, Europe, and Africa. Despite not being the most significant human exposure in the majority of countries, in the region of Central Asia, Rice consumption-related MeHg absorption accounted for 98% of the pathway where rice is the predominant dietary staple grain. It is clear that Hg trash in the soil-rice system has the potential to negatively impact the environment and human health in addition to having a significant impact on the worldwide Hg cycle. The objective of this review is to establish a scientific basis for the use of organic amendments in rice fields to reduce Hg contamination, improve soil quality, and guarantee food security [62].

b. **Effect on soil microbial activity of plants**

In accordance with [7, 14], the foremost effect of mercury (Hg) on plants is a reduction in growing rate resultantly its impact on the metabolism of plants through decrease in photosynthesis and respiration rate, drops in the absorption of nutrients and water, and amendments in antioxidant enzyme activity.

The makeup and activity of the soil microbial community can be directly impacted by Hg [32] classified the soil's basal respiratory (SBR) bioindicator as highly sensitive for recognizing reactions in the microbial community due to the strong influence of Hg, which is obvious in the decline in SBR. Furthermore, the adaptability of the microbiota following exposure to low levels of Hg contamination was highlighted by [33, 34, 64].

c. **14-Species pre-trial**

In our 14-species pre-trial, oat and common beans were the most vulnerable species, while lucerne and carrot may be more prone. In experimental settings, [97] discovered that Hg had detrimental effects on the common bean species. In their experiment, they found that when plants were subjected to higher levels of mercury (0.001, 0.04, 0.1, and 1 mmol L1) in the Hoagland nutritional solution, Hg hindered germination and lowered physiological markers including protein and chlorophyll.

The concentration of soil clay, the kind and variety of organic matter, pH, the number and shape of the micro populace, sulphur binders, and the potential for redox reactions are all frequently connected with the retention of mercury in soils. Additionally, certain organic acid discharges produced by particular plant species can have an impact on the retention of Hg. Because they choose and favour the organisms that will take part in the biogenic procedure for Hg transformation, these acids are particularly sensitive with metals in soils. The research of [50, 66], the exudates produced could also acidify the rhizosphere, which can impact Hg absorption by plants and uptake of Hg by soil colloids. In fact, in a comparable soil type, ordinary beans had a greater impact on Mercury levels than oats.

When soil mercury background values are high, crops' mercury content increases [70]. According to pertinent studies, under the influence of soil mercury pollution, cabbage and rice had mercury contents that were noticeably higher than the 0.01 mg kg^{-1} limit for vegetables set by the Chinese Food Safety Standard Contaminant Limits in Food [59, 84]. Additionally, Hg^{2+} in soil is easily changed into the more dangerous form of methylmercury [112]. As a result of digestion, migration, transformation, etc., methylmercury builds up in plants, and is continuously supplemented in the food chain (up to 10,000 times) before being consumed by humans and causing chronic poisoning [68]. Notably, despite being a non-essential element and only being present in trace amounts in plant cells, mercury is among the most harmful heavy metals [9]. According to numerous research [2, 119], heavy metals cause phytotoxicity, which inhibits plant growth, alters how vital nutrients are absorbed and stored by plants, and has a detrimental effect on fruit quality and organ development. Therefore, reducing mercury's phytotoxicity while enhancing the quality and safety of cultivated products is a priority for the entire human race.

In addition to being very medicinal ginger (Zingiber officinale Rosc.), a significant table condiment, is also rich in culinary value [3]. It is placed among the top 20 herbal increments in the United States by [11]. In China, solanaceous vegetables, green vegetables, and vegetables with rhizomes had the lowest average mercury concentrations [116].

d. ***Effect on ginger***

Naturally, ginger is a rhizome vegetable that readily absorbs mercury. By reason of the lengthy growth age of ginger, heavy fertilizer and pesticide applications are frequently made to improve yield while planting, which increases the danger of mercury leaching into edible rhizomes from the soil. Rare researches have examined the lenience, transportation, and improvement properties of mercury in vegetables, even though numerous studies have reported on the risks of mercury accumulation and toxicity in plants [54]. Mercury toxicity reduction and risk evaluation for ginger have not yet been documented. According to numerous research [20, 65], silicon has a variety of beneficial impacts on plant growth. It can also help plants cope with the pressure of harmful heavy metals [52]. On ginger, the impact of silicon on mercury increase and dispersal is still unknown. This study investigates whether mercury contamination offers a substantial risk to human health by first examining how and why mercury distresses the yield, growth and quality of ginger as well as the mechanism by which silicon reduces mercury toxicity on ginger. Furthermore, we have made a ground-breaking discovery that Grandson-ginger that is freshly cultivated has a smaller amount of mercury residue. This information can be used as a scientific basis for the development and execution of an agrarian superiority and safety assessment system. The amount of mercury that accumulates in the lower and upper parts of ginger is significantly influenced by the quantity of mercury in the soil. Additionally, there are observable variations in mercury enrichment in various organs, which might be connected to variations in plant characteristics and transport pathways [37].

Only a small amount of the soil-based mercurial could get transferred uphill into the leaves and stems of the ginger plant, where it would be amplified. This took place while the ginger leaf had apparently assimilated the soil's mercury. The majority of investigations regarding additional plants' results [26, 87] agreed that the accumulation of mercury in several organs increased as time passed, regardless of the ground. The exceptionally high ability to improve mercury in the rhizome considerably increased the risk of mercury exposure, especially since it's a product organ.

5.3 Hg Impact on a Plant's Transpiration

Hg harmful effect on plants may result from the inhibition of enzymes caused by its coupling to sulphydryl groups [30]. The sprouting and growing of embryos are stuck by such a failure of the normal -SH system since these tissues are chiefly rich in SH-groups [80]. According to [30], the reduction of biomass, photosynthetic activity, total chlorophyll, and the concentrations of nitrogen, phosphorus, and potassium are the main consequences of Hg toxicity on marine and worldly plants. In contrast, malondialdehyde levels, peroxidase activity, and the initiation of thiol-improved compounds are all rised in response to Hg contact [4, 39, 72].

According to [80], this effect may be related to disturbances in the plasma membrane shape and function, which change how permeable it is to water and nutrient intake. *B. juncea* may have had less difficulty absorbing water as a result of $HgCl_2$ exposure to cell membranes.

5.4 Root System Retention of Mercury

The majority of the Hg absorbed by *Triticum durum* roots after 25 days of experience was located on cell walls at the exterior layers of the cortical chamber, according to experiments using 203 Hg-spiked solutions. The mobility of soluble Hg multiplexes in the apoplast, a consistent system that encompasses very cell walls in the root cortex, may be related to the buildup of root Hg in *B. juncea* plants [94]. Hg tolerance in plants is knowingly unfair by the obligatory of Hg to particular places onto the cell wall, which force limit conveyance to sensitive areas in shoots [16].

Simple water mass flow and diffusion would have carried Hg to the plant branches. Hg may have been taken up and transported upward along the symplast pathway through the plasma membrane. Given that a variety of metals, including Hg, can stimulate the production of phytochelatin [86], Hg most likely moved through the symplast in the form of Hg-thiol complexes.

6 Legacy Hg Pools

The legacy Hg from watershed soils has likely been slowly and continuously mobilized, which has led to the ongoing increased Hg intake from watersheds. This legacy Hg might be prone to further rises or declines in retort to land-use alteration or a drizzlier/dryer environment. By mismanaging land use, such as through deforestation or farming, nutrients, polluted soil pools, and soil Hg can all be mobilized down to streams, wetlands, and coastal waters, which can worsen soil deterioration. Keeping an eye on the Arctic, we can observe a confluence of rapid climate change and high legacy Hg concentrations [5].

High levels of the mercury have built up in Artic soils for an extent of 0–30 cm; [60, 96]. Although the thicker surface organic matter levels are predominantly the consequence of pollution from the atmosphere after the 1960s, deep mineral and frozen soils contain mercury that has collected through centuries. It is essential to comprehend the extent of the frozen Hg reservoir in order to forecast its outcome [79].

6.1 Technology to Clean up Mercury-Contaminated Soil

The degradation of trace elements differs from that of organic pollutants. Instead, the element will be relocated from one location (such as a hazardous site) to another (such as a landfill). Alternately, they can be stabilized in place and immobilized [56, 74]. Withdrawal is typically done to remove mercury from the soil, lessen the amount of polluted soil, or subordinate the absorption of bioavailable mercury in soil to tolerable planes. In contrast, immobilization relies on encapsulating and stabilizing mercury in the soil to prevent adverse possessions on people or the other living things [22, 82, 109].

6.2 Cleaning the Soil

In order to lower Hg contents in soil, soil wash is essentially a physical separation (PS) technique. The idea behind the approach is that most pollutants have a tendency to adhere to particles [98, 103]. In order to isolate particles in soil washing, procedures like hydro classification, gravity concentration, attrition polishing, and fizz flotation are frequently used [107]. While the Hg linked to the adequate particles is focused for future handling, the coarse-grained soil is left quite unpolluted and doesn't need any more work. The washing fluid is either recycled in the procedure or discarded [10, 36]. Physical separation enables recovering of the cleaned soil, and the procedure normally takes only a short to medium amount of time [23]. This approach offers the following benefits, the amount of soil that needs to be extra processed is significantly condensed, a number of completely portable units are accessible for on-site treatment, the remediation systems are easily modular, the technologies are well-established and operational costs are often minimal. Though, Physical Separation is challenging or impractical in the next circumstances; when there are substantial concentrations of insoluble humic compounds or clay minerals, which cause Hg to be tightly bonded to soil particles [1], the polluted soil contains Hg in whole fractions of particle size [111], when the soils have a silt/clay percentage more than 30–50% and when Hg-behavior and clean particles do not differ significantly in density or surface characteristics. Chemical extraction (CE), a type of soil washing in which Hg is removed from soil using chemicals, can be utilized in conjunction with PS. Hg is solubilized during chemical extraction using alkaline substances or chelating substances are examples of solvents.

While acids and alkalis are often used to break down Hg molecules or/and elements in soil that collect Hg, chelating substances transport Hg by forming soluble Hg complexes [22]. A combination of solids and liquids is separated using hydrocyclones, and the particles are subsequently subjected to a water-rinse process to remove contaminants and reagents. To remove Hg, commercially available precipitants like lime, sodium hydroxide, or other combinations are combined with the removal fluid and washing water. The precipitated solids can require further processing or landfill disposal. One of the few methods that can completely remove mercury from soil, its performance can surpass 99% [27, 28, 102], Universal [105, 106].

In general, the main benefit of CE over PS is the ability to remove Hg that has been absorbed into the soil but is not water soluble [23]. However, the widespread adoption of CE is hampered by significant obstacles. The employment of chemical agents greatly raises dispensation costs, and recycle and water treatment are made more difficult by the presence of specific compounds in cleaning fluid. It might not be appropriate to dump the treated soil directly on site or to replant it [38]. Higher soil organic matter concentrations may powerfully hold Hg in the soil and impede removal [12]. Greater than 50% (of the dry weight of the soil) clay/silt content may necessitate longer contact times, decreasing efficiency [36]. Great quantities of the cations of important elements, such Fe and Ca, may prevent Hg from being chelated [27]. High soil heterogeneity influences extracting fluid compositions and may call for several processing steps [27].

Chemical leaching or physical separation are two ways that soil washing can perpetually eliminate pollutants from the soil [108]. The dangers of contaminant transfer during off-site treatment can be avoided by in situ soil washing, but one must ensure that complete leachate is improved and not seeped out into nearby soil and groundwater [43]. Though water only can be used as the washed solution, chemicals like HCl, HNO_3, EDTA, and $Na_2S_2O_3$ are frequently additional to increase the Hg removal efficiency during the washing process [91]. Han et al. [42] investigated the leak behaviour of sodium thiosulfate-treated Hg-tainted soil and discovered that, according to BCR analysis, more than 90% of the mercury that was weakly reducible and acid soluble could be recovered. Typically, sandy or silty soils with high levels of permeability are good candidates for soil washing. However, the drawbacks includes; unsuitable soils with low permeability (like clay). high water usage during washing and since, the extracted mercury gets into the washing solution, the water must be treated before being discharged.

6.3 Stabilization

The method of stabilizing mercury involves transforming it into chemical forms that are highly insoluble and stable over a variety of pH and redox circumstances in soil [103, 117]. For ex situ, the required excavation raises the cost even more, but for in situ, the verification of the process effectiveness becomes crucial.

6.4 Interventions

By creating mobile Hg species, some ions and soluble organic components in soil might hinder Hg stabilization. It has been claimed that using activated carbon (AC) powder as a technique to address this issue would allow for the simultaneous capture of mercury and a reduction in the interference from dissolved organics.

Furthermore, it has been proposed that pre-treating AC with sulphide impregnation will improve its capacity to absorb Hg [57]. In an experiment conducted by [117], the concentrations of 2273 mg/kg mercury and 5940 mg/L chloride were used. Hg was first stabilized using an inexpensive AC powder that had been sulphide-impregnated, and then solidified using normal Portland cement. As shown by toxicity-specific leaching methods and regular pH leaching trials, the approach effectively immobile Hg in a cement matrix. Despite being attacked by a great deal of chloride, the matrix only leached a little amount of mercury.

6.5 Electrokinetic Method

Electrokinetic remediation uses electrodes in the polluted soil as a physical way of removing mercury [25]. In the existence of an electric field, the cations will travel towards the cathode, whereas the anions would flow towards the anode [108]. The ability of dissolving mercury can be increased in soil, where it is somewhat less soluble, to optimize the efficacy of removal [90]. When the Hg-EDTA complex was present, the mercury started to flow in the direction of the anode [73]. The mixture of mercury while iodine/iodide (HgI4 2) gets stronger when HgS is oxidized to remove mercury, and the negative combination flows in the direction of the anode. As the amount of iodide diminishes, the positive isotope of mercury moves to the cathode as a result of the stability of the mercury aggregates [99]. Mercury can also be eliminated by mixing nitric acid with it [55].

The mercury and chloride mixture considerably enhances mercury removal and its movement to the anode [67]. According to Kabata-[49], one of the most adaptable agents for complexities mercury is thought to be the chloride. Mercury removal is made simpler by the electrokinetic approach [115] and can be utilized to break down contaminants in soils [45]. Mercury was successfully removed from the soil using the electrokinetic approach by the bacteria *Lysinibacillus fusiformis* [8].

6.6 Vitrification

According to [58], vitrifying is one of the most promising methods for treating polluted soil and solid waste from industries. In this process, polluted dirt is heated to the point of melting, followed by cooling to form a vitreous material, which has the appearance of glass. The vitreous bulk, where the contaminants are immobilized [13]. This technique is used to calculate the Hg contamination of soil [85]. Mercury is further removed from mining waste by means of vitrifying [75]. There are two sorts of vitrifying processes: (a) In situ vitrifying, according to [51], which transforms buried waste and polluted soil into highly durable glass and harmless debris crystal at where they originally were. Ex situ vitrifying technique is intended to deactivate or incapacitate contaminants far from contaminated landfills for trash or soil [76], although it is less expensive and environmentally detrimental than ex situ approaches [81].

6.7 Thermal Treatment

Hg becomes more volatile as the temperature rises, so contaminated soil is thermally treated to extract the metal. After combustion, mercury vapors turn into mercury liquid [89, 107]. This efficient method is then used to extract Hg from the soil's matrix [44]. In order to detoxify the Hg, $FeCl_3$ is used in a thermal treatment process while maintaining the soil original characteristics. The combination of these two also adds to the rate of combustion. Other tactics, using coal to eliminate mercury by a light thermal treatment, are also workable [18].

7 Conclusion

Mercury (Hg) is a heavy metal that is naturally present in the environment and can be found in a variety of forms, comprising elemental, organic, or inorganic forms. It is a widespread pollutant, both human activities and natural processes have built up its toxicity in the atmosphere, which caused severe consequences to ecosystems and public health. The issue again becomes worse when Hg accumulates in plant tissues and poisons the food chain poses a serious hallmark on the health of living biota and causes genotoxic, neurotoxic, and carcinogenic impacts. Higher concentrations of Hg can harm plants as well, interferes with their physiological and metabolic processes. Cleaning services, solidification, thermal treatment, and biological treatments are the effective techniques used to overcome soil contamination. Elimination of Hg pollution is essential for protecting both the environment and the well-being of humans. To reduce mercury emissions and to safeguard the food chain and human health from the harmful impacts of this hazardous metal, effective measures for its remediation requires an integrated strategies that includes scientific research, policy creation, and public awareness.

References

1. Abumaizar RJ, Smith EH (1999) Heavy metal contaminants removal by soil washing. J Hazard Mater 70(1–2):71–86
2. Adrees M, Ali S, Rizwan M, Ibrahim M, Abbas F, Farid M, Zia-ur-Rehman M, Irshad MK, Bharwana SA (2015) The effect of excess copper on growth and important food crops: a review. Environ Sci Pollut Control Ser 22(11):8148–8162
3. Ali BH, Blunden G, Tanira MO, Nemmar A (2008) Some phytochemical, pharmacological and toxicological properties of ginger (Zingiber officinale Roscoe): a review of recent research. Food Chem Toxicol 46(2):409–420
4. Ali MB, Vajpayee P, Tripathi RD, Rai UN, Kumar A, Singh N, Behl HM, Singh SP (2000) Mercury bioaccumulation induces oxidative stress and toxicity to submerged macrophyte Potamogeton crispus L. Bull Environ Cont Toxicol 65:573–582
5. AMAP (2021b) AMAP Assessment 2021b: Mercury in the Arctic. Tromso: AMAP
6. AMAP/UNEP (2013) Technical background report for the global mercury assessment 2013. Arctic Monitoring and Assessment Programme, Oslo, Norway/UNEP Chemicals Branch, Geneva, Switzerland
7. Azevedo R, Rodriguez E (2012) Phytotoxicity of mercury in plants: a review. Aust J Bot 2012:1–6
8. Azhar ATS, Nabila ATA, Nurshuhaila MS, Zaidi E, Azim MAM, Farhana SMS (2016) Assessment and comparison of electrokinetic and electrokineticbioremediation techniques for mercury contaminated soil. IOP Conf Ser Mater Sci Eng 160(1):12077
9. Beckers F, Rinklebe J (2017) Cycling of mercury in the environment: sources, fate, and human health implications: a review. Crit Rev Environ Sci Technol 47(9):693–794
10. BioGenesis Enterprises, Inc., Roy F. Weston, Inc. BioGenesisSM sediment washing technology. Full-scale, 40 Cy/Hr, sediment decontamination facility for the NY/NJ HarborFinal Report on the Pilot Demonstration Project; 1999
11. Blumenthal M (2011) Herb sales continue growth-up 3.3% in 2010. HerbalGram 90:64–67
12. Bollen A (2011) Mercury extraction from contaminated soils by L-cysteine: species dependency and transformation processes. Water Air Soil Pollut 219(1–4):175–189
13. Busto Y, Cabrera X, Tack FMG, Verloo MG (2011) Potential of thermal treatment for decontamination of mercury containing wastes from chlor-alkali industry. J Hazard Mater 186(1):114–118
14. Calgaroto NS, Castro GY, Cargnelutti D, Pereira LB, Gonçalves JF, Rossato LV, Antes FG, Dressler VL, Flores EMM, Schetinger MRC, Nicoloso FT (2010) Antioxidant system activation by mercury in Pfaffia glomerata plantlets. Biometals 23:295–305. https://doi.org/10.1007/s10534-009-9287-3
15. Carlon C (ed) (2007) Derivation methods of soil screening values in europe: a review of national procedures towards harmonisation: a report of the ENSURE action. Brussels, Belgium: EUR-306
16. Cavallini A, Natali L, Durante M, Maserti B (1999) Mercury uptake, distribution and DNA affinity in durum wheat (Triticum durum Desf) plants. Sci Tot Environ 233(234):119–127
17. Che´telat J, McKinney MA, Amyot M, Dastoor A, Douglas TA, Heimbu¨rger-Boavida L-E, Kirk J, Kahilainen KK et al (2022) Climate change and mercury in the Arctic: Abiotic interactions. Sci Total Environ 824:153715
18. Chmielniak T, Słowik K, Sajdak M (2017) Mercury removal by mild thermal treatment of coal. Fuel 195:290–298
19. Clarkson TW, Magos L (2006) The toxicology of mercury and its chemical compounds. Crit Rev Toxical 36(8):609–662
20. Coskun D, Deshmukh R, Sonah H, Menzies JG, Reynolds O, Ma JF, Kronzucker HJ, Bélanger RR (2019) The controversies of silicon's role in plant biology. New Phytol 221(1):67–85
21. Coumou D, Di Capua G, Vavrus S, Wang L, Wang S (2018) The influence of Arctic amplification on mid-latitude summer circulation. Nat Commun 9:2959

22. Dermont G (2008) Metal-contaminated soils: remediation practices and treatment technologies. Pract Period Hazard Toxic Radioact Waste Manag 12(3):188–209
23. Dermont G, Bergeron M, Mercier G, Richer-Lafl'eche M (2008) Metal-contaminated soils: remediation practices and treatment technologies. Pract Period Hazard Toxic Radioact Waste Manag 12(3):188–209
24. Doherty RM, Orbe C, Zeng G, Plummer DA, Prather MJ, Wild O, Lin M, Shindell DT et al (2017) Multi-model impacts of climate change on pollution transport from global emission source regions. Atmos Chem Phys 17:14219–14237
25. Essa MH, Mu'azu ND, Lukman S, Bukhari A (2015) Application of box-behnken design to hybrid electrokinetic-adsorption removal of mercury from contaminated saline-sodic clay soil. Soil Sediment Contam Int J 24(1):30–48
26. Esteban E, Moreno E, Penalosa J, Cabrero JI, Millán R, Zornoza P (2008) Short and long-term uptake of Hg in white lupin plants: kinetics and stress indicators. Environ Exp Bot 62(3):316–322
27. Federal remediation technologies roundtable (FRTR) (2001) Chemical extraction. Federal remediation technologies reference guide and screening manual, version 4.0
28. Federal Remediation Technologies Roundtable (FRTR) (1995) Technology cost and performance—soil washing at the King of Prussia Technical Corporation Superfund Site http://costperformance.org/profile.cfm?ID=125andCaseID=125
29. Feinberg A, Jiskra M, Borelli P, Biswakarma J, Selin NE (2023) Evaluating atmospheric mercury (Hg) uptake by vegetation in a chemistry-transport model. Earth
30. Ferrara R (1998) Atmospheric mercury sources in the Mt. Amiata area, Italy. Sci Total Environ 213(1–3):13–23
31. Ferrara R (2000) Volcanoes as emission sources of atmospheric mercury in the Mediterranean basin. Sci Total Environ 259(1–3):115–121
32. Frey B, Rieder SR (2013) Response of forest soil bacterial communities to mercury chloride application. Soil Biol Biochem 65:329–337. https://doi.org/10.1016/j.soilbio.2013.06.001
33. Frossard A, Donhauser J, Mestrot A, Gygax S, Bååth E, Frey B (2018) Long- and shortterm effects of mercury pollution on the soil microbiome. Soil Biol Biochem 120:191–199
34. Frossard A, Hartmann M, Frey B (2017) Tolerance of the forest soil microbiome to increasing mercury concentrations. Soil Biol Biochem 105:162–176
35. Gao S, Ou-yang C, Tang L, Zhu JQ, Xu Y, Wang SH, Chen F (2010) Growth and antioxidant responses in Jatropha curcas seedling exposed to mercury toxicity. J Hazard Mater 182(1–3):591–597
36. Griffiths RA (1995) Soil-washing technology and practice. J Hazard Mater 40(2):175–189
37. Gruba P, Błońska E, Lasota J (2014) Predicting the concentration of total mercury in mineral horizons of forest soils varying in organic matter and mineral fine fraction content. Water, Air, Soil Pollut 225(4):1924
38. Guo XX, Liu CQ, Zhu ZZ, Wang ZL, Li J (2011) Evaluation methods for soil heavy metals contamination: a review. Chin J Ecol 30(5):889–899
39. Gupta M, Tripathi RD, Rai UN, Chandra P (1998) Role of glutathione and phytochelatin in Hydrilla verticillata (i.f.) Royle and Vallisneria spiralis L. under mercury stress. Chemosphere 37(4):785–800
40. Gustin MS, Bank MS, Bishop K, Bowman K, Branfireun B, Chételat J et al (2020) Mercury biogeochemical cycling: a synthesis of recent scientific advances. Sci Total Environ 737:139–619
41. Gworek B, Dmuchowski W, Baczewska-Dąbrowska AH (2020) Mercury in the terrestrial environment: a review. Environ Sci Eur 32(1):1–19
42. Han C, Wang H, Xie F, Wang W, Zhang T, Dreisinger D (2019) Feasibility study on the use of thiosulfate to remediate mercury-contaminated soil. Environ Technol 40:813–821
43. He F, Gao J, Pierce E, Strong PJ, Wang H, Liang L (2015) In situ remediation technologies for mercury-contaminated soil. Environ Sci Pollut Res 22:8124–8147
44. Huang YT, Hseu ZY, Hsi HC (2011) Influences of thermal decontamination on mercury removal, soil properties, and repartitioning of coexisting heavy metals. Chemosphere 84(9):1244–1249

45. Iman T, Azadeh N (2012) Soil electerokinetic remediation and its effects on soil microbial activity-a review. Afr J Microbiol Res 6(10):2233–2238
46. IPCC (2021) Summary for policymakers. In: Climate change 2021: the physical science basis. contribution of working group i to the sixth assessment report of the intergovernmental panel on climate change. IPCC
47. Jiskra M, Sonke JE, Obrist D, Bieser J, Ebinghaus R, Myhre CL, Pfaffhuber KA, Wängberg I, Kyllönen K, Worthy D, Martin LG (2018) A vegetation control on seasonal variations in global atmospheric mercury concentrations. Nat Geosci 11(4):244–250
48. Kabata-Pendias A (2010) Trace elements in soils and plants. CRC Press
49. kabata-Pendias A (2011) Trace elements in soils and plants. CRC Taylor Fr Group, London New York
50. Kabata-Pendias A, Szteke B (2015) Trace elements in abiotic and biotic environments. CRC Press, New York
51. Kaushik CP (2014) Indian program for vitrification of high level radioactive liquid waste. Procedia Mater Sci 7:16–22
52. Keller C, Rizwan M, Davidian JC, Pokrovsky OS, Bovet N, Chaurand P, Meunier JD (2015) Effect of silicon on wheat seedlings (Triticum turgidum L.) grown in hydroponics and exposed to 0–30 μM Cu. Planta 241(4):847–860
53. Khush GS (2005) What it will take to feed 5.0 billion rice consumers in 2030. Plant Mol Biol 59(1):1–6
54. Kim YO, Bae HJ, Cho E, Kang H (2017) Exogenous glutathione enhances mercury tolerance by inhibiting mercury entry into plant cells. Front Plant Sci 8:683
55. Kirkelund GM, Jensen PE, Villumsen A, Ottosen LM (2010) Test of electrodialytic upgrading of MSWI APC residue in pilot scale: focus on reduced metal and salt leaching. J Appl Electrochem 40(6):1049–1060
56. Kumpiene J, Lagerkvist A, Maurice C (2008) Stabilisation of As, Cr, Cu, Pb and Zn in soil using amendments—a review. Waste Manag 28(1):215–225
57. Kwon S (2000) Evaluation of two sulfur impregnation methods on activated carbon and bentonite for the production of elemental mercury sorbents. Environ Eng Sci 17(6):303
58. Li C, Zhou K, Qin W, Tian C, Qi M, Yan X, Han W (2019) A review on heavy metals contamination in soil: effects, sources, and remediation techniques. Soil Sediment Contam: Int J 28(4):380–394
59. Li P, Feng X, Qiu G, Shang L, Wang S (2008) Mercury exposure in the population from Wuchuan mercury mining area, Guizhou. China Sci Total Environ 395(2–3):72–79
60. Lim AG, Jiskra M, Sonke JE, Loiko SV, Kosykh N, Pokrovsky OS (2020) A revised northern soil Hg pool, based on western Siberia permafrost peat Hg and carbon observations. Biogeosciences 2020:1–35
61. Lindqvist O, Johansson K, Bringmark L, Timm B, Aastrup M, Andersson A et al (1991) Mercury in the Swedish environment—recent research on causes, consequences and corrective methods. Water Air Soil Pollut 55(1–2):xi–261
62. Liu M, Zhang Q, Cheng M, He Y, Chen L, Zhang H, Cao H, Shen H, Zhang W, Tao S, Wang X (2019) Rice life cycle-based global mercury biotransport and human methylmercury. Nat Commun 10(1):5164
63. Liu S, Wang X, Guo G, Yan Z (2021) Status and environmental management of soil mercury pollution in China: a review. J Environ Manag 277:111–442
64. Liu Y-R, Wang J-J, Zheng Y-M, Zhang L-M, He J-Z (2014) Patterns of bacterial diversity along a long-term mercury-contaminated gradient in the paddy soils. Microb Ecol 68:575–583
65. Ma JF, Yamaji N (2006) Silicon uptake and accumulation in higher plants. Trends Plant Sci 11(8):392–397
66. Marchi G, Guilherme LRG, Chang AC (2011) Plant availability of trace elements in sewage sludge-treated soils: methodology. Rev Bras Ciênc Solo 35:1453–1460
67. Marrugo Negrete JL, Lopez´ Barboza E (2013) Electrokinetic Remediation of MercuryContaminated Soil, from the Mine El Alacran-San Jorge River Basin, CordobaColombia. Revista Facultad de Ingeniería Universidad de Antioquia

68. Martín JAR, Nanos N (2016) Soil as an archive of coal-fired power plant mercury deposition. J Hazard Mater 308:131–138
69. Mason RP, Choi AL, Fitzgerald WF, Hammerschmidt CR, Lamborg CH, Soerensen AL et al (2012) Mercury biogeochemical cycling in the ocean and policy implications. Environ Res 119:101–117
70. Meng B, Feng X, Qiu G, Liang P, Li P, Chen C, Shang L (2011a) The process of methylmercury accumulation in rice (Oryza sativa L.). Environ Sci Technol 45(7):2711–2717
71. Mondal NK, Das C, Datta JK (2015) Effect of mercury on seedling growth, nodulation and ultrastructural deformation of Vigna radiata (L) Wilczek. Environ Monit Assess 187(5):1–14
72. Moreno-Jimenez E, Gamarra R, Carpena-Ruiz RO, Mill´an R, Pe´nalosa JM, Esteban E (2006) Mercury bioaccumulation and phytotoxicity in two wild plant species of Almaden area. Chemosphere 63:1969–1973
73. Mu'azu ND, Usman A, Jarrah N, Alagha O (2016) Pulsed electrokinetic removal of chromium, mercury and cadmium from contaminated mixed clay soils. Soil Sediment Contam Int J 25(7):757–775
74. Mukherjee AB, Zevenhoven R, Brodersen J, Hylander LD, Bhattacharya P (2004) Mercury in waste in the European Union: sources, disposal methods and risks. Resour Conserv Recycl 42(2):155–182
75. Navarro A, Cardellach E, Canadas I, Rodríguez J (2013) Solar thermal vitrification of mining contaminated soils. Int J Miner Process 119:65–74
76. Novaković V, Tomić A, Nikolić N, Petrović D (2018) Zagađenje I Zaˇstita Zemljiˇsta I Podzemnih Voda. Pollution And Protection of Soil and Groundwater (Novi Sad)
77. Obrist D, Agnan Y, Jiskra M, Olson CL, Colegrove D P, Hueber J et al (2017) Tundra uptake of atmospheric elemental mercury drives Arctic mercury pollution. Nature 547(7662):201–204
78. O'Connor D, Hou D, Ok YS, Mulder J, Duan L, Wu Q, Wang S, Tack FM, Rinklebe J (2019) Mercury speciation, transformation, and transportation in soils, atmospheric flux and implications for risk management: a critical review. Environ Int 126:747–761
79. Olson C, Jiskra M, Biester H, Chow J, Obrist D (2018) Mercury in active-layer tundra soils of Alaska: concentrations, pools, origins, and spatial distribution. Global Biogeochem Cycles 32:1058–1073
80. Patra M, Sharma A (2000) Mercury toxicity in plants. Bot Rev 66(3):422–479; Rauser WE (1995) Phytochelatins and related peptides. Plant Physiol 109:1141–1149
81. Pei SL, Chen TL, Pan SY, Yang YL, Sun ZH, Li YJ (2020) Addressing environmental sustainability of plasma vitrification technology for stabilization of municipal solid waste incineration fly ash. J Hazard Mater 398:122–959
82. Petruzzelli G, Pedron F, Rosellini I, Barbafieri M (2013) Phytoremediation towards the future: focus on bioavailable contaminants. In: Plant-based remediation processes. Springer, Berlin, pp 273–289
83. Pirrone N, Costa P, Pacyna JM, Ferrara R (2001) Mercury emissions to the atmosphere from natural and anthropogenic sources in the Mediterranean region. Atmos Environ 35(17):2997–3006
84. Qiu G, Feng X, Wang S, Shang L (2005) Mercury and methylmercury in riparian soil, sediments, mine-waste calcines, and moss from abandoned Hg mines in east Guizhou province, southwestern China. Appl Geochem 20(3):627–638
85. Randall P, Chattopadhyay S (2004) Advances in encapsulation technologies for the management of mercury-contaminated hazardous wastes. J Hazard Mater 114(1–3):211–223
86. Rauser WE (1995) Phytochelatins and related peptides. Plant Physiol 109:1141–1149
87. Rellán-Álvarez R, Ortega-Villasante C, Álvarez-Fernández A, Del Campo FF, Hernández LE (2006) Stress responses of Zea mays to cadmium and mercury. Plant Soil 279(1–2):41–50
88. Rice KM, Walker EM Jr, Wu M, Gillette C, Blough ER (2014) Environmental mercury and its toxic effects. J Prev Med Public Health 47(2):74
89. Richter RB, Flachberger H (2010) Soil washing and thermal desorption: reliable techniques for remediating materials contaminated with mercury. BHM Berg-und Hüttenm¨ annische Monatshefte 155(12):571–577

90. Robles I, Lozano MJ, Solís S, Hernandez G, Paz MV, Olvera MG, Bustos E (2015) Electrokinetic treatment of mercury-polluted soil facilitated by ethylenediaminetetraacetic acid coupled with a reactor with a permeable reactive barrier of iron to recover mercury (II) from water. Electrochim Acta 181:68–72
91. Rodriguez O, Padilla I, Tayibi H, Lopez-Delgado A (2012) Concerns on liquid mercury and mercury-containing wastes: a review of the treatment technologies for the safe storage. J Environ Manag 101:197–205
92. Rothenberg SE, Feng X (2012) Mercury cycling in a flooded rice paddy. J Geophys Res 117:G03003
93. Rothenberg SE, Windham-Myers L, Creswell JE (2014) Rice methylmercury exposure and mitigation: a comprehensive review. Environ Res 133:407–423
94. Salisbury FB, Ross CW (1992) Plant physiology, 4th edn. Wadsworth Publishing Company, California, USA, p 682
95. Schartup AT, Soerensen AL, Angot H, Bowman K, Selin NE (2022) What are the likely changes in mercury concentration in the Arctic atmosphere and ocean under future emissions scenarios? Sci Total Environ 836:155477
96. Schuster PF, Schaefer KM, Aiken GR, Antweiler RC, Dewild JF, Gryziec JD, Gusmeroli A, Hugelius G et al (2018) Permafrost stores a globally significant amount of mercury. Geophys Res Lett 45:1463–1471
97. Sharma J, Shrivastava S (2014) Physiological and morphological responses of Phaseolus vulgaris caused by mercury stress under lab conditions. Recent Adv Biol Med 01:7
98. Sierra C, Menéndez-Aguado JM, Afif E, Carrero M, Gallego JR (2011) Feasibility study on the use of soil washing to remediate the As–Hg contamination at an ancient mining and metallurgy area. J Hazard Mater 196:93–100
99. Suér P, Allard B (2003) Mercury transport and speciation during electrokinetic soil remediation. Water Air Soil Pollut 143(1):99–109
100. Sun R, Jiskra M, Amos HM, Zhang Y, Sunderiand EM, Sonke JE (2019) Modelling the mercury stable isotope distribution of Earth surface reservoirs: implications for global Hg cycling. Geochim Cosmochim Ac 246:156–173
101. Teng D, Mao K, Ali W, Xu G, Huang G, Niazi NK et al (2020) Describing the toxicity and sources and the remediation technologies for mercury-contaminated soil. RSC Adv 10(39):23221–23232
102. Twidwell LG, Thompson RJ (2001) Recovering and recycling Hg from chlor-alkali plant wastewater sludge. JOM 53(1):15–17
103. United States Environmental Protection Agency (US EPA) (2007) Treatment technologies for mercury in soil, waste, and water
104. United States Geological Survey (USGS) (2012) USGS minerals yearbook. Online at: http://minerals.usgs.gov/minerals/
105. Universal Dynamics (2004) The REMERC™ process. Online at: http://udl.com/systems/remerc_x.html
106. USEPA (1997a) Mercury study report to congress. EPA-452/R-97–004
107. Vik EA, Bardos P (2003) Remediation of contaminated land technology implementation in Europe: prepared by Working Group" Remediation Technologies" of the Concerted Action" Contaminated Land Rehabilitation Network for Environmental Technologies (CLARINET)
108. Wang J, Feng X, Anderson CW, Wang H, Zheng L, Hu T (2012) Implications of mercury speciation in thiosulfate treated plants. Environ Sci Technol 46(10):5361–5368
109. Wang J, Feng X, Anderson CW, Xing Y, Shang L (2012) Remediation of mercury contaminated sites—a review. J Hazard Mater 221:1–18
110. Wang X, Yuan W, Lin CJ, Zhang L, Zhang H, Feng X (2019) Climate and vegetation as primary drivers for global mercury storage in surface soil. Environ Sci Technol 53(18):10665–10675
111. Xu J, Kleja DB, Biester H, Lagerkvist A, Kumpiene J (2014) Influence of particle size distribution, organic carbon, pH and chlorides on washing of mercury contaminated soil. Chemosphere 109:99–105

112. Xu X, Meng B, Zhang C, Feng X, Gu C, Guo J, Bishop K, Xu Z, Zhang S, Qiu G (2017) The local impact of a coal-fired power plant on inorganic mercury and methylmercury distribution in rice (Oryza sativa L.). Environ Pollut 223:11–18
113. Xun Y, Feng L, Li Y, Dong H (2017) Mercury accumulation plant Cyrtomium macrophyllum and its potential for phytoremediation of mercury polluted sites. Chemosphere 189:161–170
114. Yang L, Zhang Y, Wang F, Luo Z, Guo S, Strähle U (2020) Toxicity of mercury: molecular evidence. Chemosphere 245:125–586
115. Yang Z, Zhang Z, Chai L, Wang Y, Liu Y, Xiao R (2016) Bioleaching remediation of heavy metal-contaminated soils using Burkholderia sp. Z-90. J Hazard Mater 301:145–152
116. Yu H, Li J, Luan Y (2018) Meta-analysis of soil mercury accumulation by vegetables. Sci Rep 8(1):1261
117. Zhang J, Bishop PL (2002) Stabilisation/solidification (S/S) of mercury-containing wastes using reactivated carbon and portland cement. J Hazard Mater 92(2):199–212
118. Zhou J, Obrist D (2021) Global mercury assimilation by vegetation. Environ Sci Technol 55:14245–14257
119. Zia-ur-Rehman M, Sabir M, Nadeem M (2015) Remediating cadmium-contaminated soils by growing grain crops using inorganic amendments. Soil Remediat. Plants: Prospect. Chall. 367–396

Mercury Toxicity and Health

Health Risks of Mercury

Ayesha Zafar, Sadia Javed, Nadia Akram, and Syed Ali Raza Naqvi

Abstract Mercury is a very toxic volatile chemical pollutant and non-essential metal in the human body. Mercury is widely distributed in the environment, present in natural products and various types of medicines. There are three forms of mercury, i.e., elemental (or metallic) mercury, inorganic mercury compounds, and organic mercury compounds. The organic form of mercury is most toxic, causing neurological and developmental damage. Inorganic mercury compounds are water soluble and on ingestion, can cause gastrointestinal symptoms. They are mainly accumulated in the kidneys and are the major reason for kidney damage. In contrast, human exposure to elemental mercury is mainly by inhalation, followed by rapid absorption and distribution in all major organs. The primary target organs of elemental mercury are the brain and kidney. Elemental mercury is lipid soluble and can cross the blood–brain barrier. It may also enter the brain from the nasal cavity through the olfactory pathway. Emission of mercury from volcanoes, industrial processes, artisanal small-scale gold mining (ASGM), fossil fuels combustions and biomass burning accumulates in soil and water, where it's ingested by plants and sea animals. This can cause the bioaccumulation and biomagnification of mercury in the food chain. This chapter discusses the mercury toxicity, health risks of mercury and their routes of exposure in the food chain.

Keywords Bioaccumulation · Biomagnification · Environmental exposure · Health risks · Neurological effects · Elemental

---

A. Zafar · S. Javed (✉)
Department of Biochemistry, Government College University, Faisalabad 38040, Pakistan
e-mail: sadiajaved@gcuf.edu.pk

N. Akram · S. A. R. Naqvi
Department of Chemistry, Government College University, Faisalabad 38040, Pakistan

N. Kumar (ed.), *Mercury Toxicity Mitigation: Sustainable Nexus Approach*, Earth and Environmental Sciences Library, https://doi.org/10.1007/978-3-031-48817-7_3

1 Introduction

Mercury, also known as Quicksilver, a chemical element that exists in a liquid state under normal condition. The detrimental impact of mercury on human health is a matter of concern due to it toxicity, sustainability and recyclability. Even in small doses, mercury and the other metals in the hazardous category are poisonous and harmful. Each year, 250,000 people with intellectual disabilities are caused by the heavy, neurotoxic chemical mercury, which can enter diverse ecosystems through a variety of sources [46, 47]. Three chemical forms of mercury i.e., Elementary mercury typically found in thermometer, fluorescent, light bulbs, can be inhaled as vapors causing the respiratory problem, kidney damage and neurological disorder. It has been reported by WHO that 84% of mercury exposure can be attributed to amalgamated teeth. Every time a tooth with amalgam is chewed and brushed, extremely small amounts of elemental mercury are emitted into the atmosphere, and we aren't even aware of it [72]. Inorganic mercury in the form of cinnabar and metacinnabar, mostly found in medicines. The majority of human exposure to methylmercury (third form) from eating fish happens in coastal fisheries around the world [17, 43] Methylmercury (MeHg), can cross the blood brain barrier leading to cognitive impairments, memory loss and difficulties with motor coordination. Various studies have consistently the use of mercury persists in cosmetic product across the different countries, such as Mexico, the united state, Africa and Asia. Mercury (Hg) is a very hazardous substance which is dangerous for both the environment and human health [42, 49]. Some examples of organic mercury compounds that are more harmful compared to organic ones include dimethylmercury, ethyl mercury and phenylmercury [45]. Methylmercury has a significant affinity for sulfhydryl (thiol, –SH) molecules. Hg disrupts neurons, obstructs neurotransmitter activity, and stops the body from producing proteins. Mercury's neurotoxic effects interfere with gene expression, cell signaling pathways, protein phosphorylation, calcium ion balance, and cell growth. The linking of methylmercury to Cys can be a mediator for the various negative effects of this metal, including its inhibition of enzyme activity. The world Health organization (WHO) and European medical agency (EMA) have established permissible limits for various heavy metals ions, spanning from ppt to ppm. As of June 1, 2020, the WHO website recently released information identifying arsenic (As), cadmium (Cd), lead (Pb) and mercury (Hg) as four of the ten major compounds that pose significant public health risks. There would be a leak of around 2000 tons of mercury into the environment each year, which has become a public concern. With 128 nations' signatures in October 2013, the Minamata Convention on Mercury is an international agreement that has been in effect since 2017. Its goal is to cut down on mercury emissions worldwide [40, 78, 80]. The hazardous substances have the potential to impair biological systems, alter biochemical processes, and have severe, even deadly, consequences. According to data, distinctive properties of mercury such as toxicity, stability, bioavailability and bioaccumulation has received the particular consideration in ecotoxicology.

2 Sources of Mercury Exposure

Despite numerous findings regarding the harmful health impacts of heavy metals, exposure to heavy metals is rising, especially in developing and underdeveloped nations. Construction and building materials can be significant threats to public health because they can release pollutants into the air [56] (Fig. 1).

2.1 Traditional Medicines and Cosmetics

Mercury (Hg) can be found in cosmetic products as an impurity or intentionally added as a preservative in the form of organic mercury compounds like Thiomersal and phenylmercury salts (commonly with borate). Contamination with mercury in cosmetics can occur during the manufacturing process or due to the use of inadequately cleaned raw materials. It is important to note that, except for volpar and Thiomersal, the presence of mercury in cosmetic products is completely restricted. Even trace amounts of mercury in these products can indicate contamination. Eyewear can only contain thiomersal and phenylmercuric salts, with a maximum concentration of 0.007% as measured in Cosmetics. In the United States, the use of Hg in preparations is forbidden, with the exception of eye medications, where its concentration must not exceed 65 parts per million (ppm). Moreover, it is important

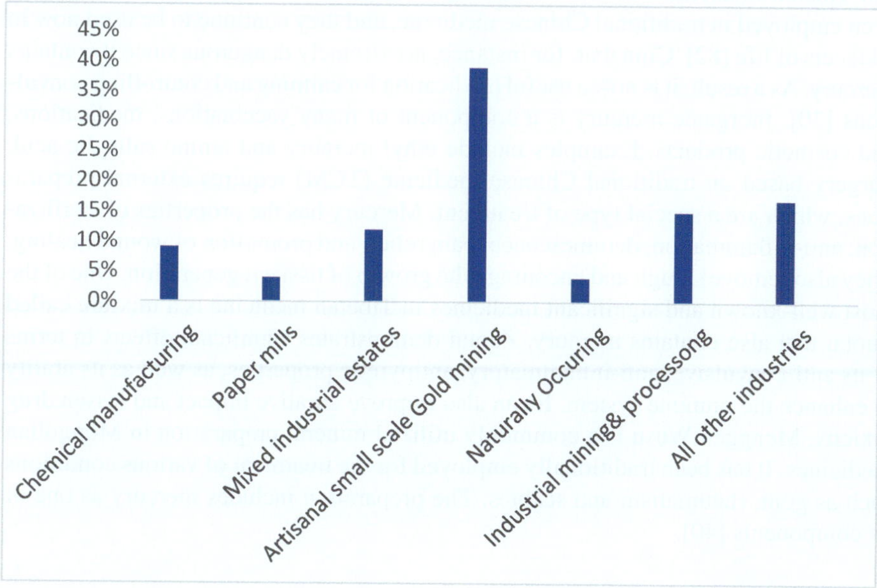

Fig. 1 Prevalence of mercury toxicity

to note that the presence of mercury in cosmetics products is strictly regulated. Hg is permitted in other cosmetics items, albeit in extremely low quantities of less than 1 ppm. However, this allowance is only applicable if it cannot be entirely eliminated from the products following the guidelines of good manufacturing practices (GMP). The highest concentration for creams from Jamaica was 17,547 mg/kg in one report [69]. In several countries including China, Japan, Sri Lanka, Taiwan, Thailand and united states, approximately 6% of the sample analyzed exhibited mercury concentrations exceeding 1000 mg/kg. In a Thai retail store, a remarkable product caught attention due to its significant value. This particular item happened to be a whitening and lifting cream, which had an impressive concentration level of 45,622 ± 322 mg/kg. In accordance with available research findings, it has been observed that skincare products utilized for skin lightening frequently contain notable quantities of mercury (Hg). Hg salts prevent the production of melanin, which lightens the skin tone. According to a study by [16], HgCl2 can directly inhibit tyrosinase, which explains how it works and why Hg is harmful [16, 62]. A guidance on preventing exposure to Hg from various sources, such as creams, has also been released by the United States Environmental Protection Agency (US EPA). These sources include creams that may bear the terms "anti-aging" or "skin lightening. "United States Environmental Protection Agency. Throughout the history of medicine and alchemy, various civilizations such as prehistoric Greece, India, Persia, Arabia and China extensively documented the utilization of mercury, specifically liquid metal mercury for medical purpose. Even in present times, mercury continues to hold a prominent position in traditional Asian medicines, including Ayurveda, Unani, Siddha, traditional Chinese medicine and Tibetan medicines. In the last 3,500 years ago, various types of mercury have been employed in traditional Chinese medicine, and they continue to be used now in all facets of life [82]. Cinnabar, for instance, is extremely dangerous since it contains mercury. As a result, it is now a useful medication for calming and controlling convulsions [30]. Inorganic mercury is a component of many vaccinations, medications, and cosmetic products. Examples include ethyl mercury and amino salicylic acid. Surgery based on traditional Chinese medicine (TCM) requires external preparations, which are a special type of treatment. Mercury has the properties of sterilization, anti-inflammation, detumescence, pain relief, and promotion of wound healing. They also remove slough and encourage the growth of tissue regeneration. One of the most well-known and significant medicines in Tibetan medicine is a mixture called Zuotai that also contains mercury. Zuotai demonstrates significant effects in terms of its anti-convulsive, anti-inflammatory, antipyretic properties, as well as its ability to enhance the immune system. It can also improve curative impact and lessen drug toxicity. Menggen Wusu is a commonly utilized mineral preparation in Mongolian medicines. It has been traditionally employed for the treatment of various conditions such as gout, rheumatism and scabies. The preparation includes mercury as one of its components [40].

2.2 Mercury Dental Amalgam

Dental amalgam, a substance utilized for cavity fillings caused by dental decay, consists of a combination of metal alloy and liquid mercury. Low-copper amalgams contain traces of other metals and 50% mercury. Due to its low cost, simplicity of use, and durability over time, amalgam has become the material of choice for dental restorative procedures. According to preliminary research results, mercury poisoning can occur even in extremely small doses. Very little amounts of elemental mercury are discharged into the atmosphere every time an amalgamated tooth is chewed and brushed, and we aren't even aware of it. The World Health Organization (WHO) believes that amalgamated teeth are responsible for around 84% of mercury exposures. According to Maqbool et al., the daily mercury emission from dental amalgam is roughly 30 μg per cubic centimeter.

Our mobile phones and other electronic devices have created electric and magnetic fields that are prevalent all around us in the world in which we live. The exposure to mercury from amalgamated teeth is increased by these electric and magnetic fields. Humans may get severe neurological diseases as a result of these mercury releases. Children whose mothers had amalgam fillings may have a much higher prevalence of autism spectrum disorders [72]. According to the World Health Organization, breathing in mercury vapors may affect the nervous, immunological, pulmonary, and renal systems [48]. Amalgam fillings consisting of 50% metallic mercury by weight, have been employed in American dentistry for over 150 years. The US Food and Drug Administration (FDA) stated that certain patients may experience harm from the mercury vapor released by amalgam fillings [24]. Numerous researches hypothesized that mercury from the filling may eventually seep into the mouth [3].

2.3 Fish and Sea Food

All trophic levels of aquatic food chain are directly contaminated from the emission of heavy metals. The link between heavy metal poisoning of the environment and seafood has been one of the most explored concerns in recent years [31]. Children who swallowed methyl-Hg-tainted Minamata Bay fish in 1956 had signs of mercury poisoning. A local chemical plant had dumped 75–150 tons of mercury into this little body of water, poisoning it. The fish consumed by the locals carried extremely high amounts of mercury (50 mg Hg/kg), which proved deadly to many people who were severely exposed [41]. Several research studies have extensively investigated the levels of heavy metals in fish and seafood, their toxicity, and the mechanisms by which they accumulate in the aquatic food chain. Evaluating the bioavailability of specific metals and their various forms is crucial due to the significant variation in metal toxicity based on their physicochemical characteristics. Additionally, it is important to note that not all metals are readily accessible to humans. Hg pollution has recently become a serious problem for both human health and seafood contamination. Approximately 65 years ago, an infamous incident of mercury accumulation and sea food contamination was identified in Minamata Bay, Japan. This incident led to severe illness and fatalities among individuals who consumed the polluted fish and shellfish. The previous data calculated the quantity of mercury (Hg) carried by fishing from the ocean to the land. Based on the calculations, Asian countries were responsible for majority of the 13 tons of mercury (Hg) that were transferred from the ocean to land in 2014. In addition, they estimated the amount of Hg that each person would be exposed to through fish intake and discovered that 38% of the 175 nations studied had populations that had this exposure. The people of the Maldives have the highest weekly methyl mercury (MeHg) level 23.1 5.3 g of MeHg/kg of body mass per week. The dominant types of mercury Hg found in fish include inorganic Hg (II) and methylmercury (MeHg), as well as dimethylmercury (Me2Hg) and ethyl mercury (EtHg). MeHg accumulates and increases in concentration as it moves up the aquatic food chain, a process known as bioaccumulation and biomagnification. MeHg in consumed fish is absorbed in about 90–95% of cases through the GI tract, where it eventually enters the central nervous system and placenta. Hence, it is assuring to acknowledge that scientific studies and institutional endeavors focused on addressing this pressing concern are duly acknowledging the presence and buildup of of mercury in seafood, along with its detrimental impact on human health. Most health regulations and legislations mandate compliance with the prescribed maximum allowance concentrations (MPC) of mercury to safeguard individuals consuming fish and shellfish.

2.4 *Mercury in Water*

The human body organs affect by contaminated drinking water and children are particularly susceptible to poisoning [20]. The toxicity of a substance depends on whether it contains elemental (EM), inorganic mercury (IM), or organic mercury (OM) compounds. Consequently, the exposure scenarios for these different forms of Hg vary significantly, posing challenges in toxicity assessment Microorganisms present in water and soil have the capability to convert EM and IM into methyl mercury, an OM that accumulates in food chain. Methylmercury has the ability to readily traverse both the placental and blood brain barriers, leading to well documented neurotoxicity and adverse effects on human development. Pregnant women should therefore take extra precautions. Approximately 80% of the vapors from EM that are inhaled during exposure are retained in the lungs. These vapors have the potential to cross the blood brain barrier, leading to neurological damage or even fatality. Additionally, while IM and EM are present in notable quantities in cosmetics, cleaning products and traditional medicines, the primary source of IM exposure for individuals is through their diet. Up to 80% of the EM vapors inhaled during exposure is retained in lung tissue before crossing the blood–brain barrier and causing neurological harm or possibly death. In addition, while IM and EM can be found in significant levels in complexion enhancement, detergents, and conventional medicines, individuals are mostly exposed to IM through their foods [29].

Several dangerous metals, such as Pb, Cd, and Hg, damage brain and neurological processes and cause skin lesions, weight loss, and other environmental health concerns, according to the WHO and USEPA. These pollutants are classed as inorganic compounds due to their high degree of hazard [36].

3 Pathways of Heavy Metals

Through inhalation, the skin, or the digestive system, heavy metals can penetrate into the human body. Heavy metals due to their toxicity Pb, Hg, Cd etc. have shown to pose a serious hazard to human health systems, mostly due to their capacity to harm DNA and membranes, as well as to interfere with protein and enzyme performance [81] Marine, freshwater, and starfish fish are the main sources of mercury intake, according to reports [54]. However, compared to 80% when inhaled, only 7–15% of mercury is absorbed through the digestive tract. This comparison emphasizes the risks associated with inhaling mercury vapors, which can arise from activities such as manual gold mining, handling silver ore, and performing amalgam dental work. Mercury has detrimental effects on the neurological system and impairs kidney function. Increased levels of blood mercury have also been associated with infertility. Mercury exists in three distinct chemical forms: metallic (Hg°), mercuric (Hg (II)), and organic compounds. While metallic mercury is minimally absorbed through the intestines, its vapors are rapidly absorbed through the lungs and can pass through both the placenta and the blood–brain barrier (BBB). Like the aforementioned metals, divalent mercury (elemental mercury) has also been demonstrated to induce oxidative stress, which can disrupt calcium homeostasis [75]. $HgCl_2$ is still included in some nations' lightening creams as the main ingredient. By substituting the copper cofactor, it permanently reduces the activity of the melanin-producing enzyme tyrosinase [16]. It is crucial to acknowledge that the accumulation of detrimental impacts resulting from the exposure to diverse xenobiotics such as heavy metals, insecticides, and other toxins can manifest [9]. Pathway of heavy metal exposure is illustrated in Fig. 2.

4 Chemical Form and Properties of Mercury

The chemical form of mercury compound is directly related to its toxic properties, biological behavior, toxicokinetic and clinical manifestations [77].

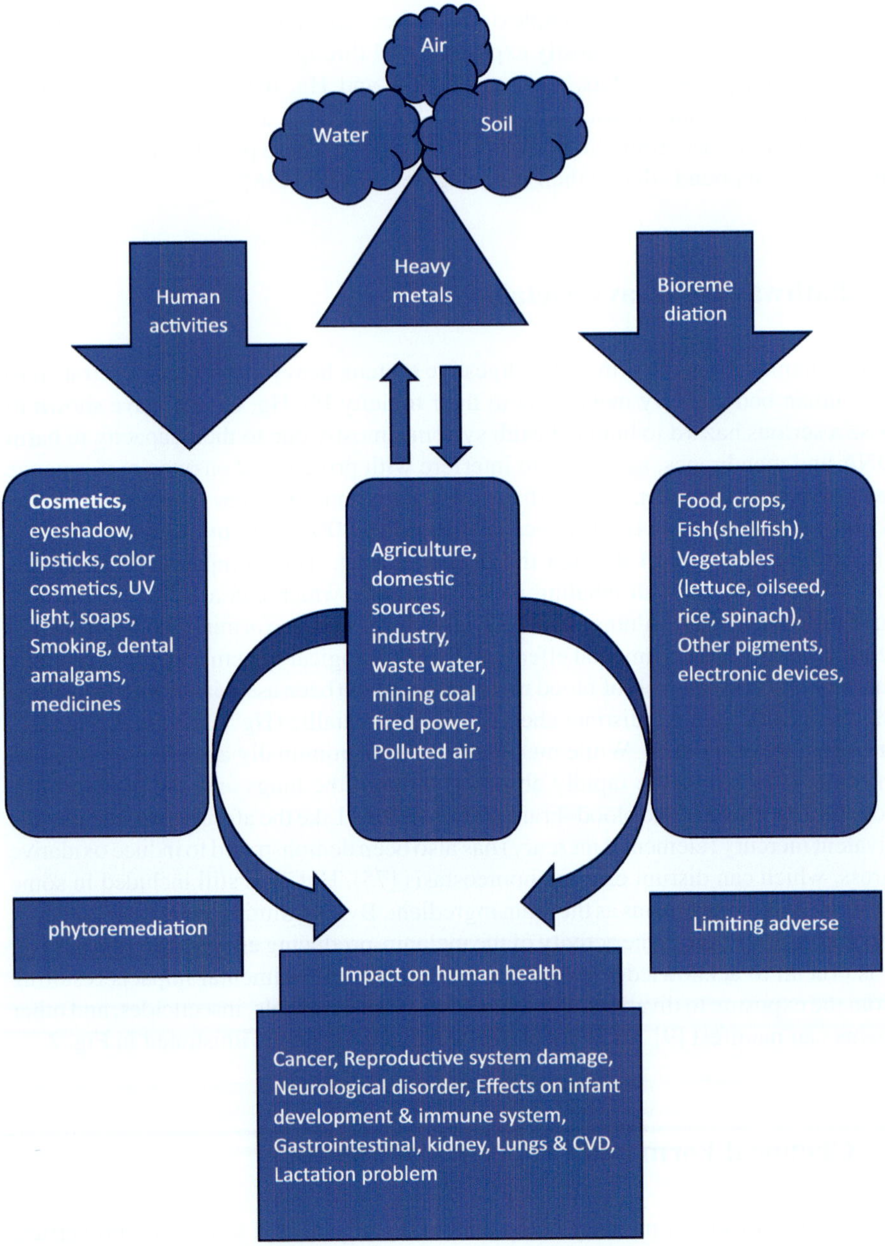

Fig. 2 Pathway of heavy metal exposure

4.1 Elemental Mercury

Elemental mercury, also known as "quicksilver," is one of the rare metals that exists in a liquid state at room temperature. It possesses a density 13 times greater than water. The term "small" mercury leak is often used to describe a quantity of mercury equivalent to or less than the amount found in a typical fever thermometer, which is approximately 0.7 g. Areas that are inadequately ventilated, confined, or situated at lower elevations have an increased potential for mercury vapor accumulation, thus posing a greater risk of poisoning. Small mercury spills are not usually linked with high mercury vapors concentrations. When elemental mercury is spilled, it condenses into small beads that spread swiftly and hard to remove. Using a typical household vacuum to remove elemental mercury bead pollution might exacerbate the condition. Elemental mercury may also seep into permeable surfaces and are problematic to extract from carpets, textiles, untreated wood, and upholstered furniture. To remove mercury from hard surfaces, use cardboard, masking tape, and eyedroppers on small spills. To remove elemental mercury from hard surfaces, you may also use commercially available powdered Sulphur. Larger mercury releases need thorough investigation and specialized remediation.

The process of in vivo oxidation occurs when absorbed elemental mercury is converted into inorganic ions Hg^{+} and Hg^{+2}. These ions then interact with sulfhydryl groups, leading to the inactivation of enzymes and alterations in cell membrane permeability. The proximal convoluted tubule is the main site of deposition for inorganic mercury, and the primary mode of excretion is through the renal system. Elemental mercury vapors exposure can have an effect on a variety of organ systems, although it most frequently affects the kidneys and the brain. Children are more vulnerable to the adverse impacts of mercury vapors compared to adults due to their smaller size and proximity to the ground, where concentrated mercury vapors tend to accumulate. Children also breathe more slowly than adults, which increases their exposure to mercury vapors. Although occurrences of elemental mercury-induced poisoning are infrequent in medical practice, its impact on public health can be severe, leading to detrimental health effects [32].

4.2 Inorganic Mercury

Inorganic mercury is widely distributed in nature, most frequently in the mineral forms of cinnabar and metacinnabar. In the meanwhile, when elemental mercury combines with any element other than carbon (because contact with carbon results in organic mercury), inorganic mercury compounds and salts are produced. Mercuric oxide, mercuric sulphide, and mercuric chloride are some of the inorganic mercury compounds that are most frequently discovered. These often take the shape of a white crystalline powder and can travel considerable distances in water, leach into soil following erosion of rocks harbouring them, or be released into the air when mining ores that house them. Other sources of emission include factories or wastewater treatment facilities that handle mercury as part of their operations, coal-fired power plants, incinerating medical and municipal waste, and manufacturers. It is a well-liked fungicide, disinfectant, and wood preservative, as well as a key component in many skin creams and soaps. On the other hand, mercuric sulphide is frequently utilized as a coloring agent in paints or tattoo ink. https://www.envirotech-online.com/news/health-and-safety/10/breaking-news/what-is-inorganic-mercury/57483.

Elemental form which is highly volatile, which when exposed to causes major health issues like harm to the brain system and lungs as well as the possibility of birth problems in pregnant women, the inorganic form of mercury is extremely toxic [76]. Hg is primarily present in inorganic [8]. However, a bacterial mechanism can convert this inorganic mercury into an organic form (for instance, methylmercury, or MeHg). MeHg possesses the capability to undergo bioabsorption and biomagnification within the aquatic food chain, starting from plankton and extending to the largest predatory fish [7]. Furthermore Hg found in fish muscle tissue is MeHg, that has high affinity for the sulphydryl group found in protein.

4.3 Organic Mercury

Methylmercury (CH_3HgX), ethyl mercury, merbromin, phenylmercuric are the example of organic mercury. Mercury enters the food chain when it is typically transformed from elemental mercury through biomethylation. Due to bioaccumulation and biomagnification, species such as top predatory fish, have the greatest mercury concentrations. Inorganic mercury is less hazardous to living things than organic mercury. The main source of organic mercury exposure in people due to ingestion is by far fish intake. Hunter-Russell syndrome and Minamata sickness are just two of the illnesses that mercury exposure can cause. Due to its potential to cause severe neurological complications and birth defects, mercury poses a significant risk to both developing fetuses and pregnant women. Infants can also be affected as methylmercury is excreted into breast milk and absorbed through the digestive system. The effects of organic and inorganic mercury differ significantly. Organic mercury, which readily crosses the blood–brain barrier and leads to neurological harm, is more hazardous compared to its inorganic counterpart [74].

5 Mechanisms of Mercury Toxicity

Inhibition of enzymatic activity disrupt the normal biochemical reaction, impairing cellular function and metabolic process. MeHg's neurotoxicity is caused by a number of recognized processes, including oxidative stress, mitochondrial toxicity, and disturbances in calcium homeostasis [23, 33]. Recent research indicates that epigenetic controls are involved in the toxicity mechanisms brought on by ambient levels of mercury [10, 13, 57, 67] which are also important factors in MeHg's effects on subsequent generations [34, 55, 65]. Heavy metal poisoning is a complex phenomenon that involves various detrimental effects on metabolic processes and cellular structures. The toxicity of heavy metal ions can be attributed to several mechanisms: 1. Interaction with specific functional groups in proteins, such as thiol, histidine, and carboxyl groups, as well as small molecules like glutathione (GSH). This interaction can result in the loss of protein activity, disruption of protein structure, and perturbation of signaling and regulatory pathways. 2. Inactivation of proteins by replacing essential metal ions, particularly those present in the active sites of specific enzymes. 3. Similarity to functional groups found in biochemical molecules, such as phosphate groups. 4. Generation of reactive oxygen species (ROS) through autooxidation and Haber's process. By addressing these mechanisms, we can gain a better understanding of the harmful effects associated with heavy metal poisoning.

5.1 *Oxidative Stress*

When heavy metals accumulate in the body, they cause considerable harm to the majority of organs. The majority of the time, they induce oxidative stress in the cells, which boosts the levels of reactive oxygen species, a few oxidants, and antioxidants and, through a number of processes, resulting in cell death. The liver is the first organ to be impacted. Work in an industry related to heavy metals or other factors may be to blame for the introduction of these heavy metals [68]. ROS levels that are too high injure cell components and disturb redox balance. The word "oxidative stress" refers to a condition in which ROS are created in excess. In cells, many oxidation states of redox-active heavy metals are present, where they directly interact with ROS to convert less dangerous ROS into more detrimental ones.

5.2 *Reactive Oxygen Species*

"ROS, or Reactive Oxygen Species, is generated as a byproduct during aerobic metabolism". They exist in both radical and nonradical forms, and both are capable of rapidly interacting with organic molecules to induce cell component damage. Aerobic species have evolved a range of antioxidant defenses, but they have also worked out how to utilize ROS as signaling molecules and in pathogen defense. Under stressful circumstances, excessive ROS production frequently happens as a result of disturbed metabolism. In its ground state, atomic oxygen exhibits a unique arrangement of electrons. Due to the presence of two unpaired electrons in two antibonding orbitals, *2p, it is diradical. It possesses three energy levels in the external magnetic field, hence the name triplet oxygen ($3O_2$). Because most chemical compounds contain paired antiparallel electrons in their orbitals, this configuration renders $3O_2$ less reactive. One of the unpaired electrons undergoes spin reversal as a result of the excitation of

$3O_2$, which results in the creation of singlet oxygen ($1O_2$). O_2 exists in two singlet states: 1 g + O_2 and $1gO_2$, both of which have paired electrons in a *2p orbital and opposite spin electrons that are still in distinct orbitals, respectively. The 1 g+ state transforms into the 1 g state, which has lower energy, relatively quickly. The latter can interact with other molecules since its lifespan (4 s in water) is sufficient. $1O_2$ is significantly more reactive than $3O_2$ due to its paired electrons. When it combines with substances like photosynthetic pigments that contain unsaturated bonds, cycloadducts, hydroperoxides, and endoperoxides are produced. Since these bonds are typically present in membrane lipids, $1O_2$ results in lipid peroxidation. Additionally, it turns sulphides into sulfoxides. In terms of proteins, the amino acid residues Trp, Tyr, His, Met, and Cys are sensitive to oxidation by 1O2, although this ROS mostly oxidises guanine in nucleic acids. The primary reactive oxygen species (ROS) responsible for leaf damage and the reduction of PS II activity under light exposure is $1O_2$. Oxygen molecules can also be decreased. One water molecule is produced by full four-electron reduction, while all of its intermediates are ROS. The initial step, the one-electron reduction, needs energy, but the subsequent processes might happen on their own. Superoxide anion ($O_2\cdot$) is recognized for its detrimental effects on Fe–S clusters within enzymes. It can also react with Cys thiol groups and decrease transition metals like Fe^{3+} and Cu^{2+}. His, Met, and Trp are other amino acids that are highly vulnerable to $O_2\cdot$. Hydroperoxides are created when molecules containing double bonds interact with $O_2\cdot$. The highly oxidizing peroxynitrite (ONOO) is produced when nitric oxide (NO) and oxygen ($O_2\cdot$) combine. $O_2\cdot$ is protonated to the hydroperoxide radical ($HO_2\cdot$) at low pH levels. Due to its neutral charge, $HO_2\cdot$ can permeate across biological membranes and start the oxidation of lipids.

Hydrogen peroxide (H_2O_2) is a relatively stable byproduct of the two-electron reduction of oxygen, but because of its lower reactivity. H_2O_2, which has no electrical charge, can diffuse through membranes. H_2O_2 can react with various functional groups, including thiols, indoles, imidazole, phenols, thioesters, and methionyl groups. Additionally, it harms PS II and hemo group Mn clusters. The Fenton reaction is a chemical process involving the reaction between hydrogen peroxide H_2O_2 and ferrous ions (Fe^{2+}). This reaction is part of a larger cycle known as the Haber–Weiss reaction, which describes the overall mechanism of hydroxyl radical ($OH\cdot$) formation in the presence of both iron ions and hydrogen peroxide. In the chloroplasts of Cr-treated soybean, an analogous reaction was proposed to take place with Cr ions serving as catalysts. Free metal ions have a propensity to adhere to proteins' and DNA's surfaces, where they can take part in the production of $OH\cdot$. The only factor limiting the rate of $OH\cdot$'s reaction with any nearby molecule is diffusion, making it the most reactive ROS. As a result, the harmful effect of $OH\cdot$ is essentially contained to the location where it is formed.

Ozone (O_3) and compounds such alkoxy radicals ($RO\cdot$), peroxy radicals (ROO ·), and hydroperoxides (ROOH) that are created when any of the aforementioned forms of ROS react with organic molecules are other types of ROS.

5.3 *Epigenetic Changes*

Attention-deficit hyperactivity disorder (ADHD) is a neurodevelopmental condition that affects both adults and children. It is estimated that approximately 2.5% of adults and 5% of children worldwide experience ADHD. The neurotoxic effects of methylmercury (MeHg) pose a significant risk to the development of fetuses. When asymptomatic mothers have internal exposure to MeHg, it can have a detrimental and long-lasting impact on the neurodevelopment of the fetus. Cysteine, an amino acid, can form a complex with MeHg. Methionine, an amino acid, is mimicked structurally by the MeHg-cysteine complex. As a result, the compound can enter the brain without restriction via methionine transporters. MeHg has the ability to alter the cellular redox equilibrium, resulting in a series of harmful effects. Exposure to methylmercury (MeHg) in the environment has been found to impact DNA methylation levels, which is a critical process involved in the epigenetic regulation of gene expression. However, the exact implications of consuming fish on neurobehavioral functions remain uncertain. The dopaminergic neurotransmission system plays a crucial role in motor control and emotion regulation within the brain [28]. The efficacy of dopamine neurotransmission stimulants in treating attention deficit hyperactivity disorder (ADHD) suggests that individuals with ADHD may experience abnormal dopaminergic neurotransmission. The production of dopamine involves several steps, one of which includes the rate-limiting enzyme tyrosine hydroxylase (TH). Synaptic vesicles store dopamine intracellularly and release it into the synaptic cleft for neurotransmission [61]. Dopamine synthesis takes several steps, among one of them is tyrosine hydroxylase (TH) that is rate-limiting enzyme to produce dopamine. Synaptic vesicles contain intracellular dopamine and discharge it into the synaptic cleft for neurotransmission. Dopamine transporters (DATs) regulate extra synaptic dopamine levels and are able to return it to presynaptic neurons. Further research on the epigenetic impacts of MeHg will reveal additional details about the mechanisms underlying environmental causes of ADHD since epigenetic markers are particularly vulnerable to environmental influences.

6 Overview of Hg Concentration in Food Items

Pollution from human activities such as fossil fuel burning, industry, and small-scale gold mine raises mercury levels in the environment. The issue of certain elements in food items, such as metalloids like arsenic (As) and metals like lead (Pb), mercury (Hg), and cadmium (Cd), is a growing concern. These substances have been found to undergo biomagnification in the food chain. Their presence in food has been shown to have adverse effects on public health, as evidenced by research conducted by Wei et al. in 2019. According to the 2015–2020 Dietary Guidelines for Americans, it is recommended that adults consume a minimum of 8 oz or 2 servings of seafood per week. Seafood is known to be a good source of heart-healthy omega-3 fatty acids, such as eicosapentaenoic acid (EPA) and docosahexaenoic acid (DHA). However, some individuals choose to limit their seafood intake due to concerns about mercury exposure. It is important to note that methylmercury, a neurotoxin, can accumulate in fish as they age [50]. Recent field studies conducted in both mercury (Hg) mining and non-mining regions have demonstrated that rice plants possess a greater capacity to accumulate Hg, particularly methylmercury (MeHg), when compared to other cereal crops. Various factors, such as environmental conditions and the specific genotypes of rice, influence the extent of Hg accumulation in these plants [40]. While the average mercury levels in most species remained below the recommended 0.2 g/g/ww threshold for subsistence fisheries, it is concerning that 70% of lake trout exceeded this limit, reaching 0.35 g/g/ww. Additionally, 19% of lake trout surpassed levels that were 2.5 times higher, specifically set for commercial sale. Consequently, we strongly recommend the implementation of lake trout consumption advisories to ensure the safety of pregnant women and young children. Furthermore, we advocate for the initiation of community-based monitoring programs on a regular basis [38].

7 Dietary Hg Intake in Different Countries and Potential Health Risks

The risk associated with the consumption of total mercury in the diet by individuals of different age groups is determined by the provisional tolerable weekly intake of total mercury, set at 4 g/kg of body weight. This corresponds to a tolerable daily intake (TDI) of 0.57 g/kg of body weight, as recommended by the FAO/WHO JECFA (Expert Committee on Food Additives). Recent studies conducted in Spain have indicated that the mercury levels in the population's diet exceed the limits established by the European Food Safety Agency [73]. Based on the research conducted by Yu s á et al. findings from human biomonitoring studies and food and environmental toxin programs conducted between 2014 and 2020 indicate that a notable proportion of both children and mothers are exposed to elevated levels of mercury (Hg) [59]. Here is the revised paragraph without plagiarism.

In a study conducted by Yusà and colleagues, it was observed that the concentration of total mercury in the hair of breastfeeding mothers was six times greater than the average internal exposure of mothers in 17 other European countries (0.225 $\mu g\ g^{-1}$). Furthermore, it was found that 27% of these mothers exceeded the health-based guideline value proposed by the European Food Safety Authority (EFSA), which is 1.9 $\mu g\ g^{-1}$. Another study by Pérez et al. revealed that children in the Valencian Region exhibited hair mercury levels almost five times higher than children from 17 other European nations (0.145 $\mu g\ g^{-1}$). Additionally, approximately 13% of these children had hair mercury levels exceeding the Joint FAO/WHO Expert Committee on Food Additives (JECFA) limit of 2.3 $\mu g\ g^{-1}$, while 18% exceeded the EFSA health-based advice [73].

Tea production in 58 countries, primarily Asia, is growing rapidly, Concern over Hg contamination in tea and infusions is on the rise [15, 27, 71]. Poland ranks highly in both the world and Europe for the quantity of tea eaten [70]. The people of the UK and Ireland consume the most tea infusions (around 190 L per person annually), which is double what is consumed in other European nations [12].

Hg can build up in ecosystems and is very mobile in the environment. Hg affects neurons, interferes with the function of neurotransmitters, and prevents the creation of proteins in the body. The neurotoxic effects of mercury disrupt calcium ion homeostasis, protein phosphorylation, cell signaling pathways, gene expression, and cell proliferation. Nephrotoxic effect on health due to mercury are linked to a decrease in the activity of antioxidant enzymes [1, 25]. Mercury poisoning and mercury deposition in fish and fish products have been recorded in numerous nations over the past few decades, including Pakistan [35]. A survey of people over the age of 18 conducted in six cities between 2016 and 2017 by the Chinese Academy of Environmental Sciences found that diet contributed the most to people's exposure to mercury (61.23–99.77%), followed by drinking water, soil, and air. All types of mercury cause kidney issues, and too much mercury can cause neurological issues, sleep disturbances, hearing loss, decreased reproductive abilities, and cardiac issues.

7.1 Organ Specific Effects

7.1.1 Kidney

Domestic water quality is often subpar, especially in poorer nations, and this has contributed to a large number of water-borne illnesses. The proximal tubules are particularly concentrated in the kidneys' simple target for mercury toxicity. The main and most common environmental toxin that causes nephrotic consequences is mercury. The kidneys are easily targeted for mercury toxicity due to significant accumulation, particularly in the regions of the proximal tubules. It is also a hazardous metal of explicit concern with severe nephrotoxic effects. The previous research proved that proteinuria may also be caused by lysosomal sensitivity because the proximal tubule cells are destroyed and can enter readily via the glomerular filter. Testing for mercury in urine provides a more accurate and purposeful picture of exposure since the mercury is more concentrated. Hemoglobin levels are frequently evaluated with a simple blood test kit, and kidney function is also assessed in order to evaluate the early symptoms of mercury exposure [2].

7.1.2 Lungs

Hg^0 gas is swiftly absorbed by the lungs (around 75%), but metallic mercury is barely absorbed by the digestive system. After being introduced into the body, mercury ions (Hg^0) disperse and traverse various membranes, including those found in the blood–brain barrier, brainstem, cerebellar nuclei, spinal cord, and placenta. In the erythrocytes and other cells, catalase oxidizes Hg^0 to produce Hg^{2+} . After a few hours, the distribution and excretion of elemental mercury (Hg^0) exhibit similarities to those observed in the case of divalent mercury ions (Hg^{2+}). When Hg^0 gas is inhaled, it oxidizes to Hg^{2+} and is stored in the brain [79]. Most of the mercury that enters the body is absorbed by red blood cells. However, unbound mercury quickly circulates through the body and accumulates in vital organs such as the kidneys, heart, and brain. The kidneys are where the body's highest amounts are found. Catalase oxidizes mercury in tissues and RBCs to create divalent chemical mercury. Inadequate solubilized mercury may return to the alveoli, where it is expelled after expiration [22]. The initial indications of acute mercury inhalation involve severe irritation and coughing in the bronchial passages. Subsequently, individuals may experience symptoms such as fever, diarrhea, inflammation of the mouth, difficulty breathing, vomiting, dehydration, and shock [53]. Individuals who managed to survive the initial assault may encounter various health complications such as pulmonary dysfunction, azotemia, ulcerative colitis, and reduced urine output. Moreover, the different types of mercury tend to accumulate in the liver, kidneys, hair, and nails. Eventually, they are excreted through perspiration, tears, saliva, and breast milk [11]. The initial goal of methyl mercury poisoning is the Central Neuron System. Visual confusion, ataxia, paresthesia, slurred speech, muscular tremors, cognitive problems, movement dysfunctions, and, in severe cases, paralysis and death, are all symptoms of exposure to methyl mercury. The neurological system is still developing, but it is more sensitive to methyl mercury. MeHg exposure from the consumption of contaminated seafood is thought to be a route that may be harmful to the fetus's neurodevelopment. Fish quickly and almost entirely absorb MeHg, which readily passes through both the placenta and the blood–brain barrier.

7.1.3 Bones

Osteoporosis, a chronic condition characterized by a reduction in bone density, has become a significant global public health concern. Previous research found that one in five males and one in three females over the age of 50 had osteoporosis or osteopenia. Osteoporosis is also becoming more common each year as a result of an ageing and expanding population. Various data found that the metabolism of bones can be affected by heavy metals. Metallic and inorganic Hg are converted into organic forms, mostly methylmercury, in the aquatic environment via biochemical changes (methylation) in microorganisms (aerobic bacteria). In the human body, Hg molecules disrupt enzymatic and hormonal responses. Currently, there is a scarcity of literature regarding the impact of mercury (Hg) on human bone tissue. A study conducted by Rasmussen et al. investigated Hg concentrations in ancient human bone samples. However, due to the archaeological nature of the material, it is inappropriate to compare these concentrations to samples obtained from contemporary individuals. Today's population has distinct dietary habits, environmental pollution levels, and treatment approaches, which differ significantly from the medieval era when Hg was used to treat diseases like leprosy and syphilis. In their research, Rasmussen et al. observed varying levels of mercury concentration in both the spongy and compact bone. These findings suggest that spongy bone, characterized by a higher metabolism and a larger surface area in contact with blood vessels, may facilitate the transportation of Hg. Among males aged 17–19 and 38–45 years, the lowest Hg contents were found in the long bones of the shoulder and knee joints. The pelvis and the hip joint consist of bone structures with high porosity, which exhibited the highest levels of Hg content. Limited information is available regarding the quantities of Hg found in human bone structures. The examination of biological and environmental factors impacting Hg concentrations in bones indicated that weight, BMI, osteoporosis, dental amalgam use, and the consumption of fish and shellfish had no significant influence [18]. According to Campbell et al., lead exposure has a deleterious impact on bone mineral density (BMD). By inhibiting the P2X7/PI3K/AKT signaling pathway and modulating the ratio of receptor activator for nuclear factor-B ligand (RANKL) to osteoprotegerin (OPG), Ma et al. demonstrated that cadmium exposure is capable of inducing osteoporosis. Based on the research conducted by He et al., it was observed that cadmium exhibits a substantial increase in the expression of RANKL. However, its influence on OPG is comparatively less significant.

7.1.4 Cardiovascular System

Inhalation of metallic mercury can lead to an increase in heart rate and blood pressure. However, studies have shown that even with chronic exposure to very small doses of mercury (0–0.27 mg/m^3 in one study and 0.075 mg/m^3 in another) over a period of 0.5–7 years, there was no significant elevation in blood pressure or abnormal electrocardiograms observed. Research has also indicated that individuals with dental amalgam, which contains mercury, experienced a slight increase in systolic and diastolic pressure. However, these increases were not significantly different from those observed in individuals without dental amalgam or in the general population. Additionally, exposure to mercury (Hg) at the ubiquinone-cytochrome-b site of the mitochondrial respiratory chain has been found to generate hydrogen peroxide and cause mitochondrial dysfunction [26]. The presence of mitochondrial reduced glutathione (GSH) decreases by over 50%, while the levels of thiobarbiturate reactive substances, an indicator of increased mitochondrial lipid peroxidation, rise by 68%. Hg selenides, which are insoluble compounds of mercury with selenium, limit the availability of selenium. This is problematic as selenium is an essential cofactor for glutathione peroxidase, a crucial scavenger responsible for neutralizing H_2O_2 and lipid peroxides. Insufficient selenium levels can elevate the risk of cardiovascular disease (CVD) and cerebrovascular accidents (CVA).

7.2 Non Organ Specific Effects

7.2.1 Carcinogenicity

The intricate relationship between metal carcinogens and biological materials is complex, and assuming that various metals share identical mechanisms of action may be overly simplistic. However, a closer examination reveals three fundamental pathways that are applicable to the majority of carcinogenic metal compounds: (1) impairment of cellular redox control, resulting in DNA damage; (2) inhibition of crucial DNA repair systems, leading to genomic instability; and (3) activation of oncogenic pathways, disrupting the balance between cell proliferation and cell death. The development of cancer through oxidative stress depends on the severity and timing of free radical accumulation, which can result in mutagenesis, dysregulated cell division, and impaired repair processes. Carcinogenic metal compounds have been found as DNA repair pathway inhibitors at low concentrations due to the generation of hazardous intermediates. Metal substances that cause cancer can induce cellular proto-oncogene expression, alter the gene expression, promote mitogenic signaling pathways. This results in an unstable genome, which promotes cellular growth. Apoptotic pathways and angiogenic models are further inhibited. Finally, pre-cancerous lesions can develop and progress to malignant tumors [58]. Besides gastrointestinal malignancies, the occurrence of various cancers including liver, prostate, lung, urinary bladder, thyroid, and kidney have been predominantly associated with exposure to heavy metals. The abnormal functioning of miRNAs, or microRNAs, has been identified as a contributing factor in the development of different types of human cancers. Recent studies have demonstrated that deregulation of microRNAs contributes to the development of cancer in a variety of tissues. The hazardous effects of elemental mercury are typically limited to the inhalation of vapors, which causes "metal fume fever" in people. Diseases of the skin or the digestive system can develop with little elemental mercury intake. But because the monoatomic gas is so lipid-soluble and highly diffusible, inhalation results in fast and full absorption through the alveoli.

7.2.2 Reproductive Effects

Mercury can accumulate in female ovaries causing infertility, and ovarian failure. Experiment animal studies have revealed that higher mercury dosages increase the possible number of reproductive abnormalities such as stillbirth, congenital deformities, and spontaneous abortion. Infertile individuals with unknown causes of infertility showed higher levels of mercury in their hair, blood, and urine compared to fertile individuals. Exposure to mercury has been associated with an increased incidence of menstrual and hormonal abnormalities, as well as an elevated risk of adverse reproductive outcomes. A comparative study involving 84 patients with polycystic ovary syndrome (PCOS) and 70 healthy volunteers revealed that individuals with PCOS had elevated mercury exposure. The exposure to mercury obstructed the production of luteinizing hormone (LH) and follicle-stimulating hormone (FSH) in females, originating from the anterior pituitary gland. Menstrual disorders such as irregular bleeding, shorter or longer menstrual cycles, and discomfort have been linked to mercury exposure. In an experimental group, female mice exposed to methylmercury at doses of 0.25–1.00 mg/kg/day (administered via gavage) displayed decreased fertility and survival rates, although litter size remained unaffected. A positive correlation was observed between mercury accumulation in the ovaries and the rate of follicular atresia in laying hens (40-week-old Hy-Line Brown) fed four experimental diets containing different doses of mercury (0.280, 3.325, 9.415, and 27.240 mg/kg) [44]. Conversely, FSH and LH levels exhibited an inverse relationship with mercury dosages. The experimental groups exhibited significant reductions in catalase, superoxide dismutase, glutathione reductase, and glutathione concentration. Mercury-susceptible animals experienced uterine inflammation in the endometrium and myometrium. While the uterine endometrial area decreased compared to the control group, there was no difference in the myometrium area between the groups. In another study involving human subjects, higher levels of mercury were detected in tissue samples from pre-menopausal women under the age of 50 with typical endometrial hyperplasia, endometrial cancer, and normal endometrial tissues.

Mercury's negative impacts on male reproductive features include impaired spermatogenesis, decreased spermatozoa mobility, and an increase in pathological alterations. Mercury chloride ($HgCl_2$) was shown to be harmful to the testicles of adult male Wistar rats [4]. Histological analysis of the testes in the group exposed to mercury (40 mg/kg bw; $HgCl_2$; orally administered daily for 28 days) showed disrupted cytoarchitecture and impaired spermatozoa quality. The weights of the testes and the gonadosomatic indices were significantly lower in the mercury-treated group compared to the control group. The mercury-exposed group exhibited degradation of spermatogenic cells in the germinal epithelium, obstruction and enlargement of seminiferous tubule lumens, and irregular vacuolization of the basement membrane. Testicular atrophy induced by mercuric chloride was also observed in a separate study. Rats injected with mercury subcutaneously (5 mg/kg mercury chloride; 5 days) exhibited the formation of fibrotic histological structures in mature active seminiferous tubules. Additionally, the mercury-exposed group displayed a reduced quantity of spermatocytes. Analysis of mercury-exposed cultures revealed changes in the acrosome (anterior section of the skull), connection piece (connecting region), and mitochondrial segment, indicating an adverse reaction [52]. Mercury has negative effects on sperm at higher blood levels (40.6 mmol/L), which result in a progressive motility rate of less than 50% and a normal morphology rate of less than 14% [14, 37].

7.2.3 Immune System Effects

Trace metals have the potential to negatively impact the immune system, leading to a decrease in the population of various immune cells including neutrophils, macrophages, natural killer cells, B cells, and T cells. Recent studies focusing on human exposure to mercury have demonstrated its association with the development of autoimmune reactions and inflammation. This is believed to occur through the stimulation of proinflammatory substances like tumor necrosis factor (TNF), interleukin 1 (IL-1), interferon-gamma (IFN-), as well as the production of autoantibodies [21, 63]. The mechanism of mercury-induced autoimmunity involves a series of interconnected events. These include the activation of T-cell dependent humoral response, enlargement of draining lymph nodes accompanied by the emergence of new germinal centers, production of IgG antibodies, and the accumulation of immune complexes in glomeruli and blood vessels. Exposure to mercury triggers an inflammatory response, characterized by the expression of damage-associated molecular patterns and the activation of innate immune sensors such as nucleic acid sensing Toll-like receptors (TLRs). Consequently, inflammatory cytokines are released, leading to the initiation of chronic inflammation. Moreover, the deposition of mercury in lysosomal compartments induces inflammation and autoimmune responses, while heightened cathepsin B activity further enhances immune response activation. Prolonged tissue damage and inflammation resulting from mercury deposition contribute to the expansion of secondary lymphoid organs and the formation of ectopic lymphoid structures. It is important to remember that early inflammatory events such as cathepsin B activation,48 production of proinflammatory cytokines 48, and cellular infiltrates 50 are connected to future autoimmune reactions. Mercury dispersion in the environment is a natural occurrence in terms of human interaction. To ensure the authenticity of the content and provide plagiarism-free data, the original lines have been rewritten. In order to safeguard various aspects of human activities, it is crucial to implement rigorous monitoring measures in sectors such as gold mining, aquatic food production, mercury-free beauty products, chloralkali industry, plant cultivation, and battery manufacturing [5, 64].

7.2.4 Neurological Effects

Methylmercury is a neurotoxic pollutants, causes neurodegeneration and mental disorder [39]. The majority of central nervous system (CNS) damage attributed to MeHg is believed to arise from its ability to induce oxidative stress in neurons. This can occur through an increase in reactive oxygen species (ROS) levels or a reduction in the effectiveness of antioxidant defense mechanisms [6]. MeHg can alter the function of the mitochondria and calcium homeostasis, dysregulate the cofilin phosphorylation/dephosphorylation pathway, cause microtubule disarray that alters the cytoskeletal structure, and interfere with the metabolism of neurotransmitters like glutamate and gamma-aminobutyric acid (GABA) signaling. The astrocytes appear to control MeHg-induced neurotoxicity because they are directly implicated in these events. Although astrocytes may potentially contribute to neurotoxicity through the disruption of their typical functions or the acquisition of abnormal capabilities, several studies have indicated that the demise or dysfunction of astrocytes precedes neuronal cell death. Glutamatergic excitotoxicity induced by MeHg contribute to the specific and potent inhibition. The administration of MeHg leads to a notable decrease in the uptake of glutamate and aspartate by astrocytes, along with an increase in the release of these excitatory neurotransmitters. The mechanisms underlying MeHg toxicity seem to be intricate and can be influenced by factors such as dosage, age, species, and duration of treatment. Additional investigations are required to gain a comprehensive understanding of this correlation and develop strategies to mitigate central nervous system (CNS) harm [19].

8 Biomarkers of Hg Exposure (Urine, Nail, Hair)

Biomarkers play a crucial role in epidemiological studies as they enable the measurement of metal and metalloid exposure, providing estimates of internal dosage. These biomarkers encompass multiple sources and pathways of exposure, making them valuable tools in assessing and quantifying such exposures. Internal dose indicators, which include total Hg levels in hair, urine, and blood in the absence of speciation tests, are the most often used biomarkers of mercurial exposure. The connection between hair Hg levels and those in the brain and whole blood is 250:5:1. The ratio of mercury (Hg) levels in blood and brain, as well as blood and hair, can vary based on an individual's characteristics, including age, gender, and genetic inheritance. The amount of Hg in hair is a good biomarker for addressing MeHg exposure in a non-invasive manner. In 2008, the World Health Organization (WHO) and the United Nations Environment Programme (UNEP) set a minimum detectable threshold of 50 g/g for adverse neurological effects (such as paresthesia) in adults caused by neurotoxic substances. It has been observed that individuals who consume a significant amount of seafood may have hair mercury (Hg) levels exceeding 10 g/g. To address the issue of potential mixed sources of exposure (such as dietary methylmercury (MeHg) versus occupational elemental mercury (Hg^0)), the analysis of stable isotopic signatures of Hg in hair has emerged as a valuable method for distinguishing between different sources of mercury in the body. This approach helps to mitigate the risk of confusion arising from exposure to multiple sources. Because they are non-invasive and portable, nails are an appealing biomarker medium in distant locations. In the nailbed heavy metals are deposit by the bloodstream. Toenails have a longer exposure time since they take 12–18 months to fully develop and grow at half the pace of fingernails (1.62 mm/month vs. 3.47 mm/month). Numerous studies have indicated that those who live near environmentally hazardous areas have greater TTE concentrations in their nails. Elevated TTE concentrations in nails have been associated to high blood pressure, hypertension, and mental stress [60].

8.1 Urinary Hg Levels for Different Diseases

The preferred biomarker for assessing inorganic mercury (Hg) exposure in both the general population and occupational settings is urinary Hg. Urine Hg levels indicate the amount of inorganic Hg that has accumulated in the kidneys following acute exposure. Urine Hg measurements can reflect exposure to both inorganic and organic Hg, which is demethylated and excreted through the kidneys. However, the extent to which organic Hg is demethylated and eliminated in urine remains uncertain due to conflicting findings from previous studies. Blood samples can capture all forms of Hg exposure. Organic Hg is predominantly present in red blood cells, where it binds to hemoglobin, while inorganic Hg is present in both red blood cells and plasma. The half-life of mercury in blood is approximately 50 days. Blood Hg levels serve as a reliable biomarker for recent exposure and can also be used to diagnose long-term exposure in populations regularly exposed to this metal. Additionally, during pregnancy, maternal blood Hg concentrations can serve as a biomarker for prenatal exposure [51, 66].

9 Conclusion

Exposure to mercury can lead to various health problems, depending on the chemical form and duration of exposure. Human exposure to mercury primary occurs through the seafood (fish, shellfish). Through biomagnification, mercury concentration increase, as go to food chain leading to higher level in organisms that are top in the food web. Methyl mercury affect the central nervous system, leading to developmental impairment especially in fetus and young children. Mercury is often called "Silent killer" due to its hazardous effects on human health. In this chapter, exposure and harmful effect of mercury is well documented. By promoting global cooperation, implementing effective mitigation strategies, we can strive to reduce the mercury pollution on both human health and environment.

References

1. Abbott LC, Nigussie F (2021) Mercury toxicity and neurogenesis in the mammalian brain. Int J Mol Sci 22(14):7520
2. Abdeldayem R (2022) Domestic water and accumulating mercury toxicity in the kidney. Appl Water Sci 12(6):114
3. Alanazi LM et al (2021) The hazardous effects of dental amalgam fillings in human body: a systematic review. J Pharm Res Int 33(49B):45–54
4. Almeer RS et al (2020) Ziziphus spina-christi leaf extract attenuates mercury chloride-induced testicular dysfunction in rats. Environ Sci Pollut Res 27:3401–3412
5. Anka AU et al (2022) Potential mechanisms of some selected heavy metals in the induction of inflammation and autoimmunity. Eur J Inflamm 20:1721727X221122719
6. Arrifano GDP et al (2021) Revisiting astrocytic roles in methylmercury intoxication. Mol Neurobiol 58(9):4293–4308
7. Bairagi S et al (2021) Preserving cultural heritage through the valorization of Cordillera heirloom rice in the Philippines. Agric Hum Values 38(1):257–270
8. Bajema KL et al (2021) Estimated SARS-CoV-2 Seroprevalence in the US as of September 2020. JAMA Intern Med 181(4):450–460

9. Balali-Mood M et al (2021) Toxic mechanisms of five heavy metals: mercury, lead, chromium, cadmium, and arsenic. Front Pharmacol 227
10. Bjørklund G et al (2020) Mercury exposure, epigenetic alterations and brain tumorigenesis: a possible relationship? Curr Med Chem 27(39):6596–6610
11. Briffa J et al (2020) Heavy metal pollution in the environment and their toxicological effects on humans. Heliyon 6(9):e04691
12. Brodziak-Dopierała B, Fischer A (2022) Analysis of mercury content in various types of tea (Camellia sinensis) and Yerba Mate (Ilex paraguariensis). Int J Environ Res Public Health 19(9):5491
13. Cao M et al (2019) Identification of potential long noncoding RNA biomarker of mercury compounds in zebrafish embryos. Chem Res Toxicol 32(5):878–886
14. Chao H-H et al (2022) Comprehensive review on the positive and negative effects of various important regulators on male spermatogenesis and fertility. Front Nutr 9
15. Chen H-Y et al (2021) Cumulative tea consumption is inversely associated with colorectal adenomas in adults: a cross-sectional study in a Taiwanese population. Cancer Epidemiol 73:101945
16. Chen J et al (2020) Inhibition of tyrosinase by mercury chloride: spectroscopic and docking studies. Front Pharmacol 11:81
17. Cheng M et al (2021) Human methylmercury exposure and potential impacts in central tibet: food and traditional tibetan medicine. Bull Environ Contam Toxicol 107:449–545
18. Ciosek Ż et al (2023) Iron, zinc, copper, cadmium, mercury, and bone tissue. Int J Environ Res Public Health 20(3):2197
19. de Paula Arrifano G et al (2023) Neurotoxicity and the global worst pollutants: astroglial involvement in arsenic, lead, and mercury intoxication. Neurochem Res 48(4):1047–1065
20. Ekawanti A et al (2021) Mercury pollution in water and its effect on renal function of school age children in gold mining area Sekotong West Lombok. In: IOP conference series: earth and environmental science. IOP Publishing
21. Elblehi SS et al (2019) L-α-Phosphatidylcholine attenuates mercury-induced hepato-renal damage through suppressing oxidative stress and inflammation. Environ Sci Pollut Res 26(9):9333–9342
22. Elder A et al (2022) Exposure, dose, and toxicokinetics of metals. Elsevier, Handbook on the Toxicology of Metals, pp 55–86
23. Farina M, Aschner M (2019) Glutathione antioxidant system and methylmercury-induced neurotoxicity: An intriguing interplay. Biochimica et Biophysica Acta (BBA)-General Subjects 1863(12):129285
24. Geier D, Geier M (2022) Dental amalgam fillings and mercury vapor safety limits in American adults. Hum Exp Toxicol 41:09603271221106341
25. Gupta A et al (2020) Mercury, a silent killer to human health and environment: a review of India. Atmosphere (Pirrone 2010; UNEP 2013; Street et al. 2011) 2:7
26. Habeeb E et al (2022) Role of environmental toxicants in the development of hypertensive and cardiovascular diseases. Toxicol Rep 9:521–533
27. Heshmati A et al (2020) The concentration and health risk of potentially toxic elements in black and green tea—both bagged and loose-leaf. Qual Assur Saf Crop Foods 12(3):140–150
28. Hirano S (2021) Clinical implications for dopaminergic and functional neuroimage research in cognitive symptoms of Parkinson's disease. Mol Med 27(1):40
29. Ismanto A et al (2023) Groundwater contamination status in Malaysia: level of heavy metal, source, health impact, and remediation technologies. Bioprocess Biosyst Eng 46(3):467–482
30. Jain A et al (2019) New insights and rethinking of cinnabar for chemical and its pharmacological dynamics. Bioengineered 10(1):353–364
31. Jinadasa B et al (2021) Mitigating the impact of mercury contaminants in fish and other seafood—a review. Mar Pollut Bull 171:112710
32. Johnson-Arbor K et al (2021) Characteristics and treatment of elemental mercury intoxication: a case series. Health Sci Rep 4(2):e293

33. Ke T et al (2019) Post-translational modifications in MeHg-induced neurotoxicity. Biochim Biophys Acta (BBA)-Mol Basis Dis 1865(8):2068–2081
34. Ke T et al (2021) Developmental exposure to methylmercury and ADHD, a literature review of epigenetic studies. Environ Epigenetics 7(1):dvab014
35. Khan EA, Abbas Z (2021) A scoping review of sources of mercury and its health effects among Pakistan's most vulnerable population. Rev Environ Health 36(1):39–45
36. Khattak SA et al (2021) Potential risk and source distribution of groundwater contamination by mercury in district Swabi, Pakistan: application of multivariate study. Environ Dev Sustain 23:2279–2297
37. Kleshchev M et al (2021) Impaired semen quality, an increase of sperm morphological defects and DNA fragmentation associated with environmental pollution in urban population of young men from Western Siberia, Russia. PLoS ONE 16(10):e0258900
38. Koch I et al (2021) Correlation of mercury occurrence with age, elemental composition, and life history in sea-run food fish from the Canadian arctic archipelago's lower northwest passage. Foods 10(11):2621
39. Li X et al (2021) Mechanisms of oxidative stress in methylmercury-induced neurodevelopmental toxicity. Neurotoxicology 85:33–46
40. Li Y et al (2022) Looping mercury cycle in global environmental-economic system modeling. Environ Sci Technol 56(5):2861–2879
41. Liu C, Ralston NV (2021) Seafood and health: what you need to know? Advances in food and nutrition research. Elsevier 97:275–318
42. Liu D et al (2020) Novel carbon-based sorbents for elemental mercury removal from gas streams: a review. Chem Eng J 391:123514
43. Liu M et al (2021) Rivers as the largest source of mercury to coastal oceans worldwide. Nat Geosci 14(9):672–677
44. Ma Y et al (2018) Mercuric chloride induced ovarian oxidative stress by suppressing Nrf2-Keap1 signal pathway and its downstream genes in laying hens. Biol Trace Elem Res 185:185–196
45. Macrae CF et al (2020) Mercury 4.0: From visualization to analysis, design and prediction. J Appl Crystallogr 53(1):226–235
46. Malvandi H (2021) Assessing the potential health risk from mercury through consumption of the most popular and preferable fish species, Rutilus frisii kutum, on the Northern Coast of Iran. Biol Trace Elem Res 199(4):1604–1610
47. Malvandi H et al (2020) Assessment of mercury contamination in perch species in the Southern Caspian Sea. Arch Environ Contam Toxicol 79(1):147–155
48. Manyani A et al (2021) Dental amalgam risks in dental staff: systematic review. In: E3S web of conferences, EDP Sciences
49. Marczak M et al (2019) Investigation of subbituminous coal and lignite combustion processes in terms of mercury and arsenic removal. Fuel 251:572–579
50. Marrugo-Madrid S et al (2022) Health risk assessment for human exposure to mercury species and arsenic via consumption of local food in a gold mining area in Colombia. Environ Res 215:113950
51. Martinez-Morata I et al (2023) A state-of-the-science review on metal biomarkers. Curr Environ Health Rep 1–35
52. Massányi P et al (2020) Effects of cadmium, lead, and mercury on the structure and function of reproductive organs. Toxics 8(4):94
53. Miraj SS et al (2020) Clinical toxicology of mercury: source, toxidrome, mechanism of toxicity and management. Metal toxicology handbook, CRC Press, pp 393–404
54. Moriarity RJ et al (2020) Subsistence fishing in the Eeyou Istchee (James Bay, Quebec, Canada): a regional investigation of fish consumption as a route of exposure to methylmercury. Chemosphere 258:127413
55. Neely MD et al (2021) Single cell RNA sequencing detects persistent cell type-and methylmercury exposure paradigm-specific effects in a human cortical neurodevelopmental model. Food Chem Toxicol 154:112288

56. Oteyola AO, Ola-Oladimeji FA (2022) Heavy metal contamination from construction materials. Ecol Health Eff Build Mater 113–131
57. Pamphlett R et al (2019) Mercury in the retina and optic nerve following prenatal exposure to mercury vapor. PLoS ONE 14(8):e0220859
58. Parida L, Patel TN (2023) Systemic impact of heavy metals and their role in cancer development: a review. Environ Monit Assess 195(6):766
59. Pérez R et al (2019) Biomonitoring of mercury in hair of children living in the Valencian Region (Spain). Exposure and risk assessment. Chemosphere 217:558–566
60. Pettigrew SM et al (2022) Adult exposures to toxic trace elements as measured in nails along the interoceanic highway in the Peruvian Amazon. Int J Environ Res Public Health 19(10)
61. Pitzianti MB et al (2020) New insights on the effects of methylphenidate in attention deficit hyperactivity disorder. Front Psych 11:531092
62. Podgórska A et al (2021) Natural and conventional cosmetics—mercury exposure assessment. Molecules 26(13):4088
63. Pollard KM et al (2019) Mercury-induced inflammation and autoimmunity. Biochimica et Biophysica Acta (BBA)-General Subjects 1863(12):129299
64. Pollard KM et al (2018) Environmental xenobiotic exposure and autoimmunity. Curr Opin Toxicol 10:15–22
65. Prince LM et al (2021) Environmentally relevant developmental methylmercury exposures alter neuronal differentiation in a human-induced pluripotent stem cell model. Food Chem Toxicol 152:112178
66. Queipo-Abad S et al (2019) Concentration of mercury species in hair, blood and urine of individuals occupationally exposed to gaseous elemental mercury in Asturias (Spain) and its comparison with individuals from a control group formed by close relatives. Sci Total Environ 672:314–323
67. Rahman MA et al (2020) Emerging risk of environmental factors: insight mechanisms of Alzheimer's diseases. Environ Sci Pollut Res 27:44659–44672
68. Renu K et al (2021) Molecular mechanism of heavy metals (Lead, Chromium, Arsenic, Mercury, Nickel and Cadmium)-induced hepatotoxicity–a review. Chemosphere 271:129735
69. Ricketts P et al (2020) Mercury exposure associated with use of skin lightening products in Jamaica. J Health Pollut 10(26)
70. Rozkrut D (2019) Rocznik statystyczny rolnictwa. Główny Urząd Stat. Warszawa 300:324
71. Sanlier N et al (2018) Tea consumption and disease correlations. Trends Food Sci Technol 78:95–106
72. Sattar T (2022) Brief discussion on mercury poisoning, its sources and remedies to cure it. J Chem Health Risks 12(2):131–142
73. Silva MJD et al (2023) Determination of total mercury in Spanish samples of baby food, fast food, and daily meal. J Braz Chem Soc 34:517–526
74. So SCA et al (2021) Blood and urine inorganic and organic mercury levels in the United States from 1999 to 2016. Am J Med 134(1):e20–e30
75. Sreeshma J, Sudandiradoss C (2021) Identification of metal binding motifs in protein frameworks to develop novel remediation strategies for Hg 2+ and Cr (VI). Biometals 34:621–638
76. Subbaiah MP et al (2021) Recent advances in effective capture of inorganic mercury from aqueous solutions through sulfurized 2D-material-based adsorbents. J Mater Chem A 9(34):18086–18101
77. Teixeira FB et al (2018) Exposure to inorganic mercury causes oxidative stress, cell death, and functional deficits in the motor cortex. Front Mol Neurosci 11
78. Wang F et al (2021) Spherical-shaped CuS modified carbon nitride nanosheet for efficient capture of elemental mercury from flue gas at low temperature. J Hazard Mater 415:125692
79. Wang Y-J et al (2022) Chronic toxic effects of waterborne mercury on silver carp (Hypophthalmichthys molitrix) Larvae. Water 14(11):1774
80. Wei J et al (2022) Conversion of 2H MoS_2 to 1 T MoS_2 via lithium ion doping: Effective removal of elemental mercury. Chem Eng J 428:131014

81. Witkowska D et al (2021) Heavy metals and human health: Possible exposure pathways and the competition for protein binding sites. Molecules 26(19):6060
82. Zhao M et al (2022) Mercury and mercury-containing preparations: history of use, clinical applications, pharmacology, toxicology, and pharmacokinetics in traditional Chinese medicine. Front Pharmacol 13:510

A Review of Mercury Contamination in Water and Its Impact on Public Health

Rakesh Pant, Nikita Mathpal, Rajnandini Chauhan, Arsh Singh, and Amit Gupta

Abstract Mercury is a toxic heavy metal that can have significant health impacts on humans and other organisms. Mercury can enter the environment through natural sources such as volcanic activity or through human activities such as industrial pollution. The aim of this chapter is to focus on mercury contamination in water and its impact on public health. Mercury can enter water bodies through various sources such as natural weathering of rocks, atmospheric deposition, runoff from agricultural lands, and industrial effluents. The major sources of mercury pollution in water are mining activities, pulp and paper mills, and coal-fired power plants. Mercury can have significant impacts on public health, especially for vulnerable populations such as pregnant women, infants, and children. Exposure to mercury can cause a range of health problems such as neurological disorders, developmental delays, and cardiovascular disease. Preventing and mitigating mercury contamination in water requires a multi-faceted approach. Industrial sources of mercury pollution must be regulated and monitored to prevent contamination of water bodies. Improving water treatment technologies and practices can also help to remove mercury from contaminated water. Mercury contamination in water is a significant public health issue that can have long-lasting impacts on human health. Preventing and mitigating mercury contamination in water requires a coordinated effort from industry, government, and the public. By taking action to reduce mercury pollution, we can protect public health and preserve the integrity of our water resources.

Keywords Heavy metal · Contamination · Mercury pollution · Health impact · Water contamination

R. Pant · N. Mathpal · R. Chauhan · A. Singh · A. Gupta (✉)
Department of Microbiology, Graphic Era (Deemed to Be) University, Dehradun 248001, UK, India
e-mail: dr.amitgupta.bt@geu.ac.in; amitvsbt@gmail.com

R. Pant
Department of Quality, Eureka Forbes Ltd., Lal Tappar, Dehradun 248140, UK, India

N. Kumar (ed.), *Mercury Toxicity Mitigation: Sustainable Nexus Approach*, Earth and Environmental Sciences Library, https://doi.org/10.1007/978-3-031-48817-7_4

1 Introduction

Mercury (Hg) is a worldwide contaminant, and its organic form, methylmercury (MeHg), has been linked to cognition abnormalities in human fetuses and cardiovascular consequences in adults. At normal temperatures and pressure, mercury is a highly malleable liquid. This substance is named after the Latin word hydrargyrum, meaning "liquid silver metal" [1]. The chemical form of mercury in the air influences its persistence & spread in the environment. Although the compounds of mercury can linger in the air for more than four years, they are swiftly deposited near their source [2]. Humans are mostly exposed to MeHg through food intake. Hg may also be found in many types of goods, machinery, surroundings, and workplaces. In accordance to experts, the major causes of mercury are the combustion of contaminated fossil fuels and other solid fuels. The numerous kinds of mercury have diverse health impacts based on exposure period, amount of exposure, and mercury type. The inorganic form of mercury penetrates the body through food or via the skin rather than inhaling due to its non-volatile nature. However, the majority of inorganic mercury that goes into our systems is stored in the proximal tubule of the kidneys. Symptoms of mercury poisoning from elemental mercury vapour inhalation is illustrated in Fig. 1. Although the signs of access vary according to the kind of contact, acute high-dose exposure can result in life-threatening complications. It can, for example, blacken or fades the oral mucous membrane, causing searing chest discomfort, gastric system injury, mercurial stomatitis, and renal function problems [3]. Mercury element reaches the human body by breathing since it is often present in the form of vapor. When elemental mercury is breathed, it is promptly absorbed into the lungs and subsequently disseminated into the blood at a rate of 80%. Multiple investigations have revealed that mercury is not only harmful to human health, but it may also have an impact on industrial processes and equipment. Several investigations on mercury and its exposure, toxicity, and consequences on human health, safety, and the surroundings have been published in journals throughout the course of time. Several strategies for detecting and collecting mercury content and concentrations have also been developed and researched [1]. Mercury (Hg) is a heavy metal with unique physicochemical features that pose concerns to both the ecosystem and human health. Dental workers are still occupationally exposed to metallic mercury (Hg^0) vapor in many regions of the world while making amalgam as a filling material and when inserting, plastering, and removing amalgams via drilling [3]. Mercury (Hg) pollution in the vast Amazon Basin has been a major source of worry over the previous two decades [4].

Mercury is one of the World Health Organization's top ten substances of public health concern. International assessment programmes have established that mercury pollution sources are worldwide spread; discharge levels continue to rise over time; and these discharges are mostly the result of anthropogenic activities. Mercury and its derivatives have numerous qualities that make them helpful in civilization, but balancing these important applications against the metal's hazardous features is difficult. Our understanding of the dangers of mercury to human health is evolving. An

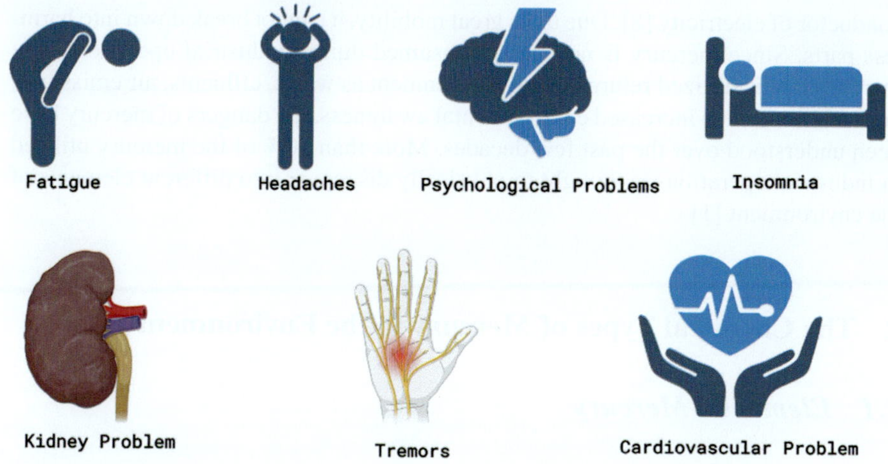

Fig. 1 Symptoms of mercury poisoning from elemental mercury vapour inhalation

early focus on workplace safety and ecological catastrophes was driven by work-related well-being, medication, and toxicology experts, and trailed by epidemiological and exposure science investigations of certain susceptible population groups [5].

Mercury is a hazardous chemical that is widely utilized in skin-lightening cosmetics. Mercury is often utilized because of its potential to generate a dramatic whitening effect, particularly when applied in large quantities. The World Health Organization recommends that mercury levels in skin-lightening cosmetics be less than one part per million (ppm) [6]. Mercury toxicity varies depending on its chemical form; therefore, symptoms and indicators change when exposed to elemental mercury, inorganic mercury compounds, or organic mercury compounds. The sources of exposure vary significantly for the various types of mercury. While it is well acknowledged that mercury and its derivatives are very toxic molecules with serious consequences, there is continuous disagreement over how dangerous these substances, particularly methylmercury, are. New results during the last decade suggest that harmful effects may occur at lower doses than previously anticipated, possibly affecting a bigger portion of the world population [7].

Because of its high toxicity and mobility in environments, mercury is currently considered an environmental contaminant posing a significant danger to human health. Mercury exposure can come from both natural and man-made sources. Mercury exposure can occur as a result of human actions such as the use of fossil fuels, chloralkaline industries, mining, trash burning, and the use of coal and petroleum. Volcanic activity, earthquakes, erosion, and the volatilization of mercury existing in the marine environment and flora are other natural sources of mercury.

Mercury offers a huge range of industrial and commercial uses. It is commonly used in electrical devices like meters, switches, batteries, etc. since it is a good conductor of electricity [8]. Due to its great mobility, it cannot break down into harmless parts. Since mercury is not really consumed during industrial operations, any mercury that is utilized returns to the environment as waste, effluents, air emissions, or products. Due to increased environmental awareness, the dangers of mercury have been understood over the past few decades. More than 90% of the mercury utilized in industrial operations is thought to physically disappear into different elements of the environment [1].

2 The Chemical Types of Mercury in the Environment

2.1 Elemental Mercury

Mercury is a bright, silver-white metal that is liquid at room temperature. It was once termed as quicksilver. It is used in older thermometers, fluorescent lights, and some electrical switches. Elemental mercury breaks up into smaller droplets when it is dropped, and these droplets may either attach to particular surfaces or fit through very small gaps. Elemental mercury that has been exposed to air can dissipate and transform into an odourless, colourless deadly vapour at room temp. When heated, it transforms into an odourless, colourless gas [8]. Elements of mercury are those that have not been exposed to other substances. Mercury combines with other substances to generate compounds like methylmercury or inorganic mercury salts. Burning coal & other fossil fuels causes it to be emitted into the atmosphere. Usually, mercury may be found in a number of different forms, including (i) soluble substances, (ii) insoluble mercury, and (iii) volatile mercury species [9].

2.2 Inorganic Mercury

Inorganic mercury is widely distributed in the ecosystem and is typically found in the mineral's cinnabar and metacinnabar as well as impurities in other minerals. Mercury is easily able to mix with other elements such as sulfur, chlorine, and oxygen to generate inorganic salts. Inorganic mercury salts can be present in the soil and are transported by water. These salts can be dispersed into the environment as a particulate matter after the mining of deposits of mercury-containing ores. The industry that uses mercury as well as coal-fired power plants and landfills are further sources of elemental and inorganic mercury emissions. Along with the discharge of contaminated water from businesses or water treatment facilities, inorganic mercury may also enter water or soil through the deterioration of rocks that include inorganic mercury salts [10].

Many skin-lightening soaps and lotions still include inorganic mercury compounds. Mercuric chloride is a topical antibacterial and sterilizer, wood preservative, and fungal that is also used in photography. Natural and professional contexts can both expose people to inorganic mercury salts. Mining, the production of electrical equipment, and the use of mercury in chemical and metal processing are among the professions that provide a greater peril of exposure to mercury and its salts. The cutaneous route of exposure to mercuric chloride in the general population can be brought on by the use of soaps, lotions, topical disinfectants, and sterilizers [8].

2.3 Organic Mercury

Any mercury compound containing carbon is referred to as organic mercury. The most prevalent type of mercury that individuals in the United States come into contact with is methylmercury, a very poisonous chemical molecule. Methylmercury is present in the bodies of almost all humans, which is a reflection of how common it is in the environment. The majority of people, however, have mercury levels in their systems that are below the range linked to potential health risks. When airborne particles may pick up inorganic mercury salts. These specks are spread throughout the earth by rain and snow. Even when mercury is dumped on land, it frequently returns to the atmosphere as a gas or in combination with other particles, where it is then dumped again. People are most frequently exposed to methylmercury when they consume fish and shellfish that contain high tissue concentrations of the toxin [3].

Mercury experiences a variety of intricate chemical and physical changes when it moves back and forth among the surroundings, terrestrial, & aquatic, various of which are still not fully understood. Mercury may combine with carbon in tiny microorganisms to change from an inorganic to an organic state. Methylmercury is a highly pervasive and harmful organic mercury element in our surroundings [11].

3 Natural Source of Mercury Contamination in Water

Almost all geological media contain tiny but varied quantities of mercury. The main sources of mercury include mineral deterioration, forest fires, and natural degassing of the earth's crust, which includes evaporation from soil and water surfaces. Due to their volatility, elemental and oxidized forms of mercury are constantly being released into the surroundings [12]. Mercury is transported and distributed throughout the environment via a number of cycles. The worldwide cycle includes the air movement of elemental mercury vapors from land-based sources to the seas, as well as the erosion and leaching of geological formations that contain mercury by rainfall, which also moves the metal to streams and lakes via surface runoff [13]. Because mercury is a strong neurotoxin that may be lethal to humans, animals, and fish, it is one of the most dangerous toxins to our country's waterways. The earth's crust naturally contains mercury, but human activities like mining and the burning of fossil fuels have caused in significant worldwide mercury contamination [14]. Mercury that is released into the air eventually sinks into the ground or is washed into the ocean. Once deposited, certain microbes have the ability to transform it into methylmercury, a very dangerous form of mercury that accumulates in fish, shellfish, and mammals that consume fish. Eating fish and shellfish that are methylmercury-contaminated is how most people get exposed to mercury. Natural phenomena including Volcanic activity, Erosion of rocks and soils, Atmospheric deposition, Degassing of the Earth's crust all produce mercury [13].

3.1 Volcanic Activity

A major source of Hg in the environment is volcanoes. However, estimates of the typical current volcanic emission rates range widely, from 75 to 700 Mg/ya. The baseline conditions that are only surpassed by big eruptions are reflected in modern volcanic emission rates. For instance, the 1991 Mount Pinatubo eruption caused the atmospheric Hg levels in the Austrian Alps to double [15]. It has also been proposed that transient spikes in Hg contents in ice core records can be used to track significant volcanic outbreaks over the last few hundred to thousands of years. Modern industrial Hg discharges to the atmosphere, which are approximately similar in ratio to geogenic sources given the harmful effects of Hg and its persistence in the surroundings, have been the focus of significant worldwide concern. As a result, Hg may not only be a possible fingerprint for enormous volcanism linked to extinction events but also contribute to atmospheric deposition. Over the past ten years, a vast amount of data has been generated due to the quickly growing interest in mercury in the geologic record. Aquatic organisms may be directly impacted by the eruption of volcanic ash and other materials. When volcanic ash is dispersed across the land and subsequently washes into rivers and lakes, it can increase the amounts of mercury in the water by releasing mercury from the subterranean rocks and soils [16].

3.2 Erosion of Rocks and Soils

The development of inorganic and organic mercury compounds is often favored by soil conditions. The amount of mercury in soil serves as a proxy for the metal's propensity to pollute surface runoff and groundwater. Direct dumping or landfilling of garbage containing mercury both have the potential to pollute the soil [17]. The presence of mercury in water body sediments may reveal previous pollution. Due to its high toxicity, bioaccumulation, and prolonged residence in the surroundings during broad travel, mercury (Hg) is a worldwide pollutant that has a negative impact on both ecosystems and people. As a result of bioaccumulation, soil poisoning with mercury might eventually harm human health by stunting crop development or killing plants. Mercury in soil, therefore, has a tight connection to both agricultural development and human survival [18]. Many researchers are interested in mercury contaminants, therefore, some studies have previously described how mercury behaves in the world's surface environment. Additionally, research on the variables influencing mercury distribution patterns in various land use types is beneficial for generating fundamental concepts for resolving probable mercury pollution in the future. Mercury is naturally distributed in water as a result of interactions between water and soil and rock during storm runoff, percolation into the earth, and movements beneath the earth where various geochemical stresses are present. The distribution of mercury in the earth's crust and a wide range of other Earth activities determine the pattern of this distribution [10]. Eliminating erosion's influence on mercury contamination of water sources can assist. This may be accomplished in a variety of ways, including by reducing the amount of land developed, using best management practices for mining and other land use operations, and implementing erosion control measures like planting plants [19]. Different factors might have an impact on soil concentrations from precipitation to potential evapotranspiration (PET). Higher leaching may result in greater Hg losses from the soil. Therefore, more precipitation results in more Hg being deposited wetly [11].

3.3 *Atmospheric Deposition*

Mercury's main source in the waters is atmospheric deposition. Three different forms of mercury enter the ocean through atmospheric deposition. Through air–water interaction, gaseous elemental mercury (Hg^0) finds its way into the ocean. Both wet and dry deposition allows the entry of inorganic mercury (Hg^{2+}/HgII) and mercury that is particle-bound (Hg(P)). Most aquatic habitats get their mercury mostly from atmospheric deposition. The major source of mercury in the environment, according to the U.S. Environmental Protection Agency (USEPA), is emissions from coal-fired power plants [20]. Pollution is wet deposition, which rain removes from the atmosphere. Pollutants from the sky that fall to the ground, trees, etc. are known as dry deposition. Pollutants are deposited as wet deposition, but rain thoroughly removes them from the atmosphere. According to estimates, the majority of inland water bodies and estuaries in the United States get 50–90% of their mercury burden via mercury wet deposition. Mercury is only ever relocated and finally buried beneath silt and soil, it is never actually taken out of the ecosystem [17].

3.4 *Degassing of the Earth's Crust*

Degassing, which is release of gases from a planet's interior, can take place directly during volcanism or indirectly through the weathering of igneous rocks on the planet's surface. By degassing the earth's surface and resuspending and dissolving previously deposited mercury particles, mercury is released into the atmosphere. Only 10% of the current emissions to the environment are thought to come from volcanoes and other natural sources. The use of mercury for shady gold mining and the generation of electricity from fossil fuels, particularly coals with high mercury concentrations, are the main anthropogenic sources; the yearly emissions from both sources have been estimated to be over 2000 tonnes [14]. Informal gold mining is a significant usage of liquid metallic mercury that has grown over the past several decades. As a result, this practice exposes the gold miner to mercury vapor and causes a significant mercury emission into the environment [21]. Volcanic activity, which has the potential to release significant amounts of mercury into the atmosphere during eruptions, is a important natural basis of degassing. Mercury may travel large distances in the atmosphere before falling to the ground as rain or by dry deposition onto land and water surfaces. This deposition may cause contamination of soil and surface water bodies [18].

4 Anthropogenic Source of Mercury Contamination in Water

Mercury emissions that are created by humans might come through the usage of fuels, raw materials, finished goods, or industrial processes. A portion of the mercury that is present in our environment now was discharged in the past. After being initially released into the environment, mercury may continually be reemitted into the atmosphere by land, water, and other surfaces. We are aware that anthropogenic emissions continue to considerably increase the amount of mercury in the atmosphere. The main manufacture of non-ferrous metals, burning of fossil fuels, and artisanal and small-scale gold mining are foremost anthropogenic sources of mercury emission. Other somewhat minor sources include the manufacture of cement, trash from consumer goods, crematories, polluted sites, and the chloralkaline industry. Recognized substantial sources of mercury leakage in the environment include industrial usage and commercial items. The main sources of mercury emissions are hazardous waste combustion, incinerators, and coal-burning power stations. Additionally, wastewater, hospitals, dental offices, municipal and industrial sites, as well as the breaking or disposal of devices containing mercury such as fluorescent lights, thermostats, and thermometers, all contribute directly to the presence of mercury in the environment [22].

4.1 Artisanal and Small-Scale Gold Mining

The use of mercury in small-scale gold mining is the main manmade source of mercury in the atmosphere. In many nations, artisanal and small-scale gold mining use elemental mercury. Gold-containing materials are combined with mercury to create a mercury-gold amalgam. The amalgam is heated to vaporise the mercury and release the gold. This procedure has severe health hazards and the potential for serious injury. Environmental deterioration, ecosystem loss, and mercury (Hg) contamination from gold processing (amalgam burning) are the primary environmental issues connected to artisanal gold mining operations in Ghana [23]. Around 38% of all anthropogenic emissions on a worldwide basis come from Hg emissions from ASGM. Large amounts of mercury are released into the air, land, and water as a result of this practise, which typically takes place in unclean and primitive conditions. As a result, populations close to artisanal mining sites may gradually build up mercury in their systems, which is particularly harmful to unborn babies, young children, and the elderly. Long-term mercury exposure can harm kidneys, produce mercury poisoning, and affect hearing, vision, and balance [24].

4.2 Power Plants Powered by Coal

Another substantial source of mercury pollution in the environment is coal-fired power stations. Despite the relatively low mercury concentration of fossil fuels like coal, burning significant amounts of them can cause pollution. When coal and other fossil fuels are burned, mercury is released into the atmosphere. Because coal contains far more mercury than other fossil fuels, coal-fired power stations frequently produce more mercury pollution than those that burn other fossil fuels. In some areas of the country, mercury pollution from coal plants is very bad.

The mercury that coal-burning power stations produce settles on soils and enters streams. There, microbes convert the relatively innocuous mercury into the dangerous organic form methylmercury. It is the only metal that is liquid at ambient temp and will evaporate into mercury vapor when heated merely moderately. As a result, when coal is burned at a power plant, the mercury that is present in it is expelled via the smokestacks as gas, where it diluted as it travelled. By contributing over 48% of all domestic mercury emissions, coal-fired power plants are the major source of mercury emissions in the U.S., according to the Environmental Protection Agency (EPA).

Depending on where it comes from, coal's mercury concentration ranges from 0.02 to 0.31 parts per million, with Indian coals from different mines having an average of 0.11 ppm. Mercury may be found in coal in a variety of forms, and when coal is burned, the mercury vaporizes. The quantity of mercury emissions produced by coal-fired power stations is thought to be rather large. More than 42% of the mercury emissions in Eastern Africa and over 40% in the former Soviet Union are caused by coal burning. Indian coal doesn't have a lot of mercury in it, but when it burns, traces of mercury that are present in the coal might produce mercury emissions. Because the nation's thermal power capacity is forecast to grow, the overall coal consumption for energy production is anticipated to rise even more; as a result, mercury emissions from coal use are likely to rise in the future.

4.3 Industrial Processes

For more than 2000 years, people have mined and utilized mercury all over the planet. Mercury is used in a various of industrial processes. Color-alkali industries, coal-fired facilities, plastics industries, pulp and paper industries, electrical appliances, medical devices, some medications, and agricultural goods all utilize and emit mercury in various ways. Since the dawn of the industrial period, mercury contamination has grown by approximately three times due to extensive emissions of mercury from mining, the manufacturing of chlorine, and fossil fuel energy sources. An estimated 2500 tonnes of mercury are emitted annually globally due to human activity. It might enter the atmosphere in a number of ways, such as via industrial processes. The largest industrial sources of mercury emissions are coal-fired power plants, incinerators, and cement factories. When coal is burnt to produce electricity, mercury from the coal is released into the atmosphere. Cement production is a significant source of mercury emissions since mercury is a component of the raw materials used to make cement.

4.4 Municipal Wastewater

Due to the fact that home and industrial sources contribute metal to wastewater treatment works (WWTWs), their discharges have the potential to increase the amount of metal in rivers to a dangerous level. The Water Framework Directive (2000/60/EC) classifies mercury as a priority hazardous substance (PHS), which necessitates the halt of emissions. Diverse sources, such as dentistry practice wastes, which may give up to fifty percent of the crude mercury concentrations in wastewater, fertilizers, landfill leachate, paints, residential waste inputs, groundwater penetration, stormwater drainage contributions, and historical sources of mercury, all contribute to the presence of mercury in the wastewater. After industrial wastes, municipal sewage has emerged as the second-most significant source for the direct discharge of manmade mercury into sea habitats. One-fourth of the total worldwide Hg emission inventory is thought to have come from sewage discharges.

The land application was used to dispose of around 29.3% of the sludge, followed by incineration (26.7%) and sanitary landfills (20.1%). Monitoring and regulating the discharge of municipal wastewater is essential to lowering the amount of mercury that enters the ecosystem. Proper treatment techniques may be able to remove the bulk of the mercury in wastewater, but more work may be needed to reduce mercury inputs into the municipal wastewater system. Initiatives aimed at raising public awareness and educating the public can also help reduce the overall amount of mercury that enters the wastewater system by persuading consumers to use fewer products that contain the metal.

4.5 Agricultural Practices

The worldwide cycle of mercury is significantly influenced by soil because it serves as a crucial conduit between the atmosphere and water. In addition to being a mercury sink, soil also acts as a mercury source, taking up mercury from the surroundings and re-emitting it into the air, water, or plants. As a result of bioaccumulation, soil poisoning with mercury might eventually harm human health by stunting crop development or killing plants. Mercury in soil, therefore, has a tight connection to both agricultural development and human survival [15].

Mercury (Hg) is regarded as a worldwide contaminant, and studies have shown that it has negative impacts on ecology and human health. Parent material and soil characteristics, such as organic matter, soil microbes, and soil pH, also human actions, like non-ferrous mining, petroleum refining, fossil fuel combustion, discharge of waste from industrial production, and fertilizer applications, all have a significant impact on the concentrations of mercury in farmland soil. Several techniques have been put out to look for potential Hg sources of heavy metal concentrations in soil, including statistical and geochemical, regulatory reference value, and GIS-based techniques. For the prevention and management of contamination in the soil-crop system, efforts to identify the sources of Hg in farming soil are of utmost importance [25].

4.6 Landfills

Municipal solid waste dumps have a significant impact on neigh boring ecosystems and human health through mercury (Hg) discharge and migration. Mercury is a dangerous heavy metal that may harm the environment and the public's health. Batteries, thermometers, fluorescent lights, and electrical equipment are just a few of the waste products that are deposited in landfills. When mercury-containing goods are broken down and dumped in landfills, mercury may leach into the environment. Additionally, bacteria in landfills may convert mercury into the extremely dangerous element methylmercury, which can contaminate soil, water, and the environment [26].

Municipal solid waste (MSW) is disposed of in landfills, which are highly varied and heterogeneous settings that are employed both in developed and developing nations. The Hg discharged from landfills comes from Hg-containing solid waste that undergoes physical, chemical, and biological changes to become volatile Hg^0 and the more hazardous volatile methylmercury. Therefore, in order to limit the emission of mercury into the atmosphere, it is essential to properly dispose of mercury-containing products and manage landfill sites. A significant source of mercury contamination may come from landfills [26].

5 Medical Source of Mercury Contamination in Water

Medical waste disposal is one potential cause of mercury contamination in water. Medical devices including thermometers, blood pressure monitors, and dental fillings frequently contain mercury. If these things are disposed of improperly, such as by putting them in the garbage, mercury may leak into the environment and contaminate water supplies [27]. Pharmaceuticals containing mercury, such as certain vaccinations and topical antiseptics, are another possible cause of mercury pollution in water. These goods can increase the amount of mercury in the water when they are inappropriately disposed of. To avoid contaminating the environment, healthcare institutions must appropriately handle and dispose of medical waste. To prevent the discharge of mercury into the environment, people should dispose of any mercury-containing devices, such as batteries and compact fluorescent light bulbs, correctly [28].

5.1 Dental Amalgams

Due to its low cost, simplicity of usage, strength, longevity, and bacteriostatic action, amalgam has been used in dentistry for over 150 years and is being used today. Because to illegal dental mercury transfers, crematoria discharges from the dead, and sewage sludge supplied to farmers, as well as artisanal and small-scale gold mining, dental amalgam is a hidden source of mercury contamination throughout the world. These important mercury sources cause pollution of the air, water, and food, which has an adverse effect on human health. Dental amalgam restorations may modestly increase these levels, although this is not clinically or practically significant. The major exposure to mercury from dental amalgam is when a restoration is put in or taken out of a tooth. Less mercury is emitted when the reaction is finished, which is far less than the existing health threshold [29].

According to the United States EPA, dental clinics are the major cause of mercury emissions to publicly owned treatment works (POTWs), which are municipal wastewater treatment facilities. Dental offices in the US discharge more than 5.1 tonnes of mercury into POTWs annually, which can eventually pollute water supplies, according to the EPA. Even while dental amalgam alone does not considerably increase the amount of mercury in water, improper disposal of the substance might be harmful to the environment. Proper management is necessary to lessen the detrimental impacts of dental amalgam waste on the ecosystem and public health [30].

5.2 Medical Waste Incineration

Hospitals, veterinary clinics, or medical research institutes all create garbage, which is burned during medical waste incineration. Both contagious ("red bag") medical wastes and non-infectious, normal household wastes are included in these wastes. The emission variables shown here include emissions from the combustion of both contagious and non-infectious trash [31].

Any mercury-containing items, including as thermometers, blood pressure monitors, and batteries, can leak mercury into the environment when medical waste is burned. Heavy metal mercury is poisonous and can have major negative effects on both human and animal health. Mercury may infiltrate streams after being discharged into the environment, where it can build up in fish and other aquatic life. Dangerous quantities of mercury may be ingested by people when they eat tainted fish, which can result in brain damage and other health issues. In order to minimize the environmental effect of medical waste incinerators, regulatory procedures can be put in place to guarantee adequate monitoring and regulation [29].

5.3 Pharmaceutical Manufacturing

Pharmaceuticals make up a sizable class of aquatic pollutants and pose a substantial hazard to the health of creatures that are not the intended targets. On a worldwide basis, no stringent legislative restrictions on drug use and disposal into bodies of water have been put in place. Mercury and preparations containing mercury have historically been utilized extensively in traditional Chinese medicine and applied in various therapeutic settings, primarily in the form of mercury sulphides [30].

Manufacturing of some medications, and cleaning and repairing equipment are some of the pharmaceutical industry's uses for mercury. The discharge of wastewater containing mercury from pharmaceutical production is one of the main causes of mercury contamination in water. Wastewater from several pharmaceutical production procedures, such as synthesis, purification, and packaging, contains mercury. Pharmaceutical production procedures occasionally require mercury, despite the fact that it is poisonous. There are, however, companies that are doing everything they can to reduce their mercury emissions and their impact on the environment [29].

6 Atmospheric Deposition Source of Mercury Contamination in Water

The atmospheric deposition of mercury is a significant factor in water contamination. Mercury may be released into the atmosphere via natural processes such as volcanic eruptions, forest fires, and the burning of fossil fuels. Most aquatic habitats get their mercury mostly from atmospheric deposition. Inorganic mercury is the main form of mercury that is deposited in the atmosphere [16]. Atmospheric deposition of mercury may cause contamination in both remote and inhabited areas. The quantity of mercury deposited in water through atmospheric deposition might vary depending on the amount of mercury released, the distance from the source of release, and the environmental parameters of the water body. Mercury in water can be contaminated significantly by air deposition, especially when there are few other sources of mercury contamination in the area [21].

6.1 Gaseous Elemental Mercury (GEM)

Three chemical and physical forms of mercury are present in the surroundings: gaseous elemental mercury, divalent reactive gaseous mercury, and particulate mercury (PHg). Research on the environment should pay close attention to the many sources of urban gaseous elemental mercury (GEM or Hg^0) emissions. The fate of mercury (Hg) in the atmosphere is significantly influenced by gaseous exchanges at the water–air interface at polluted sites [25]. Due to its high volatility and poor solubility in its elemental form (Hg^0 or GEM, Gaseous Elemental Mercury) [1], which requires long-range transit before being transformed to the oxidized, more soluble, and readily removed form (Hg^{2+}) [3, 4], Hg can stay in the atmosphere for longer than a year. Hg can so travel and have an effect on distant ecosystems that are far from the sources of emission. GEM could cause water supplies to become contaminated by air deposition [32].

Despite the fact that GEM is not regarded to be harmful in and of itself, it can contribute to mercury poisoning of water by changing into more dangerous forms of mercury, such as methylmercury, as well as by depositing on water surfaces where it can transition into other forms of mercury. Strategies to reduce GEM emissions and their environmental impact are needed to protect against the harmful effects of mercury contamination [33].

6.2 Particulate Mercury (PBM)

Due to its high dry deposition velocities and scavenging coefficients, which push mercury into both terrestrial and marine ecosystems, particulate-bound mercury (PBM) is a worry for the environment on a worldwide scale. Particulate mercury (PBM) is mercury that is found in the atmosphere as minuscule particles or droplets [34]. PBM may be transported over vast distances and deposited on water and land surfaces by air deposition. When PBM is deposited on water surfaces, it can transform into more hazardous forms of mercury, such as reactive gaseous mercury (RGM) and methylmercury. When someone consumes contaminated food or drink or breaths in polluted air, PBM exposure may result. PBM exposure has been related to a multitude of wellbeing difficulties, counting neurological damage, developmental harm, and respiratory and cardiovascular problems.

7 Mining and Industrial Source of Mercury Contamination in Water

Water pollution from mining and industrial processes is the main cause of mercury contamination. Mining ores for gold and silver frequently exposes miners to a naturally occurring element called mercury. Processing these ores releases mercury into the environment. Industrial processes such as cement manufacture, trash incineration, and coal-fired power plants also release mercury into the air, where it eventually ends up in rivers. Fish and another aquatic kinds might be exposed to toxic methylmercury when mercury is in the water. It is also possible to reduce mercury pollution in water by regulating the use and disposal of mercury-containing items such as thermometers and fluorescent light bulbs [35]. Different sources of mercury contamination are illustrated in Fig. 2.

Fig. 2 Different sources of mercury contamination

7.1 Gold Mining

Water pollution caused by gold mining is a major cause of mercury contamination. Gold is often extracted from other minerals using mercury in small-scale gold mining. Gold is left behind when miners burn off mercury mixed with ore. In this process, mercury levels in the environment, in particular rivers, are elevated. As a result, it is impossible to prevent cyanide and mercury pollution in nearby areas through the flow of rivers and groundwater. Mercury was mostly connected to suspended particles in river water.

Gold-containing materials are combined with mercury to produce a mercury-gold amalgam. The amalgam is heated to vaporize the mercury and release the gold. Large amounts of mercury are also released into the environment during the burning of the gold-mercury amalgam. More than any industrial activity, artisanal gold mining today accounts for more than 35% of all human-made mercury emissions worldwide [35].

7.2 Industrial Processes

Mercury pollution from industrialized sources and incinerators is a big problem in the environment. It is anticipated that incineration and industrialized sources would account for 97% of all artificial mercury releases in the U.S. Garbage incineration, cement production, and chlorine synthesis are among the most significant sources of mercury pollution in water. There are many processes in which mercury is used as a catalyst or a raw material. Mercury can be discharged into the environment as a gas or a liquid [7].

Methylmercury, which is extremely dangerous and can build up in fish and other aquatic species when mercury is discharged into the environment, can condense in streams and become methylmercury. Consequently, tuna and swordfish, which humans eat often, have dangerously high mercury levels [10].

The use of pollution control technology and laws on mercury emissions have been implemented to reduce mercury pollution from industrial operations. In the Minamata Convention on Mercury, an international convention aimed at minimizing mercury pollution, regulations for regulating industrial mercury emissions are included [6].

In addition to decreasing mercury pollution in water, safer, more environmentally friendly alternatives can also be used to reduce mercury use in industrial processes. Correct handling and disposal of trash from industrial operations may also prevent mercury from entering the environment [7].

7.3 Landfills

According to the Minamata Convention on Mercury, municipal solid waste (MSW) landfills are recognized as significant cause of atmospheric mercury (Hg) discharges, which may pose concerns to human health and the environment. Municipal solid waste (MSW) is disposed of in landfills, which are highly varied and heterogeneous settings that are employed both in developed and developing nations [20].

The Hg discharged from landfills comes from Hg-containing solid waste that undergoes physical, chemical, and biological changes to become volatile Hg^0 and the more hazardous volatile methylmercury. About 5% of the anthropogenic Hg emissions to the atmosphere come from landfills. Thermometers, fluorescent lights, and other mercury-containing equipment are regularly disposed of in landfills. These things could deteriorate over time and release mercury into the soil and groundwater.

7.4 Oil and Gas Production

Mercury naturally occurs in gas and oil deposits at a variety of quantities spanning six orders of magnitude, with natural gas concentrations reaching up to 5 mg/m^3 and crude oil concentrations reaching up to 600 ppm (mg/kg). Across and within reservoirs and geographical locations, as well as across crude oil and gas samples, mercury levels can vary significantly. Mercury contamination of goods and equipment, poisoning of catalysts, and extensive corrosion of technical equipment all result from the presence of mercury in hydrocarbons. These issues have an impact on both technology and the environment and increase the danger of accidents [36].

Heavy incidents result from metal mercury's quick electrochemical corrosion of aluminum alloys (such as heat exchangers) and steel's liquid metal embrittlement (LME). Regenerative or non-regenerative adsorbents can be used to remove mercury from natural gas. In both situations, hydrocarbon gas enters an adsorption tower at the top and exits the bottom for further processing or sale after passing through the adsorbent where the mercury is adsorbed [1].

Laws addressing mercury emissions and the promotion of cleaner technology have been implemented to reduce mercury emissions from oil and gas extraction. The Environmental Protection Agency (EPA) of the United States, for example, has established guidelines for mercury emissions from oil and gas facilities. Using safer, more eco-friendly alternatives to mercury in oil and gas production procedures can reduce mercury pollution in water. The proper handling and disposal of oil and gas production waste can also prevent mercury from entering the environment [36].

8 Toxic Effects of Mercury Impact on Public Health

Metals have long been understood to be potent carcinogens. The poisonous character of mercury compounds may be reflected in the larger clinical application of mercury compared to its use as an antibacterial agent. All people are subject to mercury exposure. The majority of individuals are chronically exposed to low amounts of mercury. Because mercury is so hazardous, it may be lethal to breathe in and dangerous to absorb through the skin. The neurological system, the digestive and respiratory systems, as well as the kidneys, may all be negatively impacted. Acute mercury poisoning results from transient exposure to high amounts of mercury vapor. However, some persons experience acute mercury poisoning as well as high amounts of exposure. Mercury exposure as a result of an industrial accident is an illustration of acute exposure [2].

The most significant health risk linked to mercury exposure is neurotoxicity. Methylmercury may easily breach the ordinarily protective blood–brain barrier, where it is disseminated to all tissues the brain's protective blood–brain barrier. It is especially concerning to expectant mothers and women who are nursing since it can easily pass via the placenta to growing fetuses and their developing brains. Children's learning problems and the disruption of fish-eating animals' reproductive processes have both been connected to low levels of exposure. Other harmful health impacts of mercury exposure in people include neurological, renal, gastrointestinal, genetic, cardiovascular, and developmental issues, as well as mortality [5]. For the general public's health to be protected from the harmful effects of mercury, exposure to mercury in all its forms should be limited. A number of steps can be taken to do so, including reducing industrial emissions, disposing of mercury-containing goods properly, and avoiding tainted food and water sources. By avoiding mercury-containing seafood, buying mercury-free products, and properly disposing of mercury-containing household appliances, people can also reduce their own exposure to mercury [3].

8.1 Mercury Exposure and Long-Term Health Effects

Inhalation is typically assumed to be the source of the negative health consequences of prolonged exposure to elemental mercury. Nevertheless, a little quantity of mercury liquid and vapour is also absorbed via the skin, and this mode of exposure might add to the total exposure. Elemental or inorganic mercury poisoning is more likely to occur in isolated cases. The metal is visually appealing, and there are several documented instances of kids playing with it to the harm of the entire family. Because methyl mercury is lipid soluble, it can reach bodily tissues like the brain. Recent research suggests that methylmercury enters the brain and other cells through the neutral amino acid transporters and has a structure that mimics methionine when paired with cysteine [11]. For methylmercury poisoning, the Central Nervous System

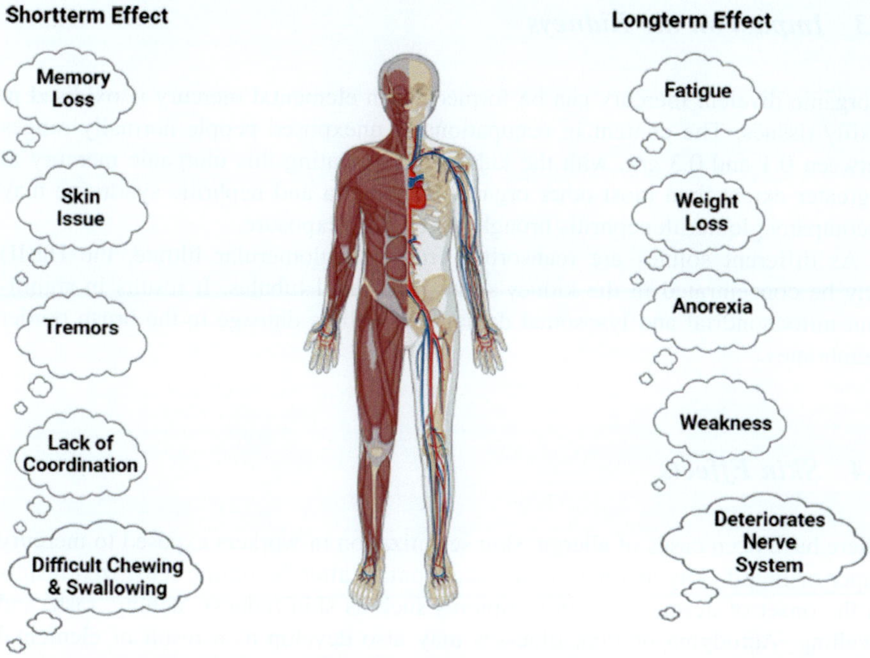

Fig. 3 Mercury exposure short-term and long-term health effects

(CNS) is the target organ. Methyl mercury readily passes the placenta and has terrible consequences on the developing brain. Additionally, it prevents the polymerization of microtubes, which is necessary for neuronal migration and cell division, both of which are critical to the healthy growth of the developing foetal brain. Both adults and foetuses are susceptible to the harmful effects of mercury, although foetuses are 5–10 times more vulnerable. Mercury exposure short-term and long-term health effects are described in Fig. 3.

8.2 Neurological Effects

Strong neurotoxic mercury is. It can harm the human central nervous system permanently, even at very low exposure levels. Mercury poisoning commonly manifests as loss of feeling in the fingers and toes, tremors, hazy eyesight, uncoordinated muscles, and brain cell death.

The neurological system is very toxic to methylmercury. Adult central nerve systems are less vulnerable to methyl-mercury than developing ones. Studies on a population that was exposed to methyl-mercury from fish also point to a link between an increase in cardiovascular disease incidence.

8.3 Impact on the Kidneys

Inorganic divalent mercury can be formed when elemental mercury is oxidised in bodily tissues. The content in occupationally unexposed people normally ranges between 0.1 and 0.3 g/g, with the kidney accumulating this inorganic mercury to a greater extent than most other organs. Proteinuria and nephritic syndrome may accompany glomerulonephritis brought on by high exposure.

As different solutes are reabsorbed from the glomerular filtrate, the Hg(II) may be concentrated in the kidney's proximal renal tubules. It results in significant mitochondrial and lysosomal damage as well as damage to the brush border membranes.

8.4 Skin Effects

There have been cases of allergic skin sensitization in workers exposed to mercury vapor or liquid. Any amount of chemical contact after becoming sensitized results in the onset of dermatitis with symptoms such as skin redness, itching, rash, and swelling. Acrodynia or Pink illnesses may also develop as a result of elemental mercury exposure. Acrodynia, often known as pink illness, is a symptom of mercury poisoning that some people experience and may be brought on by sympathetic nervous system excitation. Calomel ("sweet mercury"), a form of inorganic mercury, was historically often used to cure a variety of illnesses, such as syphilis, typhus, and yellow fever. Calomel-based teething powders contributed to a widespread mercury poisoning epidemic known as "pink disease" or acrodynia in newborns and young children before their toxicity was discovered in the 1940s [20].

9 Mercury-Related Diseases

From a toxicological perspective, metallic, divalent inorganic, and methylmercury molecules are the main types of mercury. The high vapour pressure of elemental mercury makes its vapours potentially poisonous when inhaled. Extreme shyness, excitability, memory loss, and other behavioural and personality abnormalities are all signs of mercurial erethism. In recent years, it has become clear that mercury may have a role in a variety of conditions with nebulous symptoms, including multiple sclerosis and Alzheimer's disease [15]. The majority of casualties have been workers in chemical manufacturing companies, farmers, and members of their families who have unintentionally consumed dressed seeds since organic mercury compounds have been utilised predominantly for seed dressing [34]. Within hours of exposure, high mercury vapour concentrations largely impact the lungs, producing pleuritic discomfort, shortness of breath, fever, lethargy, nausea, and blurred vision. Bronchitis with interstitial filtrates is visible in the lungs, and pulmonary function tests reveal restrictive alterations with decreased diffusion capacity. Mercuric chloride produces severe necrosis throughout the gastrointestinal tract, cardiovascular shock, and abrupt renal failure. It is very toxic to all metabolic systems [28].

10 Conclusion

It is well acknowledged on a national and international level that mercury is one of the most dangerous heavy metals and negatively affects human health. Due to the rising demand for mercury on the local market, both its import and criminal trade has dramatically increased during the past few years. Mercury poses a major health danger due to its usage in several industrial, occupational, domestic, and healthcare settings as well as from local and international sources. Mercury handling regulations and safeguards are still insufficient. In addition to existing laws, regulatory controls must be reinforced in order to address the issue of mercury in the environment [30].

The limitation and prohibition of the use of elemental mercury and mercury compounds through disincentives like high taxes and import duties, and the provision of incentives (lower or zero taxes or import duties) for commercial uses of mercury alternatives. Regulations on the recovery and recycling of mercury from various items as well as restrictions on the disposal of mercury-containing products in landfills and waste are also in place. Effective mercury waste management should be used to significantly decrease and/or remove mercury in waste. Mercury emission requirements must be improved, as must appropriate, affordable environmental monitoring. Replace mercury-containing devices or chemicals with safer substitutes as you phase them out [31]. Promoting clean energy, eliminating the use of mercury in gold mining, ceasing mercury mining, and phasing out non-essential mercury-containing items are just a few measures to prevent negative health impacts. The central nervous system, kidneys, cardiovascular system, and immunological system are particularly vulnerable to the negative effects that mercury pollution may have on people's health. Diverse factors, such as tainted food and water, industrial emissions, and inappropriate disposal of mercury-containing objects, can expose people to mercury. It is critical to decrease exposure to mercury in all of its forms through a mix of preventative efforts and individual behaviour in order to safeguard public health from the hazardous effects of mercury. This entails cutting back on industrial emissions, disposing of mercury-containing items appropriately, avoiding tainted food and water sources, and taking precautions to limit individual exposure to mercury. We may contribute to the protection of the health of present and future generations by taking steps to decrease mercury exposure [22].

References

1. Raju S, Otitolaiye VO, Mahfud R, Al Rawahi M (2022) Impacts of mercury exposure on human health, safety and environment: literature review and bibliometric analysis (1995 to 2021). Int J Occup Saf Health 12(4):336–352
2. Azevedo BF, Furieri LB, Anha FMP, Wiggers GA, Vassallo PF, Sim˜oes MR, Fiorim J, Priscila Rossi de Batista PRD, Fioresi M, Rossoni L, Ivanita Stefanon I, Alonso MJA, Mercedes Salaices M, Vassallo DV (2012) Toxic effects of mercury on the cardiovascular and central nervous systems. J Biomed Biotechnol 1–11
3. Bjorklund G, Hilt B, Dadar M, Ulf Lindh U, Jan Aaseth J (2019) Neurotoxic effects of mercury exposure in dental personnel. Basic Clin Pharmacol Toxicol 124:568–574
4. Passos CJS, Mergler D (2008) Human mercury exposure and adverse health effects in the Amazon: a review. Carlos J. S. Passos 1 Donna Mergler, Cad. Saúde Pública. 24(4):S503–S5020
5. Basu N, Bastiansz A, Dórea JG, Fujimura M, Horvat M, Shroff E, Weihe P, Zastenskaya I (2023) Our evolved understanding of the human health risks of mercury 52(5):877–896
6. Ramli FF (2021) Clinical management of chronic mercury intoxication secondary to skin lightening products: a proposed algorithm. Bosn J Basic Med Sci 21(3):261–269
7. Ripoll L, Javier Rayos J, Miguel Ángel Aguirre MA, Lorena Vidal L, Antonio Canals A (2005) Natural deep eutectic solvent-based microextraction for mercury speciation in water samples. Anal Bioanal Chem 545–557
8. Chang LW, Tjalkens RB (2010) Comprehensive toxicology, 2nd edn. Neurotoxicology Metals 13:483–497
9. Grasby SE, Them II TR, Chen Z, Yin R, Ardakani OH (2019) Mercury as a proxy for volcanic emissions in the geologic record. Earth-Science Rev 196(102880):1–20
10. Qu R, Han G, Liu M, Li X (2019) The mercury behavior and contamination in soil profiles in mun river basin, Northeast Thailand. Int J Environ Res Public Health 16(4131):1–20
11. Qu R, Han G, Liu M, Li X (2019) The mercury behaviour and contamination in soil profiles in mun river basin, Northeast Thailand. Int J Environ Res Public Health 16(4131):1–16
12. Hargreaves AJ, Vale P, Whelan J, Constantino C, Dotro G, Cartmell E (2016) Mercury and antimony in wastewater: fate and treatment. Water Air Soil Pollut 227(89):1–17
13. Liu J, Lin L, Wang K, Ding K, Xie Z, Zhang P (2022) Concentrations and species of mercury in municipal sludge of selected Chinese cities and potential mercury emissions from sludge treatment and disposal. Front Environ Sci 10:1–10
14. Wang X, Mao Y (2019) Mercury in municipal sewage and sewage sludge. Bull Environ Contam Toxicol 102:643–649
15. Gworek B, Dmuchowski W, Baczewska A, Bartoszewski P, Kałabun OB, Jakubowska JW (2017) Air contamination by mercury, emissions and transformations—a review. Water Air Soil Pollut 228(4):1–31
16. Han N, Zhang J, Hu Z, Ma Y, Duan Y, Han Y, Chen X, Zhou Y, Cheng J, Wang W (2018) Particulate mercury in ambient air in Shanghai, China: Size-specific distribution, gas–particle partitioning, and association with carbonaceous composition. Environ Pollut 238:543–553
17. Salam LB, Shomope H, Ummi Z, Bukar F (2019) Mercury contamination imposes structural shift on the microbial community of an agricultural soil. Bull Natl Res Cent 43(163):1–13
18. Morosini PC, Terzaghi E, Raspa G, Zanardini E, Anelli S, Armiraglio S, Petranich E, Covelli S, Guardo AD (2021) Mercury vertical and horizontal concentrations in agricultural soils of a historically contaminated site: Role of soil properties, chemical loading, and cultivated plant species in driving its mobility. Environ Pollut 285(117467):1–20
19. Zhong T, Chen D, Zhang X (2016) Identification of potential sources of mercury (Hg) in farmland soil using a decision tree method in China. Int J Environ Res Public Health 13(11):1–10
20. Tao Z, Dai S, Chai X (2017) Mercury emission to the atmosphere from municipal solid waste landfills: a brief review. Atmos Environ 170:303–311

21. Sun L, Zhang X, Zheng J, Zheng Y, Yuan D, Chen W (2021) Natural stable isotopic compositions of mercury in aerosols and wet precipitations around a coal-fired power plant in Xiamen, southeast China. Atmos Environ 173:72–80
22. Grandjean P (2017) Mercury. In: International encyclopedia of public health, 2nd edn, pp 110–116
23. Veiga MM, Telmer KH (2009) World emissions of mercury from artisanal and small-scale gold mining. Mercur Fate Transp Glob Atmos 131–172
24. Yevugah LL, Darko G, Bak J (2021) Does mercury emission from small-scale gold mining cause widespread soil pollution in Ghana? Environ Pollut 284:1–14
25. Fu XW, Feng X, Dong ZQ, Yin RS, Wang, JX, Yang ZR, Zhang H (2010) Atmospheric gaseous elemental mercury (GEM) concentrations and mercury depositions at a high-altitude mountain peak in south China. Atmos Chem Phys 10:2425–2437
26. Tao Z, Deng H, Li M, Chai X (2020) Mercury transport and fate in municipal solid waste landfills and its implications. Biogeochemistry 148:19–29
27. Nguyen LSP, Hien TT, Minh Tri Truong MT, Chi NDT, Sheu GR (2022) Atmospheric particulate-bound mercury (PBM10) in a Southeast Asia megacity: sources and health risk assessment. Chemosphere 307(135707):1–12
28. Hejna M, Kapuścińska D, Aksmann A (2022) Pharmaceuticals in the aquatic environment: a review on eco-toxicology and the remediation potential of algae. Int J Environ Res Public Health 19(7717):1–40
29. Rathore M, Singh A, Pant VA (2012) The dental amalgam toxicity fear: a myth or actuality. Toxicol Int 19(2):81–88
30. Zho M, Yi L, Wang Z (2022) Mercury and mercury-containing preparations: history of use, clinical applications, pharmacology, toxicology, and pharmacokinetics in traditional chinese medicine. Front Pharmacol 13(807807):1–23
31. Denton GR, Trianni MS, Bearden BG, Houk PC, Starmer JA (2011) Impact of a medical waste incinerator on mercury levels in lagoon fish from a small tropical island in the Western Pacific. J Toxicol Environ Health A 74(13):823–827
32. Castro MS, Moore CW (2016) Importance of gaseous elemental mercury fluxes in western maryland. 7(110):1–13
33. Schiavo B, Bermea OM, Martinze ES, Martinze RG, Alvarez EH (2022) Health risk assessment of gaseous elemental mercury (GEM) in Mexico City. Environ Monit Assess 456:1–19
34. Floreani F, Acquavita A, Barago N, Klun K, Faganeli J, Covelli S (2022) Gaseous mercury exchange from water-air interface in differently impacted freshwater environments. Int J Environ Res Public Health 19(13):1–19
35. Esdaile LJ, Chalker JM (2018) The mercury problem in artisanal and small-scale gold mining. Chem Eur J 24:6905–6916
36. Mashyanov N (2020) Mercury in gas and oil deposits: corrosion problem. E3S Web Conf. Corros. Oil Gas Ind. 225(01009):1–4

The Effects of Mercury Exposure on Neurological and Cognitive Dysfunction in Human: A Review

Arti Chamoli and Santosh Kumar Karn

Abstract Mercury (Hg) is a naturally occurring noxious and volatile heavy metal found in the environment in the various forms. In the environment mercury is released via natural (weathering of rocks, volcanic eruptions) and anthropogenic activities (mining, coal fired power stations, waste incineration and industrial processes. Mercury occurs in inorganic and organic form; inorganic form consists of metallic or elemental mercury (Hg) whereas organic form covers Hg bound compounds like methyl, ethyl and phenyl mercury. By the assistance of certain microorganisms mercury can be transformed into its toxic form, namely methylmercury (MeHg). Methylmercury can act as a neurotoxin and bioaccumulates in the tissues of aquatic plants and animals and can affect the health of individuals who eat these animals and fishes. High exposure to mercury engenders complications like changes in the central nervous, digestive and immune system besides this, it also have toxic effects on liver, lungs and kidneys. Neurological and cognitive and motor dysfunction may be noticed after ingestion, inhalation or by any kind of exposure to mercury compounds. Patients with high mercury exposure show symptoms like nausea, irritability, tremors, headache, hypertension, hallucinations and even death in certain cases. Resourceful approach should be taken to evaluate the risk of occupational exposure and also to consuming fish with regard to human and animal health. Current chapter will discuss all the problem findings and recent advances of mercury poisoning and their effect on neurological functions.

Keywords Mercury-toxicity · Cognitive dysfunction · Neurological disorders · Methylmercury

---

A. Chamoli · S. K. Karn (✉)
Department of Biochemistry and Biotechnology, Sardar Bhagwan Singh University, Balawala, Dehradun 248001, India
e-mail: santoshkarn@gmail.com

N. Kumar (ed.), *Mercury Toxicity Mitigation: Sustainable Nexus Approach*, Earth and Environmental Sciences Library, https://doi.org/10.1007/978-3-031-48817-7_5

1 Introduction

Mercury is a pernicious heavy metal found naturally in the environment. It is a silvery odorless metal, slowly tarnishing in moist air. It is the only metal that is liquid at normal temperatures. Mercury is naturally introduced into the environment by volcanic fumes, forest fires, weathering of rocks and through soils. Anthropogenic activities primarily; mining, combustion (burning of fossil fuels), waste incineration, coal burning power plants and industrial processes contributes to unnatural introduction of mercury. Mercury is found in its organic and inorganic form: inorganic form which includes metallic or elemental mercury which can later transform into mercury vapor (Hg); and organic form, which includes mercury bound compounds (methyl, ethyl, phenyl, or similar groups) [1]. Loads of mercury is mostly emerged from combustion of coal which can be dispersed in the air to long before being settled on the land surface [2]. All types of mercury have been acknowledged as toxic, as it has no biological benefit from it [3].

The possible threats related to health are associated with mercury vulnerability have led to the adoption of alternative substances and practices in many industries. Regarding the use of mercury as an antiseptic and medical preservative, the medical field has moved away from its use due to the availability of safer alternatives. Antiseptics are now primarily based on compounds such as alcohol, iodine, hydrogen peroxide, and chlorhexidine, which are effective without the same risks posed by mercury. As for dental amalgam, it is a mixture of metals, including silver, tin, copper, and mercury. Dental amalgam has been used for filling cavities for many years due to its durability and cost-effectiveness. However, concerns have been raised about the potential release of mercury vapor from dental amalgam fillings. The amount of vapor released can vary depending on factors such as chewing, teeth grinding, and exposure to high temperatures from hot food or drinks. Mercury vapor is highly volatile and lipid soluble, which means it can cross biological barriers such as the blood–brain barrier and lipid cell membranes. Once in the body, mercury can be accumulated in its inorganic forms within cells. However, the overall health risks associated with dental amalgam fillings and mercury vapor exposure are a subject of ongoing scientific debate. Mercury might potentially have noxious repercussions on the central nervous system, digestive system and immune system also. According to WHO, mercury is considered to be one of the top ten chemicals that possess considerable health problems. It can also affect the lungs, kidneys, skin and eyes. Food that usually contains mercury includes fish [4].

Mercury absorption depends on the type of mercury. As the route of absorption of metallic or elemental mercury (Hg^0) is mostly by lungs (80%) via inhalation, although lesser is absorbed by the gastrointestinal tract. Hg^0 is absorbed in the lungs and converted into Hg^{2+} in red blood cells [5]. After setting foot into the circulation it is promptly dispersed to the tissues, but accumulates profusely in the kidneys. Elemental mercury can be in the body for a long term (from weeks to months). Because of its lipophilic physicochemical property it can easily cross the membrane barriers.

2 Chemical Properties and Availability

Mercury has a comparatively high vapor pressure. Mercury has two cationic states; Hg^{2+} (mercuric) and Hg^{1+} (mercurous) while elemental mercury (Hg) has no charge. Hg^{2+} can be associated with inorganic molecules (mercuric chloride, cinnabar mineral, oxygen and hydroxyl ions) and organic molecules (carbon based) like dimethylmercury, which is way more harmful than inorganic forms. However Hg^{2+} with inorganic molecules is more stable. Mercuric cation is also found in carbon based compounds. Mercury existing in nature as an element that is non-biodegradable, meaning it cannot be broken down or decomposed by living organisms. However, it can undergo various transformations and transport processes in the environment.

Mercury (Hg) can exist in many physical and chemical forms in the environment, having different transport and deposition properties and effects on ecosystems. Apart from natural and direct, anthropogenic environmental changes like oxygen depletion, acidification and draining of water-drenched areas may mobilize metals apart from availability to organisms. Mercury generally exists in three forms namely elemental or metallic mercury (H^0), inorganic mercury (Hg^{2+}) and organic mercury (MeHg or methyl mercury). Mercury (Hg) is an element that exhibits unique chemical properties. It can form various compounds and amalgams with different elements like salts of mercury: Mercury can form salts with oxygen (O), sulfur (S), and chlorine (Cl); For example, mercuric sulfide (HgS), mercuric oxide (HgO) and mercuric chloride ($HgCl_2$).vapor and liquid form is the inorganic form of mercury. When monoatomic gas is released from volatile liquid Hg, it is called Hg vapors. Hg vapors play predominant role in global cycling of heavy metals because of its existence as mercurous (1+) or mercuric (2+)cation [6].

Inorganic mercury runs in a cycle between air, land and water. The salts of inorganic mercury can attach to air particles which are later deposited on land via rain or snow or it can returns to the atmosphere in the form of gas with the particles and redeposits elsewhere. During all this process inorganic mercury undergoes a series of physical and chemical transformations. These transformations are facilitated by certain microorganisms and converting it into an organic form of mercury such as methylmercury; which is highly toxic. Commonly methylmercury exposure happens to fish consuming populations, since fish have high concentration of methyl mercury in their tissues. Overexposure and consumption of methyl mercury containing products can lead to severe problems which may include: lack of movement coordination Loss of peripheral vision; "Pins and needles" feelings, mostly in the feet, hands and the mouth; hearing, speech, walking disability; and/or Muscle frailty [7].

3 Toxigenicity and Mechanism

Mercury enters the body through vapors. It can affect nervous system, enzyme system and causing cognitive dysfunction. Mercury toxicity in humans depends and differs with the degree of exposure, dose and form of mercury. Vapor inhalation can cause severe pneumonitis and can be fatal in extreme cases. After inhalation of mercury vapors primarily it aims the brain. Mercurous and mercuric salts predominantly can cause impairment of the gut lining and kidney; while methylmercury is broadly distributed all over the body. Methyl mercury can bioaccumulate and biomagnify through the aquatic food web [8].

Methylmercury can induce cell damage through various mechanisms; Disruption of Calcium Homeostasis: Methylmercury can interfere with calcium signaling and disrupt the balance of calcium ions within cells. Calcium ions play pivotal roles in various cellular processes, including signal transduction, enzyme activation, and neurotransmitter release. By disrupting calcium homeostasis, methylmercury can impair these cellular functions, leading to cellular dysfunction and damage Induction of Oxidative Stress: Methylmercury can induce oxidative stress within cells. Oxidative stress occur when there is a variance between the production of reactive oxygen species (ROS) and the cell's potential to detoxify them or restore the resulting damage. Methylmercury can either directly generate ROS or interfere with the cell's antioxidant defenses, such as glutathione, which help neutralize ROS. The accumulation of ROS can cause damage to cellular components, including lipids, proteins, and DNA, accelerating cell dysfunction and death. Interactions with Sulfhydryl Groups: Methylmercury has a high affinity for sulfhydryl (–SH) groups, which are present in many cellular proteins and enzymes. It can form strong covalent bonds with sulfhydryl groups, disrupting the structure and function of important proteins and enzymes. This interaction can lead to the inhibition of enzymatic activities, impair cellular processes, and disrupt the overall functioning of cells. These mechanisms collectively contribute to the toxicity of methylmercury and its potential to begin cellular damage, particularly in the central nervous system, where methylmercury tends to accumulate and exert its most severe effects [9].

Once mercury gets inside the body, it undergoes the processes of absorption, distribution, metabolism, and excretion (ADME). Mercury can be absorbed through the respiratory system when mercury vapors are inhaled or through the gastrointestinal tract when ingested. It can also be absorbed through the skin, although to a lesser extent. Once absorbed, mercury can circulate through the bloodstream, where it can reach various organs and tissues. It has a particular affinity for the central nervous system, including the brain, where it can accumulate over time. Mercury can undergo metabolic transformations in the body. The main form of mercury encountered in the environment is elemental mercury (Hg^0), which can be converted into inorganic mercury (Hg^{2+}) by certain bacteria or through oxidative processes. Inorganic mercury can further undergo metabolic conversions in the body, such as being transformed into methylmercury (MeHg) by certain microorganisms, particularly in aquatic environments. The excretion of mercury from the body occurs mainly through urine and

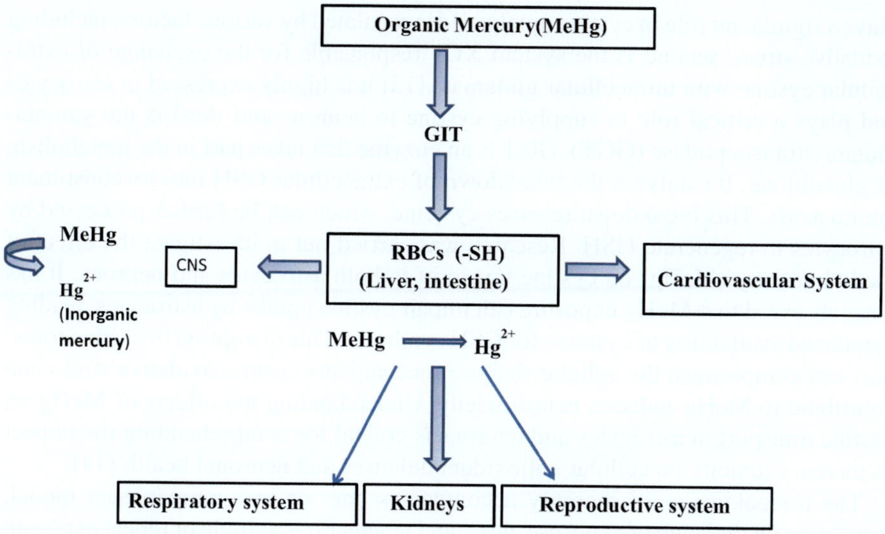

Fig. 1 Schematic representation of mercury distribution after entering the body [11]

feces. Inorganic mercury and methylmercury can be eliminated through these routes, although methylmercury has a prolonged half-life in the body in comparison to inorganic mercury [10]. A Schematic representation of mercury distribution after entering the body is illustrated in Fig. 1.

The toxicity of mercury arises from its capability to attach to sulfhydryl groups in enzymes and proteins, leading to the disruption of their normal function. This can result in extensive neurological symptoms and conditions, including neurobehavioral changes, cognitive impairment, motor dysfunction, and sensory disturbances. Methylmercury, in particular, is noted to have an intense affinity for the central nervous system and can promptly cross the blood–brain barrier, leading to its accumulation in brain tissues [12].

It should be noted that the toxicokinetics of mercury may differ depending on its chemical form. For example, elemental mercury vapor is primarily associated with neurological effects, while methylmercury, commonly found in contaminated seafood, is known for its neurotoxicity. Other forms of mercury, such as inorganic mercury compounds, can also contribute to toxic effects but may affect different organ systems. Overall, the toxic effects of mercury are multifaceted, and its ADME processes play a crucial role in determining its toxicokinetics and potential harm to the body, particularly in relation to the development of neuropathological conditions [10].

To support cellular defense against damage from free radicals, it is necessary to maintain the optimal intracellular concentrations of glutathione (GSH). Glial cell namely astrocytes, in the central nervous system, play a vital role in providing GSH precursors, particularly cystine, to neurons. Astrocytes take up cystine from the extracellular space through various transport mechanisms, namely; System XAG which plays a significant role in cystine uptake and is regulated by various factors, including oxidative stress, second is the system XC^- responsible for the exchange of extracellular cystine with intracellular glutamate [13] it is highly expressed in astrocytes and plays a critical role in supplying cystine to neurons and third is the gamma-glutamyltranspeptidase (GGT). GGT is an enzyme that takes part in the metabolism of glutathione. It catalyzes the breakdown of extracellular GSH into its constituent amino acids. This breakdown releases cysteine, which can be further processed by astrocytes to regenerate GSH. Research was carried out to investigate the effect of methylmercury (MeHg) on cystine transport in both astrocytes and neurons. It has been observed that MeHg exposure can impair cystine uptake by astrocytes, leading to reduced availability of cysteine for GSH synthesis. This disruption in cystine transport can compromise the cellular defense mechanisms against oxidative stress and contribute to MeHg-induced neurotoxicity. Understanding the effects of MeHg on cystine transport in astrocytes and neurons is crucial for comprehending the impact of mercury toxicity on cellular antioxidant defenses and neuronal health [14].

The toxicokinetics of mercury accompanies one- or two-compartment model, depending on the route of exposure, dose, and whether it is a single or repeat exposure which results in accumulation of merury in the body. In the case of single exposures, mercury can be described using a one-compartment model, which assumes that the distribution and elimination of mercury occur in a single homogeneous compartment within the body. However, with repeat or continuous exposure to any form of mercury, the body's handling of the metal becomes more complex. In such cases, a two-compartment model is often used to describe the toxicokinetics of mercury. This model considers the body as having two compartments: a central compartment (e.g., blood) and a peripheral compartment (e.g., tissues) [15]. Studies have demonstrated that continuous exposure to mercury can lead to the accumulation of the metal in the body over time. Mercury can be distributed throughout various tissues, including the brain, and may persist in these tissues even after exposure has ceased. This accumulation of mercury in the brain can have long-term effects on neurological function. During the initial few days after mercury exposure, metal levels in the blood are closely related to the overall retention of mercury in the body. However, after this initial period, the amount of mercury in the blood declines more rapidly compared to the whole-body load. This indicates that mercury is being redistributed from the blood to other tissues, including organs such as the brain. It is important to note that the toxicokinetics of mercury may differ with specific form of mercury (e.g., elemental, inorganic, or organic) and the route of exposure (e.g., inhalation, ingestion, or dermal contact) [16]. Among all the mercury species methylmercury is the highly toxic form. It is highly volatile and can pass through the biological membranes. After getting access into the brain, mercury is oxidized by the hydrogen peroxidase–catalase pathway and is easily converted into inorganic divalent mercury. Primarily absorbed metallic mercury is eliminated in the urine, about 10% and feces, some passes in the milk (5% and in small quantity exhaled out. Gaseous form of mercury when reaches brain it is transformed into oxidized form.

Inorganic mercury absorption via GI tract is 10–40% also distributed to different organs and majorly accumulates in the kidneys. A study on female Sprague–Dawley rats showed the distribution of mercury across the different parts of the body when dosed with a single dose of mercuric chloride (7.4 or 9.2 mg Hg/kg, p.o.). 12.6 and 18.9 ppm mercury was detected in the kidneys with traces in the liver, brain and serum in addition [17]. Although these compound do not cross the membrane barriers easily.

However the absorption of organic mercury (e.g. methylmercury) via GI tract is about 95% which later distributed to other organs from the circulation. Alike metallic mercury methyl mercury is concentrated majorly in the brain and fetus. Also, it has the ability to transform it into inorganic divalent cation in the brain region (tissues) and can stay there for a long term. However in comparison with metallic mercury the conversion is less [18]. Organic mercury is excreted in the form of feces for several months, although some of it is excreted via urine and milk also.

4 Mercury Toxicity

4.1 Humans

Various studies indicated that even exposure to low dosage of inorganic mercury can cause neurological and cognitive dysfunctions. The hippocampus in brain is peculiarly at risk to the harmful effects of mercury; hippocampus has a role in learning, spatial navigation and memory encoding. Neurological damage, inflammation and oxidative stress are the possible outcomes when inorganic mercury accumulates in the hippocampus region [19]. In skin-lightening creams inorganic mercury is used because it inhibits melanin formation and also acts as a bactericide and fungicide [20]. The nervous system is vulnerable to metallic mercury. Elevated levels metallic mercury exposure can cause brain, lungs and kidney impairment and can also acutely affect the growing fetus. Post exposure symptoms may include: nausea, vomiting, coughing, diarrhea, elevated hear rate or blood pressure, skin allergies and eye irritation. Mentioned symptoms occur when exposed to higher levels however, prolonged exposure to lower levels would cause indefinite repercussions which includes coordination problems, irritability, trouble sleeping, tremors, memory problems and complications in vision and hearing. All these symptoms from a continued exposure to low levels of mercury are improvable, once the exposure is ceased and the mercury has been removed from the body [21]. Acrodynia, also known as pink disease or Swift's disease, is a condition primarily affecting children with high levels of mercury vapor exposure. The symptoms of acrodynia include reddening and tenderness of the palms of the hands and soles of the feet, followed by peeling of the skin in these areas. Children with acrodynia may also exhibit worsen irritability, mood swings, struggling sleeping, and muscle or joint pains. In some cases, exposure levels that cause acrodynia can lead to coughing or chest pain [22].

Generally acrodynia is associated with urine mercury concentrations of 100 μg (or higher) of mercury per liter of although it's worth noting that not all cases may meet this specific threshold. When pregnant women are exposed to mercury, the mercury can cross the placenta and reach the developing fetus. Additionally, nursing infants can be exposed to mercury through breast milk. Hence after suspicion to exposure to high levels of mercury, it is recommended to seek medical attention for proper evaluation, diagnosis, and appropriate management [23].

4.2 Animals

Persistent contact with little dosage of mercury chloride ($HgCl_2$) has been shown to have detrimental effects on hippocampus in animal models, resulting in hippocampal dysfunction. The hippocampus is a brain region critical for learning and memory processes. Impairments in spatial learning and memory, as well as alterations in synaptic plasticity, have been observed in animal studies following chronic exposure to low doses of inorganic mercury. Altogether inorganic mercury such as $HgCl_2$ may cause cognitive loss. Although more investigation is required to fully acknowledge the basic mechanisms behind and also to find out safe levels of exposure [24]. Another study by [25] exhibited how mercury significantly affected the growth of mice when administered with different concentration of mercury (0–160 mg/l $HgCl_2$). Drop in weight gain was observed along with some pathological changes in cecum tissues. Moreover, malondialdehyde (MDA) content is increased by the exposure of mercury and also significantly decreases superoxide dismutase (SOD) activity and glutathione peroxidase level which eventually increases the oxidative stress in mice. It also affects the gut microbiota of the mice. Several past studies validate that the breakdown of all forms of mercury is analogous for humans and animals. The absorption and breakdown of mercury occurs in the tissues and red blood cells via redox reaction that occurs in intestinal microflora. The elimination half-life refers to the time it takes for half of the mercury concentration in the blood to be removed from the body and it supposedly varies with species, sex, dosage, etc. like the elimination half-life in the blood of monkeys receiving inorganic and organic mercury was found to be 26 days. Another study showed that highest concentration of total mercury was observed in mesenterial lymph nodes supervened by liver and kidneys of sled dogs fed on methylmercury-loaded meat and from predatory marine animals. It also exhibits that the lymphatic system may take part majorly in carrying mercury to aimed organs. demethylation occurs in all organs, excluding the skeletal muscles. Demethylation of methylmercury was considered to be slighter in the brain in comparison to other organs [16]. A study by [26] on mice illustrated the acute stages when subcutaneously treated with methylmercury. Methylmercury was elicited to induce cerebellar toxicity. Also after being exposed to Hg^{2+} a decrease in renal and hepatic nonprotein sulfhydryl (NPSH) content was observed. Cellular oxidative stress can be elicited by Lipid peroxidation, which act as a marker and has been observed after Hg^{2+} exposure. Hg^{2+} facilitates the formation of H_2O_2 in the mitochondria. Studies have illustrated that in this overall process selenium plays a crucial role in transporting mercury. This concept was accepted by observing that when rats were

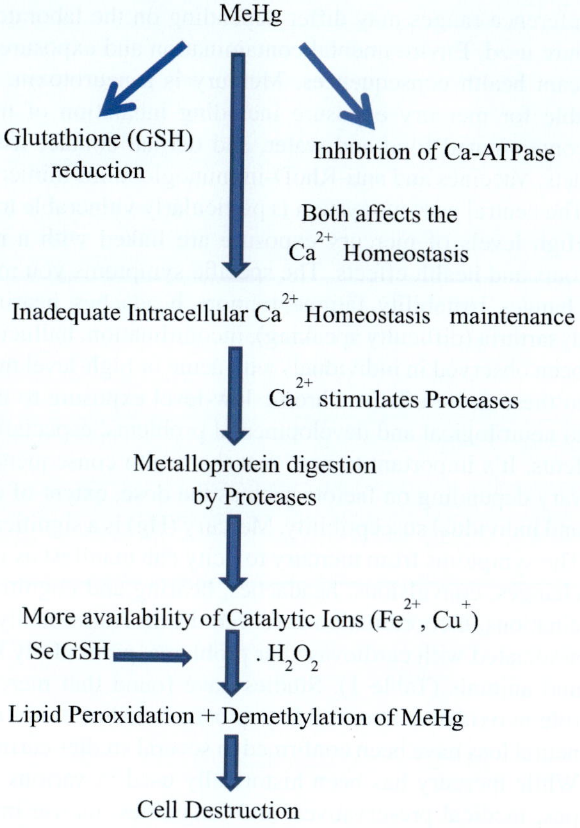

Fig. 2 Reactions of Methylmercury

administered with selenium, five days later it was detected as Se-cysteine. Hence presumed that methylmercury is chemically trapped as methylmercury-SH-cysteine or as methylmercury-SeH-cysteine complexes (Fig. 2) [27].

5 Mercury Toxicity in Humans

For individuals who have not been exposed to significant amounts of mercury, the typical concentration is about 2 ppm for hair. However, hair mercury levels can vary with the factors such as age, gender, diet, and geographic location. The normal reference range for mercury in blood is commonly reported in range of 1–10 μg/dL. For urine the reference range is around 2–20 μg/L. However, it's worth noting that these reference ranges are approximate and as with other biological samples, reference ranges may differ depending on the laboratory and the analytical procedure used. Environmental contamination and exposure to mercury can have significant health consequences. Mercury is a neurotoxin. Various routes are accountable for mercury exposure including inhalation of mercury vapors, ingestion of contaminated food and water, and dermal contact with mercury-containing products. Vaccines and anti-RhoD-immunoglobulins thimerosal is used as a preservative. The central nervous system is particularly vulnerable to the toxic effects of mercury. High levels of mercury exposure are linked with a range of neurological symptoms and health effects. The specific symptoms you mentioned, such as behavioral changes, irritability, fatigue, tremors, headaches, hearing and cognitive impairment, dysarthria (difficulty speaking), incoordination, hallucinations, and even death, have been observed in individuals with acute or high-level mercury poisoning. In addition to these acute effects, chronic low-level exposure to mercury has also been linked to neurological and developmental problems, especially in children and developing fetus. It's important to note that the health consequences of mercury exposure can vary depending on factors such as the dose, extent of exposure, age, overall health, and individual susceptibility. Mercury (Hg) is a significant concern for human health. The symptoms from mercury toxicity can manifest as irritability, fatigue, behavioral changes, convulsions, headaches, hearing and cognitive ailment, dysarthria, hallucinations, incoordination and even death. Additionally, mercury exposure has been associated with cardiovascular problems, particularly hypertension, in both humans and animals (Table 1). Studies have found that mercury exposure has significant role in oxidative stress and apoptosis. Besides, motor and cognitive destruction and neural loss have been confirmed in several studies carried out on animal models [28]. While mercury has been historically used in various applications such as antiseptics, medical preservatives, and fungicides, its use in these areas has significantly decreased in recent years due to concerns about its toxicity.

In the 1950s organic mercury was discarded into Minamata Bay which later became the cause of minamata bay epidemic. Around 20,000 people were poisoned by consuming fish from this contaminated water body. Neurological problems along with other symptoms like difficulty in coordination, abnormal reflexes, seizures and speech problems were seen in 7% of the children born to mothers who consumed adulterated fish. In adults tremors, nausea, weakness, loss of hearing, depression, loss

Table 1 Mercury toxicity and its effects

	Neurological	Motor	Renal	Cardiovascular
Adult	Memory loss, Dementia, Tremors, Incoordination, Hallucination	Retard fine motor function, Tiredness, Reduced muscular strength	Increased Plasma, Creatinine levels	Alter normal cardiovascular homeostasis
Children	Lesser memory and Attention Score, cerebral palsy	Movement abnormality, late walking	Increased Plasma, Creatinine levels	Alter normal cardiovascular homeostasis

of appetite and problematic vision was observed. In total 46 people died because of mercury exposure [29]. Another accident took place in Iraq, 1972, where approximately 459 people lost their lives and 6,500 people fell ill after consuming wheat bread containing mercury-based fungicide. Marine food is the major constituent of the diet to the people of Faroe Islands (North Atlantic). A study in 1984 evaluated how children born to mothers who ate mercury contaminated whale meat were affected. Investigators observed that children born to mothers with a 10–20 ppm mercury concentration had lesser memory, language and attention capability in comparison to the children born to mothers with lower mercury levels [30].

6 Effect of Mercury on Neurological Functions

Various mechanisms has been put forward to understand how mercury affects nervous system and kills neurons which includes protein inhibition, mitochondrial disruption, affecting ion exchange in a neuron, obliteration of structural framework of neurons and neurotransmitters and affecting cognitive functions. Generally neurons are targeted and killed by methyl mercury in the specific region of the nervous system counting Cerebellum, Visual cortex and Dorsal root ganglia [9]. In the main, developing babies are more vulnerable to methylmercury, as it is the highly toxic mercury type and can make way across the membranes and blood barriers in the body. After crossing the placental barrier mercury get accumulated in fetus's brain. methylmercury (MeHg) is one of the most toxic compound of mercury that has been considerably investigated for its detrimental consequences on the nervous system. MeHg is known to bioaccumulate in the food chain, particularly in fish and seafood, making it a significant concern for human exposure [31]. The neurological alterations engendered by methemoglobin toxicity have been well-demonstrated in both human population and experimental animals. After crossing the blood–brain barrier MeHg accumulate in the brain, where it brings out its noxious effects. The exact mechanism by which MeHg causes neurotoxicity is not completely known, but it is presumed that oxidative stress plays a crucial role. Oxidative stress occur when there is an imbalance between the production of reactive oxygen species (ROS) and the ability of the body's antioxidant defense systems to neutralize them. ROS, such as free radicals and peroxides, are highly reactive molecules that can cause damage to cellular structures, including lipids, proteins, and DNA. They can start a chain reaction of oxidative damage, leading to cellular dysfunction and death. MeHg exposure has shown raised ROS production in the brain. It can directly generate ROS through redox cycling or by inhibiting the activity of antioxidant enzymes that help neutralize ROS. The increased ROS production disturbs the antioxidant defense mechanisms, leading to oxidative stress [32]. Oxidative stress has been involved in the pathogenesis of various neurodegenerative diseases including Parkinson's disease, amyotrophic lateral sclerosis (ALS) and Alzheimer's disease. In these diseases, there is evidence of increased oxidative damage and impaired antioxidant defense systems in affected brain regions. However, the exact mechanisms by which oxidative stress contributes to the development and progression of these diseases are still under investigation [33]. It is noteworthy that while oxidative stress is believed to be involved in the neurotoxicity of MeHg and the pathogenesis of neurodegenerative diseases, it is likely that multiple mechanisms contribute to the overall toxicity. Other mechanisms that have been proposed include disruption of calcium homeostasis, mitochondrial dysfunction, and interference with neurotransmitter systems [34].

Additionally, more research is required to thoroughly explain the complex mechanisms of MeHg-induced neurotoxicity and its relationship to neurodegenerative diseases. Understanding these mechanisms could provide valuable insights for developing strategies to prevent or mitigate the harmful effects of MeHg exposure and potentially shed light on the underlying processes of neuro-degeneration in general [35]. Methylmercury is mainly risky to growing babies. Because of quick metal absorption, mercury is accumulated inside the brain of the growing fetus and is not excreted proficiently. Children exposed to mercury may be born with symptoms parallel to cerebral palsy, spasticity and different movement abnormalities, convulsions, visual problems and atypical reflexes. The brains of children died due to mercury poisoning exhibit neuron loss within the cerebellum and all around the cerebral cortex. Mercury also seems to affect brain growth by means of stopping neurons from locating their suitable region within the brain [36]. Many of the effects and complications are mentioned in the Table 2. The prime mechanisms taking part in MeHg neurotoxicity currently being studied includes intracellular calcium homeostasis impairment [37], oxidative stress and the alteration of glutamate homeostasis.

Synaptic transmission, a neurobiological process can be disturbed by mercury doing neuronal damage via hyperactivation of the receptors of NMDA (N-methyl-D-aspartate). In rat cortical neurons overactivation of postsynaptic NMDA receptors was elicited, after exposure to inorganic mercuric chloride ($HgCl_2$). $HgCl_2$ alters the membrane excitability and neuronal cytoskeletal proteins disassembly and increases the amount of intracellular Ca^{++} that enters through NMDA receptors [43].

Table 2 Characterization of mercury (Hg) and its effects

Form of mercury	Organ/Body part involved	Acute/Chronic effects	References
Elemental Hg	Peripheral Nervous system, lungs, skin	Amyotrophic Lateral Sclerosis (ALS) Erethism, Hyperirritability, Lungs-Chemical Pneumonitis Skin-erythmatous and Pruritic Skin Rash, Acrodynia, Dermatitis	[38] [39] [40]
Inorganic Hg	Gastrointestinal tract	Tachycardia, Hypertension, Necrosis of Intestinal Mucosa, Ulceration of Mouth, Tongue and Lips	[40]
Organic Hg	Brain	Cerebellar Ataxia, Dysarthria, and Constriction of the Visual Fields, Hypoplasia of Bone Marrow, and Atrophy of Lymph Nodes	[41] [42]

Methylmercury toxicity can truely exhibit various neurological signs, which can vary depending on the amount and duration of exposure. Some of these signs may include: paresthesia (abnormal sensations or a "pins and needles" sensation in the skin), ataxia (lack of muscle coordination); leading to unsteady movements and difficulties with balance, tremors (involuntary rhythmic movements); hearing impairment is a common neurological symptom associated with methylmercury toxicity and Sensory disturbances; methylmercury can disturb sensory functions, leading to problems in vision, taste, and smell. Methylmercury has a confusing trait that there can be a notable delay between exposure and the appearance of symptoms. This latency period in humans can range from a few weeks to several months; with some cases reporting symptoms may not appear for as long as 150 days after exposure. This delayed onset can make it challenging to identify the source of exposure and link it to the symptoms. However symptoms and their onset can vary among individuals due to characteristics such as the amount of exposure, individual susceptibility, and other factors that may influence the metabolism and elimination of methylmercury from the body. Most of the fish species to a certain extent contains methylmercury or its compounds and is present in higher levels in predatory marine fish like swordfish and shark, with relatable levels in other predatory fish such as large tuna. A joint US EPA/FDA issued recommendatory that pregnant women and nursing mothers should kept away from consuming high-mercury fish [44].

Although, the molecular mechanism behind the toxicity of methylmercury is still unclear. With the help of synchrotron X-ray absorption spectroscopy (XAS) [45] studied more about the forms of mercury that can be present in the human brain tissue. In the research subjects/individuals were poisoned with high levels of methylmercury, subsequently individuals exhibited increased cortical selenium along with nanoparticles of mercuric selenide, inorganic mecury and methylmercury bound to organic sulfur. This concentration of mercuric selenide and methylmercury cysteineate was much lesser in the individuals consuming high fish diet intheir lifetime. It was also noticed that selenium levels were not disturbed by the exposure of mercury. These outcomes explicate a lead to detoxification pathway in the central nervous system and helps in further biological monitoring.

Mercury compounds, including methylmercury (MeHg), have been found to inhibit the Na^+–K^+–ATPase (adenosine triphosphatase activated by Na^+ and K^+) in various tissues. Na^+–K^+–ATPase is an enzyme responsible for maintaining the electrochemical gradient of sodium and potassium ions across the cell membrane, which is essential for proper cellular function. Studies have shown that mercury compounds, including MeHg, can inhibit both the active cation flux (movement of ions) and the Na^+–K^+–ATPase activity itself [46]. The inhibition of Na^+–K^+–ATPase by mercury compounds occurs with a time delay after exposure, and it is not affected by non-penetrating mercurials, suggesting that the critical site of inhibition is within the cell membrane. Although, the inhibitory effect on Na^+–K^+–ATPase is not solely to mercury. Other metals such as lead, zinc, aluminum, copper, iron, and cobalt are also known to hinder on Na^+–K^+–ATPase. These metals can interfere with the normal functioning of the enzyme and disrupt the ion transport processes in cells. It should be noted that the effects of mercury and other metals on Na^+–K^+–ATPase can have significant implications for cellular and organ function. Disturbance in Na^+–K^+–ATPase activity can affect various physiological processes, including nerve cell signaling, muscle contraction, and fluid balance [47]. According to [48, 49] estimated methyl mercury neurotoxicity thresholds range from 50 to 200 mg/L of blood [48, 49].

7 Effect on Cognitive Functions

Cognitive function includes varied mental processes and abilities that allow individuals to interact with and understand the world around them such as perception, memory, learning, problem solving, attention, decision making, language abilities etc. Cognitive dysfunction refers to a broad range of impairments in cognitive functions and processes. Mercury toxicity can affect these cognitive functions in several ways: Neurotoxicity: Mercury has a neurotoxic effect, meaning it can damage nerve cells in the brain. This can lead to cognitive impairments, including difficulties with memory formation and retrieval, decreased attention span, and reduced processing speed. Inhibition of neurotransmitters: Mercury can disrupt the normal functioning of neurotransmitters, acting as chemical messengers and are involved in transmitting signals between nerve cells. This interference can disrupt communication between brain cells and contribute to cognitive dysfunction. Oxidative stress: Mercury can induce oxidative stress in the brain, which occurs when there is an imbalance between the production of reactive oxygen species (ROS) and the body's ability to neutralize them. Oxidative stress can damage brain cells and contribute to cognitive impairments. Inflammation: Mercury exposure can trigger an inflammatory response in the brain. Chronic inflammation has been linked to cognitive decline and neurodegenerative disorders. It is noteworthy to mention that the severity of cognitive dysfunction related to mercury exposure can vary depending on factors such as the dose and duration of exposure, individual susceptibility and the specific form of mercury (e.g., elemental, inorganic, or organic). Methylmercury (CH_3Hg^+) is an electrophilic toxicant that exhibits a strong affinity for thiol and selenol groups. There are various critical targets of CH_3Hg^+ like thiol-containing proteins, which include various enzymes and transporters taking part in neurotransmitter transport, metabolism, and signaling, as well as antioxidant selenoenzymes like glutathione peroxidases and thioredoxin reductases [50]. When CH_3Hg^+ binds to these thiol and selenol containing proteins, it disrupts their normal function and impairs their ability to carry out essential physiological processes. This disruption can significantly affect developing synapses in the brain and may lead to the behavioral and cognitive impairment, as these processes have a crucial role in normal brain cell physiology. The deregulation of neurotransmitter-related proteins can interfere with the proper transmission and regulation of signals between neurons, affecting various neurological processes. Additionally, the targeting of antioxidant selenoenzymes can lead to increased oxidative stress and cell damage [51]. Studying the AOP associated with CH_3Hg^+ neurotoxicity can indeed provide valuable insights for the growth of in silico Physiological Based

Kinetic (PBK) models. PBK models are computational tools that simulate the distribution, metabolism, and elimination of chemicals in the body based on physiological and biochemical principles. By identifying the key events and biological pathways involved in CH_3Hg^+ neurotoxicity, researchers can develop a mechanistic understanding of how methylmercury affects the nervous system. This information can then be incorporated into PBK models to predict the concentration of methylmercury in different tissues and organs over time. PBK models can help in assessing the exposure levels at which negative impact occur, predicting the toxicokinetics of methylmercury in different populations, and informing risk assessments. They can also aid in evaluating the effectiveness of different mitigation strategies and interventions to reduce exposure and minimize neurotoxic effects. Overall, studying the AOP associated with CH_3Hg^+ neurotoxicity can serve as a foundation for developing in silico PBK models, which in turn can enhance our understanding of the toxicokinetics and potential health effects of methylmercury exposure. These models have the potential to support risk assessment and inform decision-making regarding exposure mitigation strategies [52]. A case study by [53] a patient who ingested liquid Hg^0 as a suicidal attempt by consuming 6 oz (oz) of Hg^0 with wine. After performing several cognitive (WTAR-estimated premorbid psychometric intelligence was low average to average; processing speed, cognitive flexibility, and response inhibition was poor) and mental status examination (Mild upper-extremity action tremor was evident; in terms of in rate, latency, rhythm, and articulation speech was unremarkable; volume was increased and was disoriented). Patient was observed with behavioral distress as a sign of bipolar disorder with irritable mood and pressured speech. Another case study by [54] elicit 91 years old subject who exhibited cognitive waning as Alzheimer's disease. Patient had remarkably high levels of mercury by the consumption of fish containing high mercury content. He also had dental amalgams which were decades old. Urine test demonstrated increased levels of mercury in the red blood cells. The two forms of mercury: methylmercury (from fish consumption) and inorganic mercury (from dental amalgams) were highly elevated. Further suggesting that mercury may play a role as a cofactor in the growth of Alzheimer's disease. Eddins et al. [55] analyzed the effect of mercury exposure treatment affecting cognitive function in neonate for 12 weeks. Study showed that decrease or deletion of metallothioneins (MT) shoot up the susceptibility to mercury-induced developmental neurocognitive impairment. Metallothionein effects on monoamine transmitters and this may be associated with the cognitive effect. Furthermore, it was observed that MT1/MT2-null mice showed notably cognitive impairment after mercury exposure. Exposure to mercury vapor MT1/MT2-null mice show both short term and long-term defects in locomotion in the open field maze [56]. Katamanova et al. [57] collected data from the workers who have been exposed to mercury exposure during their long work period, was based on the physiological and psycologial analysis along with the chronic intoxication of mercury. Patients with light or mild cognitive disorders exhibits lower extent of cognitive induced abilities, bad long-term memory and integrative thinking, patients with average cognitive disorders are usually identified by contract visual, long-term memory, loss of concentration, poor optic and spatial gnosis while marked cognitive disorders patients deals with fall in long-term, short-term, quaint memory, lacking intelligence, optic and spatial gnosis and associative thinking. Another work by [58] showed how prenatal mercury vapor exposure can affect the cognitive responses of mice, for which two strains (MT-null and wild type) of pregnant mice were repeatedly exposed to elemental mercury vapor at 0.50 and 0.56 mg/m^3 for 6 h/day until the 18th day of gestation. A notable decrease in locomotor activities was expressed by male MT-null mice in contrast learning disabilities and response action was found retarded in females. Results exhibited that susceptibility in MT-null mice were more susceptible to the behavioral neurotoxicity of prenatal mercury exposure in comparison to wild-type mice.

8 Conclusion

Mercury is a neurotoxin which can cross the blood brain barrier and give rise to complications. Once inside the body, mercury can be accumulated in its inorganic forms. Mercury can deeply affect human health and can have impact on central nervous system leading to cognitive dysfunction and other neurological defects. Toxicokinetics of mercury is different for its different chemical form. The prime mechanisms taking part in MeHg neurotoxicity currently being studied includes intracellular calcium homeostasis impairment oxidative stress and the alteration of glutamate homeostasis. Globally there is ample documentation of mercury and its derivatives having major negative effects on many life forms including terrestrial and aquatic. Action should be taken to limit the threats to human health and the environment posed by mercury leaks. Even at modest doses, several of the neurobehavioral tests showed enough sensitivity to distinguish between groups with varied Hg body burdens. Neurobehavioral effects with a significant dose–effect relationship, mostly associated with long-term exposure to low levels of organic mercury People working with prolong mercury exposure lead to chronic intoxication of mercury.

Acknowledgements I would like to convey my profound appreciation to my co author Dr. Santosh Kumar Karn for their invaluable support and contribution to this chapter. I am grateful for the support and resources provided by Sardar Bhagwan Singh University which greatly facilitated accessing relevant literature and data making this review comprehensive.

References

1. Bernhoft RA (2012) Mercury toxicity and treatment: a review of the literature. J Environ Public Health
2. Budnik LT, Casteleyn L (2019) Mercury pollution in modern times and its socio-medical consequences. Sci Total Environ 654:720–734
3. Hylander LD, Goodsite ME (2006) Environmental costs of mercury pollution. Sci Total Environ 368(1):352–370

4. Berlin M, Zalups RK, Fowler BA (2015) Mercury. Academic, In Handbook on the toxicology of metals, pp 1013–1075
5. Halbach, S. and Clarkson, T.W., 1978. Enzymatic oxidation of mercury vapor by erythrocytes. *Biochimica et Biophysica Acta (BBA)-Enzymology, 523*(2), 522–531.
6. Boening DW (2000) Ecological effects, transport, and fate of mercury: a general review. Chemosphere 40(12):1335–1351
7. Martin S, Griswold W (2009) Human health effects of heavy metals. Environ Sci Technol Briefs Citiz 15:1–6
8. Lescord GL, Johnston T, Branfireun BA, Gunn JM (2019) Mercury bioaccumulation in relation to changing physicochemical and ecological factors across a large and undisturbed boreal watershed. Can J Fish Aquat Sci 76(12):2165–2175
9. Ceccatelli S, Daré E, Moors M (2010) Methylmercury-induced neurotoxicity and apoptosis. Chem Biol Interact 188(2):301–308
10. Rice KM, Walker EM Jr, Wu M, Gillette C, Blough ER (2014) Environmental mercury and its toxic effects. J Prev Med Public Health 47(2):74
11. Fernandes Azevedo B, Barros Furieri L, Peçanha FM, Wiggers GA, Frizera Vassallo P, Ronacher Simões M, Fiorim J, Rossi de Batista P, Fioresi M, Rossoni L, Stefanon I (2012) Toxic effects of mercury on the cardiovascular and central nervous systems. J Biomed Biotechnol
12. Ajsuvakova OP, Tinkov AA, Aschner M, Rocha JB, Michalke B, Skalnaya MG, Skalny AV, Butnariu M, Dadar M, Sarac I, Aaseth J (2020) Sulfhydryl groups as targets of mercury toxicity. Coord Chem Rev 417:213343
13. Lewerenz J, Hewett SJ, Huang Y, Lambros M, Gout PW, Kalivas PW, Massie A, Smolders I, Methner A, Pergande M, Smith SB (2013) The cystine/glutamate antiporter system xc− in health and disease: from molecular mechanisms to novel therapeutic opportunities. Antioxid Redox Signal 18(5):522–555
14. Allen JW, Shanker G, Tan KH, Aschner M (2002) The consequences of methylmercury exposure on interactive functions between astrocytes and neurons. Neurotoxicology 23(6):755–759
15. Farris FF, Kaushal A, Strom JG (2008) Inorganic mercury pharmacokinetics in man: a two-compartment model. Toxicol Environ Chem 90(3):519–533
16. Gupta RC (ed) (2012) Veterinary toxicology: basic and clinical principles. Academic
17. Lecavalier PR, Chu I, Villeneuve D, Valli VE (1994) Combined effects of mercury and hexachlorobenzene in the rat. J Environ Sci Health B 29(5):951–961
18. Risher JF, De Rosa CT, Jones DE, Murray HE (1999) Updated toxicological profile for mercury. Toxicol Ind Health 15(5):480–516
19. Dutta SS (2019) Hippocampus functions. News-Medical.net. Updated August 20
20. Gilbert SG (2014) Mercury tragedies: incidents and effects
21. Risher J (1999). Toxicological profile for mercury
22. Dathan JG, Harvey CC (1965) Pink disease—ten years after (the epilogue). BMJ 1(5443):1181
23. Meadows-Oliver M (2012) Environmental toxicants: lead and mercury. J Pediatr Health Care 26(3):213–215
24. Aragão WAB, Teixeira FB, Fagundes NCF, Fernandes RM, Fernandes LMP, da Silva MCF, Amado LL, Sagica FES, Oliveira EHC, Crespo-Lopez ME, Maia CSF (2018) Hippocampal dysfunction provoked by mercury chloride exposure: evaluation of cognitive impairment, oxidative stress, tissue injury and nature of cell death. Oxidative Med Cell Longev
25. Zhao Y, Zhou C, Guo X, Hu G, Li G, Zhuang Y, Cao H, Li L, Xing C, Zhang C, Yang F (2021) Exposed to mercury-induced oxidative stress, changes of intestinal microflora, and association between them in mice. Biol Trace Elem Res 199:1900–1907
26. Stringari J, Meotti FC, Souza DO, Santos AR, Farina M (2006) Postnatal methylmercury exposure induces hyperlocomotor activity and cerebellar oxidative stress in mice: dependence on the neurodevelopmental period. Neurochem Res 31:563–569
27. Hansen JC, Danscher G (1997) Organic mercury—an environmental threat to the health of exposed societies? Rev Environ Health 12(2):107–116

28. Cariccio VL, Samà A, Bramanti P, Mazzon E (2019) Mercury involvement in neuronal damage and in neurodegenerative diseases. Biol Trace Elem Res 187:341–356
29. Harada M (1995) Minamata disease: methylmercury poisoning in Japan caused by environmental pollution. Crit Rev Toxicol 25(1):1–24
30. Greenwood MR (1985_. Methylmercury poisoning in Iraq. An epidemiological study of the 1971—1972 outbreak. J Appl Toxicol 5(3):148–159
31. Grandjean P, White RF, Weihe P, Jørgensen PJ (2003) Neurotoxic risk caused by stable and variable exposure to methylmercury from seafood. Ambul Pediatr 3(1):18–23
32. Sasaki S, Negishi T, Tsuzuki T, Yukawa K (2023) Methylmercury-induced reactive oxygen species-dependent and independent dysregulation of MAP kinase-related signaling pathway in cultured normal rat cerebellar astrocytes. Toxicology 487:153463
33. Singh A, Kukreti R, Saso L, Kukreti S (2019) Oxidative stress: a key modulator in neurodegenerative diseases. Molecules 24(8):1583
34. Jellinger KA (2010) Basic mechanisms of neurodegeneration: a critical update. J Cell Mol Med 14(3):457–487
35. Bridges CC, Zalups RK (2010) Transport of inorganic mercury and methylmercury in target tissues and organs. J Toxicol Environ Health, Part B 13(5):385–410; Roulet M, Lucotte M, Canuel R, Rheault I, Tran S, Gog YDF, Farella N, Do Vale RS, Passos CS, Da Silva EDJ, Mergler D (1998) Distribution and partition of total mercury in waters of the Tapajós River Basin, Brazilian Amazon. Sci Total Environ 213(1–3):203–211
36. Myers GJ, Davidson PW (1998) Prenatal methylmercury exposure and children: neurologic, developmental, and behavioral research. Environ Health Perspect 106(suppl 3):841–847
37. Sirois JE, Atchison WD (2000) Methylmercury affects multiple subtypes of calcium channels in rat cerebellar granule cells. Toxicol Appl Pharmacol 167:1–11
38. Vroom FQ, Greer M (1972) Mercury vapour intoxication. Brain 95(2):305–318
39. Weiss B (2014) Psychological indices of toxicity. Encycl Toxicol 558–567
40. Park JD, Zheng W (2012) Human exposure and health effects of inorganic and elemental mercury. J Prev Med Public Health 45(6):344
41. Eto K (1997) Pathology of minamata disease. Toxicol Pathol 25:614–623
42. Takeuchi T (1968) Pathology of Minamata disease. In: Kutsuma K (ed) Minamata disease. Kumamoto Shuhan Publishing Co., Kumamoto, Japan, pp 141–256
43. Xu F, Farkas S, Kortbeek S, Zhang FX, Chen L, Zamponi GW, Syed NI (2012) Mercury-induced toxicity of rat cortical neurons is mediated through N-methyl-D-Aspartate receptors. Mol Brain 5:1–14
44. Taueg C, Sanfilippo DJ, Rowens B, Szejda J, Hesse JL (1992) Acute and chronic poisoning from residential exposures to elemental mercury-michigan, 1989–1990. J Toxicol Clin Toxicol 30(1):63–67
45. Korbas M, O'Donoghue JL, Watson GE, Pickering IJ, Singh SP, Myers GJ, Clarkson TW, George GN (2010) The chemical nature of mercury in human brain following poisoning or environmental exposure. ACS Chem Neurosci 1(12):810–818
46. Kade IJ (2012) Mercury toxicity on sodium pump and organoseleniums intervention: a paradox. J Biomed Biotechnol
47. Aschner M, Aschner JL (1990) Mercury neurotoxicity: mechanisms of blood-brain barrier transport. Neurosci Biobehav Rev 14(2):169–176
48. National Research Council (2000) Toxicological effects of methylmercury. National Academy Press, Washington, DC
49. World Health Organization (1976) Environmental health criteria. 1. Mercury. Environmental health criteria. 1. Mercury
50. Madabeni A, Nogara PA, Bortoli M, Rocha JB, Orian L (2021) Effect of methylmercury binding on the peroxide-reducing potential of cysteine and selenocysteine. Inorg Chem 60(7):4646–4656
51. Punt A (2018) Toxicokinetics in risk evaluations. Chem Res Toxicol 31(5):285–286
52. Sturla SJ, Boobis AR, FitzGerald RE, Hoeng J, Kavlock RJ, Schirmer K, Whelan M, Wilks MF, Peitsch MC (2014) Systems toxicology: from basic research to risk assessment. Chem Res Toxicol 27(3):314–329

53. Cercy SP, Wankmuller MM (2008) Cognitive dysfunction associated with elemental mercury ingestion and inhalation: a case study. Appl Neuropsychol 15(1):79–91
54. Foley MM, Seidel I, Sevier J, Wendt J, Kogan M (2020) One man's swordfish story: The link between Alzheimer's disease and mercury exposure. Complement Ther Med 52:102499
55. Eddins D, Petro A, Pollard N, Freedman JH, Levin ED (2008) Mercury-induced cognitive impairment in metallothionein-1/2 null mice. Neurotoxicol Teratol 30(2):88–95
56. Yoshida M, Watanabe C, Horie K, Satoh M, Sawada M, Shimada A (2005) Neurobehavioral changes in metallothionein-null mice prenatally exposed to mercury vapor. Toxicol Lett 155:361–368
57. Katamanova EV, Shevchenko OI, Lakhman OL, Denisova IA (2014) Cognitive disorders in patients with chronic mercury intoxication. Med Tr Prom Ekol 4:7–12
58. Yoshida M, Watanabe C, Horie K, Satoh M, Sawada M, Shimada A (2005) Neurobehavioral changes in metallothionein-null mice prenatally exposed to mercury vapor. Toxicol Lett 155(3):361–368

Health Risk Linked to Mercury Toxicity in Food and Environment

Abhishek Pathak, Pranav Anjaria, Prakrutik Bhavsar, and Varun Asediya

Abstract Mercury is a potent neurotoxin that poses significant risks to human health. The toxicity of mercury to humans depends on the specific form of mercury, the dosage, and the rate of exposure. Inhaled mercury vapor primarily affects the brain, while mercurous and mercuric salts primarily damage the gastrointestinal lining and kidneys. Methyl mercury, on the other hand, is distributed throughout the body. In this chapter, different sources of mercury contamination in the food and environment are described. Different pathways of mercury contamination in food and effect of mercury poisoning on health is also described.

Keywords Mercury poisoning · Health · Source · Food

1 Introduction

Mercury (Hg) is a heavy metal known for its toxic properties that poses significant risks to both the environment and human health. It is found naturally in the Earth's crust and released into the environment through various industrial activities [6]. Mercury pollution is a global concern, and understanding its impact on aquatic animals, livestock, and humans is crucial for safeguarding ecosystems and public health [11]. which have caused significant public health crises in Minamata Bay, Japan, and Iraq [1]. The impact of smaller exposures to mercury remains a topic of debate within the scientific community. Mercury exists in different forms, including inorganic mercury (such as metallic mercury, mercury vapor, and mercurous or

A. Pathak (✉)
Department of Veterinary Pharmacology and Toxicology, Apollo College of Veterinary Medicine, Jaipur, Rajasthan, India
e-mail: drpathak1704@gmail.com

P. Anjaria · P. Bhavsar
College of Veterinary Science and Animal Husbandry, Kamdhenu University, Anand, Gujarat, India

V. Asediya
M. B. Veterinary College, Dungarpur, Rajasthan, India

N. Kumar (ed.), *Mercury Toxicity Mitigation: Sustainable Nexus Approach*, Earth and Environmental Sciences Library, https://doi.org/10.1007/978-3-031-48817-7_6

mercuric salts) and organic mercury (compounds where mercury is bonded to carbon-containing structures like methyl, ethyl, or phenyl groups). The biological behaviour, toxicokinetic, and clinical implications of these mercury forms vary depending on their chemical composition. There is some interconversion between the different forms of mercury in living organisms. For example, when elemental mercury vapor is inhaled, it can be easily absorbed through the mucous membranes and lungs, where it is gradually transformed into other forms, including deposition in the brain. Methyl mercury, which is absorbed through the gastrointestinal tract, can be found in various tissues but does not efficiently cross the blood–brain barrier. However, once it enters the brain, it undergoes demethylation and converts into elemental mercury. In contrast, mercury salts have low solubility, relative stability, and limited absorption.

The toxicity of mercury to humans depends on the specific form of mercury, the dosage, and the rate of exposure. Inhaled mercury vapor primarily affects the brain, while mercurous and mercuric salts primarily damage the gastrointestinal lining and kidneys. Methyl mercury, on the other hand, is distributed throughout the body. The severity of toxicity varies depending on the dose, with acute exposures to elemental mercury vapor causing severe pneumonitis, which can be fatal in extreme cases. Chronic low-level exposure to elemental mercury or other forms of mercury leads to more subtle symptoms and clinical findings.

2 Impact of Mercury in Ecosystem

Mercury poisoning in aquatic animals can disrupt ecosystems and food chains, leading to population declines and imbalances [4]. The contamination of water bodies with mercury results in its bioaccumulation and biomagnification, posing a greater threat to higher trophic level species. Livestock can be exposed to mercury through contaminated feed and water, causing health issues such as reproductive problems and neurological disorders. The presence of mercury in animal-derived food products also poses a risk to human consumers. Consuming contaminated seafood can lead to mercury accumulation in the human body, resulting in severe health problems like neurological disorders and kidney damage. Certain occupations, such as gold mining and dental work, expose individuals to elevated levels of mercury, increasing the risk of mercury poisoning [21]. Addressing mercury poisoning in aquatic animals, livestock, and humans is crucial to maintain ecosystem health and safeguard human well-being. Efforts should focus on reducing mercury emissions, implementing proper waste management practices, and raising awareness to mitigate the risks associated with mercury contamination. Understanding and addressing mercury poisoning in aquatic animals, livestock, and humans is crucial for the protection of ecosystems and human health. Efforts must focus on reducing mercury emissions and implementing proper waste management practices. Additionally, regular monitoring of mercury levels in the environment and raising awareness among communities can help mitigate the risks associated with mercury contamination, ensuring a safer and healthier future for all.

3 Historical Perspectives of Mercury Poisoning

The Minamata outbreak, which occurred in the 1950 and 1960s in Minamata, Japan, is a significant historical incident related to mercury poisoning. In the 1950s, unusual events unfolded in Minamata Bay, with dying shellfish, floating fish, stunted seaweed growth, and mysterious cat fatalities. On April 21, 1956, a young girl was admitted to the Chisso Hospital with severe limb numbness, speech impairment, and difficulty eating, leading to the official recognition of Minamata Disease (MPM). By the end of 1956, there were 54 confirmed cases, including 17 deaths since the initial outbreak in December 1953. In November 1962, the first case of congenital MPM was officially acknowledged, and the affected population expanded along the coast of the Shiranui Sea [25]. It played a crucial role in shaping environmental policies and research on mercury toxicity, leading to efforts to mitigate mercury pollution and protect public health worldwide. The Minamata outbreak serves as a stark reminder of the long-term consequences of environmental contamination and the importance of understanding the toxic effects of heavy metals on human health.

4 Forms of Mercury

Mercury, a toxic heavy metal, exists in various forms in the environment. Understanding the different forms of mercury is essential in assessing its toxicity, sources of exposure, and potential health risks. The three main forms of mercury are elemental (or metallic) mercury, organic mercury compounds, and inorganic mercury compounds. Different types of mercury forms and their distinctive features are described in Table 1.

a. **Elemental Mercury (Hg^0)**

Elemental mercury is the pure, metallic form of mercury that is a shiny, silver liquid at room temperature. It has low solubility in water and does not readily bind to organic matter. Elemental mercury is commonly found in thermometers, barometers, and some electrical switches. It can be released into the environment through natural processes such as volcanic activity, as well as human activities like burning fossil fuels and waste incineration. Inhalation of mercury vapor is the primary route of exposure to elemental mercury. Once inhaled, it can cross the blood–brain barrier and accumulate in the brain, leading to neurological effects.

b. **Inorganic Mercury Compounds**

Inorganic mercury compounds include various forms such as mercuric chloride ($HgCl_2$) and mercurous chloride (Hg_2Cl_2). These compounds can be found in industrial settings, laboratories, and certain consumer products like skin-lightening creams and disinfectants. Inorganic mercury compounds are more water-soluble than elemental mercury and can enter the environment through industrial discharges and

Table 1 Types of mercury forms and their distinctive features in toxicodynamic contexts

	Elemental mercury	Methyl mercury	Inorganic mercury
Sources	Fossil fuels, dental amalgams, old latex paint, incinerators, thermometers	Pesticides, fish, poultry	Biological oxidation of mercury, demethylation of methyl mercury by intestinal microflora
Absorption	75–85% of vapor absorbed	95–100% absorbed in intestinal tract	7–15% of ingested dose absorbed and 2–3% dermal dose absorbed in animals
Distribution	Distributed throughout the body, lipophilic, crosses blood–brain barrier and placental barrier, accumulates in brain and kidney	Distributed throughout the body, lipophilic, readily crosses blood–brain barrier as well as placental barrier, accumulates in kidney and brain	Does not cross blood–brain or placental barrier, present in brain neonates, accumulates in kidney
Excretion	Sweat, urine, faeces, and saliva	90% excreted in bile, faeces, 10% in urine	Sweat, saliva, urine and faeces
Reason for toxicity	Oxidation to inorganic mercury	Demethylation to inorganic mercury, generation of free radical, binding to thiols	

improper waste disposal. Ingestion and inhalation are the primary routes of exposure to inorganic mercury compounds. These compounds can damage the kidneys, gastrointestinal tract, and respiratory system. Unlike organic mercury compounds, inorganic forms do not readily cross the blood–brain barrier.

c. **Organic Mercury Compounds**

Organic mercury compounds are formed when elemental mercury combines with carbon-containing compounds. The most well-known and toxic organic mercury compound is methylmercury (CH_3Hg^+). Methylmercury is produced through microbial processes in aquatic environments, particularly in sediment and water bodies. It bioaccumulates and biomagnifies in the food chain, with higher concentrations found in predatory fish and marine mammals. Human exposure to methylmercury primarily occurs through the consumption of contaminated fish and shellfish. Methylmercury readily crosses the placenta, leading to potential harm to the developing foetus. It affects the central nervous system and can cause neurological damage, especially in infants and children.

5 Sources of Mercury

a. **Natural Sources**

- Volcanic Activity: Volcanic eruptions release small amounts of mercury into the atmosphere.
- Weathering of Rocks: Mercury can be naturally present in certain types of rocks and is released into the environment through weathering processes.

b. **Anthropogenic (Human) Sources**

- Artisanal and Small-Scale Gold Mining: Mercury and other metal such as manganese is used in gold mining to extract gold from ore, leading to significant mercury emissions and environmental contamination [12].
- Coal Combustion: Burning coal for energy generation releases mercury into the air, which can then be deposited into water bodies and soil [17]. Coal fire plants accounts for around 40% man made mercury emission in U.S. [19]
- Industrial Processes: Various industries, such as chlor-alkali production, cement production, and waste incineration, release mercury into the environment as a byproduct [15].
- Fossil Fuel Combustion: Burning fossil fuels in power plants, factories, and vehicles can result in the release of mercury into the atmosphere.
- Waste Incineration: Burning of municipal, medical, or hazardous waste can release mercury into the air [7].
- Chemical Manufacturing: Mercury is used in the production of certain chemicals, and improper handling or disposal can lead to mercury releases.
- Dental Amalgams: Dental fillings that contain mercury can contribute to mercury pollution through improper disposal of dental waste [13].
- Consumer Products: Some consumer products, such as thermometers, fluorescent light bulbs, and skin-lightening creams, may contain mercury. Improper disposal of these products can release mercury into the environment [2].
- Agricultural Practices: The use of mercury-containing fertilizers, pesticides, and fungicides in agriculture can contribute to mercury pollution.
- Different sources of Mercury in the environment are depicted in Fig. 1.

6 Synonyms of Mercury Poisoning

a. Minamata disease: Minamata disease is a specific form of mercury poisoning that occurred in the city of Minamata, Japan, in the mid-twentieth century. It was caused by the release of methylmercury into the local waters by an industrial facility, resulting in severe neurological symptoms and disabilities in the affected population.

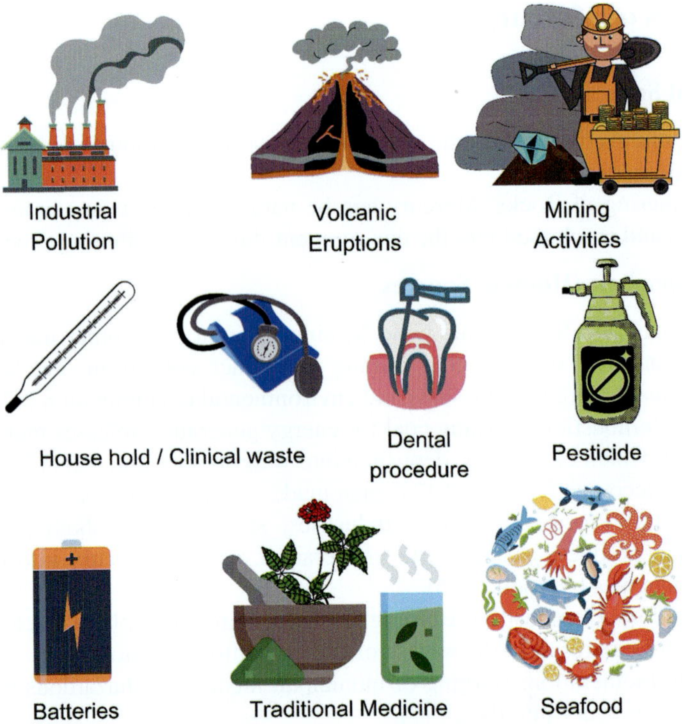

Fig. 1 Different sources of mercury

b. Quicksilver poisoning: Quicksilver is another term for mercury, and quicksilver poisoning refers to poisoning caused by mercury exposure. It refers to the toxic effects and symptoms that occur due to the ingestion, inhalation, or contact with mercury.
c. Hydrargyria: Hydrargyria refers to the condition of having a silver or grayish appearance due to mercury poisoning. It is derived from the Greek words "hydor" meaning "water" and "argyros" meaning "silver."
d. Mad hatter's disease: Mad hatter's disease is a colloquial term for mercury poisoning that originated from the historical use of mercury in hat-making processes. Hatters were exposed to mercury vapor while working with mercury-treated felt, and over time, they developed neurological symptoms, leading to the association of mercury poisoning with the term "mad hatter's disease" or "Erethism" [14].

7 Pathways of Mercury Exposure

- Ingestion: Consumption of food and water contaminated with mercury is a significant pathway of exposure for the general population. Fish and seafood are known to accumulate mercury, especially methyl mercury, from contaminated water sources. Additionally, ingestion can occur through the accidental ingestion of mercury-containing products or intentional ingestion, such as in cases of self-harm [10].
- Inhalation: Inhalation of mercury vapor is a common route of exposure, particularly in occupational settings such as industries involving mercury use or mining. It can also occur through the inhalation of mercury released from products or environmental sources, such as coal combustion or dental amalgam [23].
- Dermal Contact: Direct skin contact with mercury or mercury-containing substances can lead to absorption of mercury through the skin. This route of exposure is more common in occupational settings where individuals come into direct contact with mercury or its compounds [3].
- Maternal-foetal Transfer: Pregnant women who are exposed to mercury can transfer the toxic metal to their developing foetus through the placenta. This is particularly concerning as the developing foetus is highly vulnerable to the harmful effects of mercury [18].
- Occupational Exposure: Certain occupations, such as those involving mercury mining, production, or handling, pose a higher risk of exposure to mercury. Workers in these industries may be exposed to mercury vapor, dust, or other forms of the metal during their job tasks.
- Dental Amalgams: Dental amalgams, which contain mercury, can contribute to exposure when they degrade over time. This is more relevant for individuals with a large number of amalgam fillings or those who grind their teeth, as it can lead to the release of mercury vapor.
- Environmental Exposure: Environmental sources, such as contaminated air, water, and soil, can contribute to mercury exposure. This can result from natural processes, like volcanic activity, as well as human activities, including industrial emissions, waste incineration, and the improper disposal of mercury-containing products.

8 Mercury Poisoning and Its Types

Mercury is a toxic heavy metal that poses risks to human health. The primary sources of human exposure to mercury include the release of mercury vapor from dental amalgam and the consumption of mercury-contaminated fish and seafood. Occupational exposure to mercury vapor can also lead to mercury toxicity. The brain is particularly susceptible to the toxic effects of inhaled mercury vapor [28]. Both inorganic forms of mercury (such as mercurous and mercuric salts) and organic forms (primarily methyl mercury) can cause toxicity. Mercury, in any form, disrupts protein structure by binding to sulfhydryl groups, potentially impairing the function of various organs, with neurological functions being most commonly affected.

Methyl mercury exposure from consuming fish, shellfish, and sea mammals is particularly concerning for pregnant women and children. Numerous lakes in the United States have been closed to fishing due to mercury contamination, with predatory fish often containing higher levels of methyl mercury. It is recommended that pregnant women, women planning to become pregnant, and children avoid consuming fish known to accumulate high levels of mercury, such as shark, swordfish, king mackerel, golden bass, and snapper. Other individuals can safely consume up to 7 oz of these fish per week. Seafood like salmon, cod, flounder, catfish, crabs, and scallops may contain lower levels of mercury. Monitoring mercury levels in blood, urine, and hair can be helpful in assessing exposure. Experts from the World Health Organization (WHO) have determined that a 24 h urine mercury level exceeding 50 μg indicates excessive exposure to mercury [5].

9 Vulnerable Populations at Risk

Mercury is a potent neurotoxin that poses significant risks to human health, particularly for vulnerable populations. Certain groups are more susceptible to the adverse effects of mercury exposure due to their unique physiological characteristics, developmental stages, or dietary habits. This article focuses on four specific vulnerable populations at risk: pregnant women and developing foetuses, children and their neurodevelopment, indigenous communities relying on fish consumption, and individuals exposed to mercury in occupational settings.

9.1 Pregnant Women and Developing Foetuses

Maternal Exposure: Pregnant women are at risk of mercury exposure through contaminated food and environmental sources. Mercury can cross the placenta and accumulate in foetal tissues, including the developing brain. Prenatal exposure to mercury has been associated with adverse effects on neurodevelopment, cognitive function, and behaviour in children [10].

Neurodevelopmental Risks: Developing foetuses are particularly vulnerable to mercury's neurotoxic effects. Methylmercury, the organic form of mercury commonly found in seafood, can interfere with brain development and lead to permanent cognitive and behavioural impairments. The developing nervous system is highly sensitive to mercury exposure, and even low levels of exposure can have long-lasting consequences [20].

Fish Consumption Guidelines: Due to the potential risks associated with mercury exposure, pregnant women are advised to follow fish consumption guidelines to minimize their exposure to methylmercury. These guidelines typically recommend limiting the consumption of certain types of fish that are known to have higher mercury levels while encouraging the consumption of low-mercury fish, which provide important nutrients for foetal development.

9.2 Children and Their Neurodevelopment

Increased Susceptibility: Children are more susceptible to the adverse effects of mercury exposure due to their developing nervous systems and higher metabolic rates. Exposure during critical periods of brain development can lead to neurocognitive impairments, including deficits in attention, memory, language, and visual-spatial skills.

Behavioural and Cognitive Effects: Mercury exposure in childhood has been associated with behavioural problems, including attention-deficit/hyperactivity disorder (ADHD) symptoms and decreased IQ scores. Studies have also shown that prenatal and early-life exposure to mercury can have long-term effects on academic performance, memory, and executive functioning [26].

Educational and Intervention Programs: Recognizing the risks posed by mercury exposure to children, educational programs are implemented to raise awareness among parents, caregivers, and healthcare professionals. These programs provide information on the sources of mercury exposure, the importance of following fish consumption guidelines, and strategies to reduce exposure in homes and schools.

9.3 Indigenous Communities Relying on Fish Consumption

Cultural Practices and Dietary Habits: Indigenous communities often have cultural practices and dietary habits that involve the consumption of fish from local water bodies. Fish provide an important source of nutrition and sustenance. However, certain fish species in contaminated areas may have elevated mercury levels, posing health risks to these communities.

Health Disparities: Indigenous communities may face higher risks of mercury exposure due to their reliance on contaminated fish for sustenance. Limited access to alternative food sources, economic constraints, and geographical isolation can exacerbate the challenges of reducing mercury exposure in these populations. This can lead to health disparities and increased vulnerability to the adverse health effects of mercury.

Collaborative Approaches: Addressing mercury exposure in indigenous communities requires collaborative efforts between community members, healthcare providers, researchers, and policymakers. Initiatives aimed at assessing mercury levels in local fish, providing alternative food sources, and implementing culturally appropriate risk communication strategies are crucial in minimizing exposure and promoting the health and well-being of these communities.

9.4 Occupational Exposure Risks for Specific Industries

Industries at Risk: Certain industries and occupations are associated with increased mercury exposure risks [8]. For example, dental workers exposed to dental amalgam, artisanal and small-scale gold miners using mercury for gold extraction, and workers in industries such as chemical manufacturing, chlor-alkali production, and fluorescent light manufacturing may face occupational mercury exposure.

Health Effects on Workers: Occupational exposure to mercury can result in a range of health effects, including neurological symptoms, kidney damage, respiratory problems, and skin disorders. Workers in these industries may be exposed to mercury vapor or other forms of mercury during production, handling, or disposal processes. Proper protective measures, occupational health and safety training, and regular monitoring are essential to minimize exposure and protect workers' health.

Regulatory Measures: Governments and regulatory agencies play a crucial role in implementing regulations and guidelines to protect workers from mercury exposure in occupational settings. These measures include setting occupational exposure limits, promoting engineering controls and personal protective equipment, and conducting regular workplace monitoring to ensure compliance with safety standards.

10 Toxicodynamic of Mercury

The absorption, distribution, metabolism, and elimination of mercury are influenced by its chemical form and oxidation state. Organic forms of mercury are more easily absorbed compared to inorganic forms. The body metabolizes mercury through an oxidation–reduction cycle, and the primary routes of excretion are through urine and faeces. The elimination half-life varies depending on the form of mercury, ranging from 35 to 90 days for elemental mercury and about 40 days for inorganic salts.

While ingestion of mercury metal is generally not harmful, ingestion of inorganic salts can lead to severe gastrointestinal irritation, renal failure, and even death at high doses. Mercuric salts are typically more toxic than mercurous salts. Mercury can also trigger hypersensitivity reactions, including contact dermatitis and acrodynia (pink disease). Inhalation of mercury vapor can cause respiratory tract irritation, renal disorders, neurobehavioral changes, peripheral nervous system toxicity, renal toxicity, and even death.

Sub chronic and chronic exposure to mercury primarily affects the kidneys and/or nervous system, with the specific effects depending on the form of mercury. Organic mercury, particularly methyl mercury, rapidly enters the central nervous system, leading to behavioural and neuromotor disorders. The developing central nervous system is particularly vulnerable to methyl mercury toxicity, as evidenced by severe effects observed in regions where methyl mercury-contaminated food caused toxicity in adults and central nervous system effects in infants. The teratogenic effects of mercury exposure are not extensively documented, although there is some evidence suggesting potential disturbances in menstrual cycles and spontaneous abortions.

The reference doses (RfDs) for chronic oral exposure to methyl mercury and mercuric chloride are 0.0001 and 0.0003 mg/kg/day, respectively, based on specific health effects. The lowest observed adverse effect level (LOAEL) for mercuric chloride is 0.63 mg Hg/kg/day. Regarding inhalation exposure, the reference concentration (RfC) for inorganic mercury is 0.0003 mg Hg/m^3, which is associated with neurological disorders. The LOAELs for inhalation exposure to inorganic mercury are 0.32 mg Hg/m^3 (sub chronic) and 0.03 mg Hg/m^3 (chronic). However, the RfC for methyl mercury inhalation has not been determined [16].

11 Health Implications of Mercury Poisoning

Mercury poisoning is a condition that can affect multiple systems in the human body, leading to a wide range of health problems. The toxic effects of mercury can involve various systems, including the neurological, renal, respiratory, cardiovascular, and immune systems. In the neurological system, mercury can cause neurobehavioral changes, cognitive impairments, tremors, and even neurological disorders. The renal system may be affected by mercury-induced nephrotoxicity, leading to kidney damage and impaired renal function. Inhalation of mercury vapor can irritate the respiratory system and cause respiratory tract inflammation. Mercury has also been associated with cardiovascular effects, such as increased blood pressure and heart rate variability. Furthermore, the immune system can be impacted, leading to immunotoxicity and increased susceptibility to infections. The multisystem involvement in mercury poisoning highlights the systemic nature of its toxicity and underscores the importance of comprehensive evaluation and management of individuals exposed to mercury to minimize its detrimental effects on overall health. Effect of mercury toxicity on different system is described in Table 2.

12 Risk Assessment of Mercury Poisoning

- Risk assessment of mercury poisoning involves evaluating the potential health risks associated with exposure to mercury. Here are the key steps involved in the risk assessment process:
- Hazard Identification: This step involves gathering information on the toxicity of mercury, including its various forms and routes of exposure. Studies and scientific literature are reviewed to understand the adverse health effects of mercury on different organ systems and vulnerable populations.

Table 2 Multisystem involvement in mercury toxicity

System	Signs and symptoms
Cardiovascular	Hypertension, hypotension
Nervous system	Emotional disturbances, irritability, hypochondria, psychosis, impaired memory, insomnia, tremor, dysarthria, involuntary movements, vertigo, polyneuropathy, paraesthesia, headache
Sensory systems	Corneal opacities and ulcers, conjunctivitis, hypoacusis
Endocrine	Hyperthyroidism
Haematology	Hypochromic anaemia, erythrocytosis, lymphocytosis, neutropenia, aplastic anaemia
Mouth	Loose teeth, discoloration of the gums and oral mucosa, mouth ulcers, fetor
Gastrointestinal	Anorexia, nausea, vomiting, epigastric pain, diarrhoea, constipation
Urinary tract	Nephrotic syndrome
Skin	Tylotic eczema, dry skin, skin ulcers, erythroderma
Musculoskeletal	Acrodynia, arthritis in the legs
Reproductive system	Dysmenorrhea

- Exposure Assessment: The exposure assessment focuses on determining the extent and frequency of exposure to mercury. This includes identifying the sources of mercury exposure, such as occupational exposure, consumption of contaminated fish, or exposure to mercury-containing products. Data on exposure levels and patterns are collected and analysed.
- Dose–Response Assessment: In this step, the relationship between the dose (amount) of mercury and the resulting health effects is established. It involves evaluating the available scientific data to determine the adverse health effects associated with different levels of mercury exposure.
- Risk Characterization: The risk characterization combines the information from the hazard identification, exposure assessment, and dose–response assessment to estimate the potential health risks posed by mercury exposure. This step involves quantifying the likelihood and severity of adverse health effects based on exposure levels and the susceptibility of different population groups.
- Uncertainty Analysis: Uncertainty analysis is conducted to address the limitations and uncertainties in the risk assessment process. It involves identifying and quantifying the uncertainties associated with data gaps, variability, and assumptions made during the assessment.
- Risk Communication: The findings of the risk assessment are communicated to stakeholders, including policymakers, health professionals, and the general public. Clear and concise information about the potential health risks, recommended exposure limits, and preventive measures is provided to help individuals make informed decisions.

- By following these steps, risk assessors can evaluate the potential health risks associated with mercury exposure and provide valuable information for decision-making, policy development, and the implementation of preventive measures to protect public health.

13 Biomarkers of Mercury Exposure

Biomarkers play a vital role in identifying and quantifying mercury exposure levels.

A. **Blood and Urine Biomarkers**

1. Blood Mercury Levels: Measuring total mercury levels in blood provides valuable information about recent exposure to mercury. Blood mercury levels are commonly expressed as micrograms per liter (μg/L) or parts per billion (ppb). However, blood mercury levels only represent the current exposure and do not provide insight into past or cumulative exposure [22].
2. Urinary Mercury Levels: Urine is another common medium used to assess mercury exposure. Total mercury levels in urine can reflect recent exposure, particularly to inorganic mercury [9]. However, it should be noted that urinary mercury levels can vary widely based on factors such as kidney function, hydration status, and recent exposure. Urine mercury levels are also influenced by other sources of mercury exposure, such as dental amalgams.

B. **Hair and Nail Analysis**

1. Hair Mercury Analysis: Hair analysis is a useful tool for assessing long-term or chronic exposure to mercury [24]. Mercury is incorporated into growing hair strands, allowing for retrospective analysis. Hair samples are typically collected from the scalp or other body regions and analyzed for total mercury content. However, it is important to consider external contamination sources such as mercury-containing shampoos or hair dyes, which can affect the accuracy of results.
2. Nail Mercury Analysis: Similar to hair analysis, nail analysis provides a means to assess long-term exposure to mercury. Mercury accumulates in the nail matrix as it grows, allowing for the analysis of historical exposure levels [27]. Nail clippings or entire nails can be collected for analysis, and the mercury content can be measured. Nail analysis is particularly useful for detecting exposure to methylmercury, the organic form of mercury commonly found in seafood.

C. **Biomonitoring Studies and Their Implications**

Biomonitoring studies involve the measurement of mercury levels in various biological samples across a population or specific groups. These studies provide valuable data on the extent and distribution of mercury exposure in different populations and can help identify high-risk groups. Some key implications of biomonitoring studies include:

1. Assessing Environmental Exposure: Biomonitoring studies can help assess the overall exposure to mercury in a given population. By measuring biomarkers in different individuals, researchers can determine the extent of exposure and identify potential sources such as contaminated water or occupational settings. This information is crucial for implementing targeted interventions and regulatory measures to reduce exposure levels.
2. Evaluating Health Risks: Biomonitoring studies provide insights into the potential health risks associated with mercury exposure. By correlating biomarker levels with clinical outcomes, researchers can determine the dose–response relationships and establish threshold levels for adverse effects. This information helps in setting appropriate exposure limits and guidelines to protect public health.
3. Monitoring Effectiveness of Interventions: Biomonitoring studies also play a crucial role in evaluating the effectiveness of interventions aimed at reducing mercury exposure. By periodically measuring biomarkers in exposed populations, researchers can assess the impact of interventions such as pollution control measures or dietary guidelines. This feedback allows for adjustments and improvements in intervention strategies to further reduce exposure risks.
4. Public Awareness and Education: Biomonitoring studies raise public awareness about mercury exposure and its potential health effects. The dissemination of study findings can help educate individuals and communities about the sources of mercury exposure and the importance of adopting preventive measures. This awareness can lead to behavioural changes, such as choosing low-mercury seafood options or proper disposal of mercury-containing products.

14 Strategies for Mercury Poisoning Control

- Strategies for mercury poisoning control involve implementing measures to prevent or reduce exposure to mercury and mitigate the associated health risks. Here are some key strategies:
- Environmental Controls: Implementing strict regulations and controls on industrial processes and activities that release mercury into the environment is crucial. This includes minimizing mercury emissions from industries such as coal-fired power plants, waste incineration facilities, and artisanal small-scale gold mining. Implementing proper waste management practices to prevent mercury contamination of soil and water is also important.
- Occupational Safety Measures: Ensuring occupational safety measures in industries where mercury is used or handled is essential to protect workers. This includes providing adequate ventilation systems, personal protective equipment,

and training on safe handling and disposal practices. Regular monitoring of workplace mercury levels and health surveillance of workers can help identify and prevent exposure.

- Mercury-Free Alternatives: Promoting the use of mercury-free alternatives in various sectors is an effective strategy. Encouraging industries to adopt mercury-free technologies and processes, such as replacing mercury-containing instruments and equipment with safer alternatives, reduces the overall demand and use of mercury.

15 Dietary Interventions and Mercury Detoxification

A. **Nutritional Approaches to Mitigate Mercury Toxicity**

Mercury toxicity is a concern due to its potential adverse effects on human health. While complete elimination of mercury from the body is challenging, certain nutritional approaches can help mitigate its toxicity and support overall health. These approaches aim to enhance the body's natural detoxification mechanisms and reduce the absorption and accumulation of mercury. Here are some key nutritional strategies to consider:

- Antioxidant-Rich Foods: Including a variety of antioxidant-rich foods in the diet can help counteract the oxidative stress caused by mercury. Antioxidants, such as vitamins C and E, selenium, and various phytochemicals found in fruits, vegetables, nuts, and seeds, can neutralize free radicals and minimize their harmful effects.
- Sulfur-Containing Foods: Sulfur-containing foods, such as cruciferous vegetables (broccoli, cabbage, cauliflower), garlic, onions, and eggs, can support detoxification pathways in the body. Sulfur compounds aid in the formation of glutathione, a potent antioxidant and essential component of the body's detoxification system.
- Omega-3 Fatty Acids: Consuming foods rich in omega-3 fatty acids, such as fatty fish (salmon, sardines) and flaxseeds, can help reduce inflammation associated with mercury toxicity. Omega-3 fatty acids also support brain health and may mitigate some of the neurotoxic effects of mercury.
- Fiber-Rich Foods: Including high-fiber foods in the diet, such as whole grains, legumes, fruits, and vegetables, can facilitate the elimination of mercury through the digestive system. Fiber binds to mercury in the gut and promotes its excretion, reducing its absorption into the bloodstream.
- Hydration: Staying adequately hydrated supports the body's natural detoxification processes. Drinking sufficient water helps flush out toxins, including mercury, through urine and supports kidney function.

It is important to note that while these nutritional approaches can support overall health and potentially reduce the adverse effects of mercury toxicity, they should not be considered standalone treatments for mercury detoxification. They should be used as part of a comprehensive approach that includes minimizing exposure to mercury and seeking appropriate medical advice.

B. **Chelation Therapy and Its Effectiveness**

Chelation therapy is a medical intervention used to remove heavy metals, including mercury, from the body. This approach involves administering chelating agents that bind to the metals, forming stable complexes that can be excreted through urine. While chelation therapy has been utilized for various heavy metal poisonings, its effectiveness and safety specifically in the context of mercury detoxification are still subjects of ongoing research and debate. There are different chelating agents available for mercury detoxification, such as dimercaptosuccinic acid (DMSA), dimercapto-propane sulfonate (DMPS), and ethylenediaminetetraacetic acid (EDTA). These agents have varying affinities for mercury and can be administered through different routes, including oral and intravenous. Chelation therapy can assist in promoting the excretion of mercury from the body, thereby reducing its burden. However, the effectiveness of this therapy in addressing the long-term health effects of mercury toxicity, particularly in cases of low-level chronic exposure, remains a subject of scientific inquiry. Research on chelation therapy for mercury detoxification has produced mixed results, underscoring the need for additional studies to assess its long-term benefits and potential risks accurately. It is important to recognize that chelation therapy carries potential risks and side effects. Chelating agents can bind not only to mercury but also to essential minerals and trace elements in the body, potentially causing imbalances or deficiencies. Adverse effects, including gastrointestinal disturbances, allergic reactions, and kidney damage, have been reported in some cases. Consequently, it is imperative that chelation therapy be conducted under the supervision of qualified healthcare professionals who can monitor its safety and efficacy closely. The decision to pursue chelation therapy should be based on a comprehensive evaluation of individual circumstances, including the level of mercury exposure, presenting symptoms, and overall health. Consulting with healthcare professionals experienced in metal toxicology is crucial to assess the appropriateness of chelation therapy and explore alternative treatment options that may be available.

16 Fish Consumption Guidelines

Fish consumption advisories for mercury poisoning are guidelines issued by regulatory agencies or health authorities to inform the public about the potential risks associated with consuming fish and seafood contaminated with mercury. These advisories aim to provide recommendations on safe levels of fish consumption based on mercury levels in specific water bodies or fish species. Advice pertaining to fish consumption issued by the FDA is explained in Table 3. Here's how fish consumption advisories for mercury poisoning typically work:

Table 3 Advice pertaining to fish consumption issued by the FDA

Category	Fish
Best Choices	Anchovy, Atlantic Croaker, Atlantic Mackerel, Black Sea Bass, Butterfish, Catfish, Clam, Cod, Crab, Crawfish, Founder, Haddock, Hake, Herring, Lobster (American And Spiny), Mullet, Oyster, Pacific Chub Mackerel, Perch (Freshwater and Ocean), Pickerel, Plaice, Pollock, Salmon, Sardine, Scallop, Shad, Shrimp, Skate, Smelt, Sole, Squid, Tilapia, Trout (Freshwater), Tuna (Canned Light, Including Skipjack), Whitefish, Whiting
Good Choices	Bluefish, Buffalo Fish, Carp, Chilean Sea Bass/Patagonian Toothfish, Grouper, Halibut, Mahi Mahi/Dolphinfish, Monkfish, Rockfish, Sablefish, Sheepshead, Snapper, Spanish Mackerel, Striped Bass (Ocean), Tilefish (Atlantic Ocean), Albacore Tuna/White Tuna (Canned and Fresh/Frozen), Yellowfin Tuna, Weakfish/Sea Trout, White Croaker/Pacific Croaker
Choices to Avoid	King Mackerel, Marlin, Orange Roughy, Shark, Swordfish, Tilefish (Gulf of Mexico), Bigeye Tuna

- Predatory Fish: Large predatory fish such as shark, swordfish, king mackerel, and tilefish tend to have higher levels of mercury due to their position in the food chain.
- Freshwater Fish: Certain freshwater fish, such as largemouth bass and some species of trout, may also contain elevated mercury levels, especially in areas affected by industrial pollution or mining activities.
- Shellfish: Mercury levels in shellfish, such as shrimp, crab, and lobster, are typically lower compared to predatory fish but can vary depending on the specific species and habitat.
- Public Awareness and Education: Raising public awareness about the sources, risks, and health effects of mercury exposure is crucial. Educational campaigns, public outreach programs, and targeted information dissemination help individuals make informed decisions regarding their exposure to mercury. This includes educating communities about the risks associated with consuming mercury-contaminated fish and promoting healthier fish consumption habits.
- Monitoring and Surveillance: Regular monitoring of mercury levels in air, water, soil, and biological samples (including fish) is necessary to assess the extent of contamination and track changes over time. Surveillance systems can help identify high-risk areas and populations, enabling targeted interventions and preventive measures.
- International Cooperation: Collaboration among countries is important to address mercury pollution, as mercury can travel long distances through air and water. International agreements and frameworks, such as the Minamata Convention on Mercury, facilitate cooperation, information sharing, and coordinated actions to reduce mercury emissions and exposure worldwide.
- Regulatory Measures: Governments implement regulations to monitor and control mercury levels in fish and seafood, ensuring they meet safety standards and protecting public health.

17 Conclusion

Mercury presents naturally in the environment. When processed mercury is released, the amount of atmospheric mercury may gradually rise and enter the cycles of the atmosphere, soil, and water, where it may circulate for years. Exposure to mercury or mercury compounds can result in mercury poisoning, which can have a variety of harmful effects depending on the chemical type and method of exposure. Methylmercury (MeHg) is primarily acquired from consuming contaminated fish, seafood, and other animal products that have been exposed to mercury through consumption of contaminated lower organisms. MeHg poisoning is linked to neurological impairment in children and adults as well as harm to the nervous system in adults. Mercury that has been consumed may bioaccumulate, gradually increasing the loads on the body. The systemic pathophysiology of specific organ systems linked to mercury poisoning is discussed in this chapter.

References

1. Bernhoft RA (2012) Mercury toxicity and treatment: a review of the literature. J Environ Public Health. Hindawi Limited. https://doi.org/10.1155/2012/460508
2. Budnik LT, Casteleyn L (2019) Mercury pollution in modern times and its socio-medical consequences. Sci Total Environ 654:720–734. https://doi.org/10.1016/j.scitotenv.2018.10.408
3. Copan L, Fowles J, Barreau T, McGee N (2015) Mercury toxicity and contamination of households from the use of skin creams adulterated with mercurous chloride (calomel). Int J Environ Res Public Health 12(9):10943–10954. https://doi.org/10.3390/ijerph120910943
4. Das K, Debacker V, Pillet S, Bouquegneau J-M (2021) Heavy metals in marine mammals. In: Toxicology of marine mammals. CRC Press, pp. 147–179. https://doi.org/10.1201/9780203165577-11
5. Dasgupta A, Wahed A (2014) Common poisonings including heavy metal poisoning. In: Clinical chemistry, immunology and laboratory quality control. Elsevier, pp 337–351. https://doi.org/10.1016/b978-0-12-407821-5.00019-x
6. de Almeida Rodrigues P, Ferrari RG, dos Santos LN, Conte Junior CA (2019) Mercury in aquatic fauna contamination: A systematic review on its dynamics and potential health risks. J Environ Sci (China) 84:205–218. Chinese Academy of Sciences. https://doi.org/10.1016/j.jes.2019.02.018
7. Domingo JL, Marquès M, Mari M, Schuhmacher M (2020) Adverse health effects for populations living near waste incinerators with special attention to hazardous waste incinerators. A review of the scientific literature. Environ Res 187. https://doi.org/10.1016/j.envres.2020.109631
8. Dutta S, Gorain B, Choudhury H, Roychoudhury S, Sengupta P (2022) Environmental and occupational exposure of metals and female reproductive health. Environ Sci Pollut Res 29(41):62067–62092. Springer Science and Business Media Deutschland GmbH. https://doi.org/10.1007/s11356-021-16581-9
9. Esteban-López M, Arrebola JP, Juliá M, Pärt P, Soto E, Cañas A, Pedraza-Díaz S, González-Rubio J, Castaño A (2022) Selecting the best non-invasive matrix to measure mercury exposure in human biomonitoring surveys. Environ Res 204. https://doi.org/10.1016/j.envres.2021.112394
10. Garí M, Grzesiak M, Krekora M, Kaczmarek P, Jankowska A, Król A, Kaleta D, Jerzyńska J, Janasik B, Kuraś R, Tartaglione AM, Calamandrei G, Hanke W, Polańska K (2022) Prenatal

exposure to neurotoxic metals and micronutrients and neurodevelopmental outcomes in early school age children from Poland. Environ Res 204. https://doi.org/10.1016/j.envres.2021.112049

11. Ghotra A, Lehnherr I, Porter TJ, Pisaric MFJ (2020) Tree-Ring inferred atmospheric mercury concentrations in the mackenzie delta (NWT, Canada) Peaked in the 1970s but are increasing once more. ACS Earth Space Chem 4(3):457–466. https://doi.org/10.1021/acsearthspacechem.0c00003
12. González-Merizalde MV, Menezes-Filho JA, Cruz-Erazo CT, Bermeo-Flores SA, Sánchez-Castillo MO, Hernández-Bonilla D, Mora A (2016) Manganese and mercury levels in water, sediments, and children living near gold-mining areas of the Nangaritza River Basin, Ecuadorian Amazon. Arch Environ Contam Toxicol 71(2):171–182. https://doi.org/10.1007/s00244-016-0285-5
13. Gupta SK, Pratap A (2007) History, origin, and evolution. In: Advances in botanical research, vol 45. Academic, pp 1–20. https://doi.org/10.1016/S0065-2296(07)45001-7
14. Heck DE, Joseph LB, Murthy P, Ansehl A, Jan YH, Wahler GC, Kim HD (2020) Technology versus mercury: the metal that scars civilization. In: Technology and global public health, pp 205–218. https://doi.org/10.1007/978-3-030-46355-7_18
15. Jen YH, Yuan CS, Lin YC, Lee CG, Hung CH, Tsai CM, Tsai HH, Ie IR (2011) Partition and tempospatial variation of gaseous and particulate mercury at a unique mercury-contaminated remediation site. J Air Waste Manag Assoc 61(11):1115–1123. https://doi.org/10.1080/10473289.2011.617617
16. Kismelyeva S, Khalikhan R, Torezhan A, Kumisbek A, Akimzhanova Z, Karaca F, Guney M (2021) Potential human exposure to mercury (Hg) in a chlor-alkali plant impacted zone: Risk characterization using updated site assessment data. Sustainability (Switzerland) 13(24). https://doi.org/10.3390/su132413816
17. Liang Y, Liang H, Zhu S (2014) Mercury emission from coal seam fire at Wuda, Inner Mongolia, China. Atmos Environ 83:176–184. https://doi.org/10.1016/j.atmosenv.2013.09.001
18. Lyons K, Lowe CG (2013) Mechanisms of maternal transfer of organochlorine contaminants and mercury in the common thresher shark (Alopias vulpinus). Can J Fish Aquat Sci 70(12):1667–1672. https://doi.org/10.1139/cjfas-2013-0222
19. Martinez AI, Deshpande BK (2007) Kinetic modeling of H_2O_2-enhanced oxidation of flue gas elemental mercury. Fuel Process Technol 88(10):982–987. https://doi.org/10.1016/j.fuproc.2007.05.009
20. Mitra S, Chakraborty AJ, Tareq AM, Emran TB, Nainu F, Khusro A, Idris AM, Khandaker MU, Osman H, Alhumaydhi FA, Simal-Gandara J (2022) Impact of heavy metals on the environment and human health: Novel therapeutic insights to counter the toxicity. J King Saud Univ-Sci 34(3). https://doi.org/10.1016/j.jksus.2022.101865
21. Neghab M, Choobineh A, Zadeh JH, Ghaderi E (2011) Symptoms of intoxication in dentists associated with exposure to low levels of mercury. Ind Health 49(2):249–254. https://doi.org/10.2486/indhealth.MS1214
22. Packull-McCormick S, Ratelle M, Lam C, Napenas J, Bouchard M, Swanson H, Laird BD (2022) Hair to blood mercury concentration ratios and a retrospective hair segmental mercury analysis in the Northwest Territories, Canada. Environ Res 203. https://doi.org/10.1016/j.envres.2021.111800
23. Priyadarshanee M, Mahto U, Das S (2022) Mechanism of toxicity and adverse health effects of environmental pollutants. In: Microbial biodegradation and bioremediation: techniques and case studies for environmental pollution, pp 33–53. https://doi.org/10.1016/B978-0-323-85455-9.00024-2
24. Yasutake A, Hachiya N (2006) Accumulation of inorganic mercury in hair of rats exposed to methylmercury or mercuric chloride. Tohoku J Exp Med 210(4):301–306. https://doi.org/10.1620/tjem.210.301
25. Yokoyama H (2018) Mercury pollution in minamata. SpringerBriefs Environ Sci 341(6153)
26. Yoshimasu K (2013) 763–Childhood autism, adhd, and mercury exposures: a meta-analysis. Eur Psychiatry 28:1. https://doi.org/10.1016/S0924-9338(13)75968-6

27. Zhang W, Li F, Gao L, Sun G, Cui Z, Chen F, Li P, Feng X, Shang L (2022) Understanding the excretion rates of methylmercury and inorganic mercury from human body via hair and fingernails. J Environ Sci (China) 119:59–67. https://doi.org/10.1016/j.jes.2022.01.041
28. Zulaikhah ST, Wahyuwibowo J, Pratama AA (2020) Mercury and its effect on human health: A review of the literature. Int J Public Health Sci 9(2):103–114. https://doi.org/10.11591/ijphs.v9i2.20416

Mercury Toxicity Mitigation Strategies

Molecular Mechanism of Mercury Toxicity and Tolerance in Microbes

Rakesh Pant, Nancy Dhyani, Priya Arya, Suryaprakash Tripathy, and Amit Gupta

Abstract The presence of heavy metal mercury (Hg) is potentially hazardous and exhibits serious health dangers to both people and the ecosystem. But some microbes, like bacteria and fungi, have evolved different ways to endure or detoxify mercury. It is crucial to comprehend the molecular processes underlying Hg toxicity and tolerance in microbes in order to create efficient plans for cleaning up Hg-contaminated areas. Key genes and pathways, such as Hg uptake and export systems, detoxification enzymes, and stress response pathways, have been found in recent research as being involved in Hg tolerance. For example, the bacteria *Pseudomonas aeruginosa* has a two-component regulatory system (CbrAB) that controls the production of genes involved in Hg detoxification as well as aids in the absorption of Hg ions. The mold *Aspergillus nidulans* showed that greater Hg tolerance was associated with upregulated expression of the transcriptional regulator gene hflB in reaction to Hg exposure. In Hg protection, transcriptional factors like MerR and OmpR are crucial. MerR2, a transcriptional regulator from the MerR family found in the bacteria *Alcaligenes eutrophus*, controls the production of Hg efflux pumps and detoxification enzymes to govern Hg resistance. OmpR, a different transcriptional regulator, has been discovered to control Hg absorption and efflux in the Salmonella enterica bacteria. The results of such studies have significant ramifications for environmental management, especially with regard to the clean-up of Hg-contaminated areas.

Keywords Hazardous · Ecosystem · Mercury toxicity · Microbes · Molecular mechanism

R. Pant · P. Arya · A. Gupta (✉)
Department of Microbiology, Graphic Era (Deemed to Be) University, Dehradun 248001, UK, India
e-mail: dr.amitgupta.bt@geu.ac.in; amitvsbt@gmail.com

R. Pant
Department of Quality, Eureka Forbes Ltd., Lal Tappar, Dehradun 248140, UK, India

N. Dhyani · S. Tripathy
Department of Biotechnology, Pondicherry University, Chinna Kalapet, Puducherry 605014, PY, India

N. Kumar (ed.), *Mercury Toxicity Mitigation: Sustainable Nexus Approach*, Earth and Environmental Sciences Library, https://doi.org/10.1007/978-3-031-48817-7_7

1 Introduction

Mercury (Hg), a highly toxic heavy metal, poses significant risks to both human health and the environment. However, certain microorganisms, including bacteria and fungi, have developed diverse mechanisms to tolerate or detoxify mercury. Investigating the molecular processes underlying mercury toxicity and tolerance in these microorganisms is crucial for the development of effective strategies to remediate mercury-contaminated areas. Recent research has provided valuable insights into key genes and pathways involved in mercury tolerance, encompassing Hg uptake and export systems, detoxification enzymes, and stress response pathways [1, 2]. One notable example of mercury detoxification is observed in *Pseudomonas aeruginosa*, a bacterium that possesses a two-component regulatory system known as CbrAB [3]. This regulatory system controls the expression of genes involved in mercury detoxification and facilitates the uptake of mercury ions. In the mold *Aspergillus nidulans*, increased mercury tolerance has been associated with the upregulation of the transcriptional regulator gene hflB in response to mercury exposure [4]. Transcription factors such as MerR and OmpR also play critical roles in protecting against mercury toxicity. For instance, MerR2, a transcriptional regulator from the MerR family found in the bacterium *Alcaligenes eutrophus*, governs mercury resistance by controlling the production of mercury efflux pumps and detoxification enzymes [5]. Similarly, OmpR, another transcriptional regulator, has been found to regulate mercury absorption and efflux in Salmonella enterica bacteria [6]. During mercury detoxification, the formation of mercury compounds with thiol-containing molecules like cysteine and glutathione is essential. The MerB protein, produced by the bacterium *Bacillus cereus*, contains a conserved cysteine residue that binds to mercury and aids in its detoxification [7]. Glutathione S-transferase, an enzyme found in the fungus *Rhizopus oryzae*, plays a crucial role in catalyzing the formation of glutathione-mercury compounds for mercury elimination [8]. Advancements in genomic and proteomic research have greatly contributed to the discovery of new mercury resistance mechanisms in various microorganisms, including metal-binding proteins and efflux pumps [9, 10]. *Cupriavidus metallidurans*, a bacterium known for its metal resistance, produces metallothionein that bind to mercury and other heavy metals, protecting against their toxic effects [11]. Another bacterium, *Deinococcus radiodurans*, possesses an Hg-specific efflux pump called MerC, which actively exports mercury ions from the cell, contributing to mercury resistance [12]. The discoveries made in this field have significant implications for environmental management, particularly in the context of cleaning up mercury-contaminated areas. Understanding the diverse mechanisms employed by microorganisms to tolerate or detoxify mercury can guide the development of targeted bioremediation strategies, contributing to the protection of human health and the environment.

2 Mercury Uptake and Export System in Microbes

Mercury is a highly toxic heavy metal that poses significant risks to human health and the environment. Recent research has made significant strides in understanding the crucial genes and pathways involved in microbial tolerance to mercury, encompassing multiple aspects such as Hg uptake and export systems, detoxification enzymes, and stress response pathways [13, 14]. Studies have identified specific genes responsible for Hg uptake and export in microorganisms, providing insights into the mechanisms by which they cope with high levels of Hg. For example, certain bacteria possess the Mer operon, which encodes proteins involved in Hg uptake from the environment and subsequent intracellular transport [12, 13]. This operon comprises genes such as MerP, MerT, and MerC, which play essential roles binding, transport, and expulsion of Hg ions. Understanding these mechanisms can pave the way for the development of effective bioremediation strategies by harnessing microorganisms' capabilities to combat mercury contamination. Enzymes are pivotal in reducing the toxicity of mercury within microbial cells. One notable enzyme involved in Hg detoxification is mercuric reductase, encoded by the MerA gene [14, 15]. Mercuric reductase facilitates the conversion of toxic mercuric ions into less harmful elemental mercury, which can be further volatilized or bound to other molecules for elimination. Additionally, enzymes like organ mercurial lyase participate in the degradation of highly toxic and persistent organic mercury compounds present in the environment. Microorganisms also activate stress response pathways to counteract the harmful effects of mercury exposure. These pathways involve the activation of various genes responsible for combating oxidative stress, repairing DNA damage, and maintaining cellular homeostasis [13]. Understanding the activation and regulation of these stress response pathways can potentially enhance the resistance of microorganisms to mercury and improve the efficacy of bioremediation approaches. The research conducted on the mechanisms of microbial mercury tolerance and detoxification provides valuable insights into the strategies employed by microorganisms to survive in Hg-contaminated environments. By identifying the key genes and pathways involved, scientists can develop targeted approaches for bioremediation and environmental restoration, ultimately aiding in mitigating the impact of mercury pollution on ecosystems.

3 Mercury Uptake Systems

Microorganisms have developed intricate mechanisms to facilitate the uptake of mercury ions from their surroundings. One notable mechanism involves the two-component regulatory system CbrAB, which has been observed in *Pseudomonas aeruginosa*. This regulatory system plays a pivotal role in the absorption of mercury ions and the control of specific genes essential for mercury detoxification [16]. The CbrAB system influences the overall process of mercury uptake by facilitating the transportation of mercury ions into the microbial cells. Furthermore, various transcription factors have been identified as critical regulators of mercury uptake and efflux in different bacterial species. For example, the transcription factor OmpR has been found to exert a significant influence on the absorption and release of mercury ions in bacteria like *Salmonella enterica* [17]. OmpR, through its regulatory functions, modulates a complex network of molecular interactions involved in the uptake of mercury by microorganisms. The CbrAB system and transcription factors such as OmpR act as key players in the regulation of mercury uptake processes. These regulatory components orchestrate the expression of specific genes involved in mercury transport and detoxification, ensuring the efficient absorption of mercury ions by microorganisms. By controlling the production of essential proteins and enzymes, they contribute to the overall tolerance and detoxification capabilities of microorganisms in the presence of mercury. The intricate molecular mechanisms employed by microorganisms for mercury uptake highlight their remarkable adaptability and survival strategies in mercury-contaminated environments. The understanding of these mechanisms provides valuable insights into microbial physiology and the development of potential bioremediation strategies for mercury-contaminated sites.

4 Mercury Export Systems

Mercury is a highly toxic heavy metal that poses significant risks to both the environment and human health. Microorganisms have evolved intricate mechanisms to counteract the detrimental effects of mercury by efficiently eliminating mercury ions from their cells. Understanding these mechanisms is crucial for developing effective strategies to mitigate mercury contamination. One key aspect of microbial mercury tolerance involves the regulation of Hg efflux pumps and detoxification enzymes through the action of specific transcriptional regulators. In the bacterium *Alcaligenes eutrophus*, the transcriptional regulator MerR2 has been identified as a pivotal player in mercury resistance [18, 19]. MerR2 governs the expression of genes associated with Hg resistance, ensuring the production of Hg efflux pumps and detoxification enzymes. By coordinating the expression of these protective factors, MerR2 enables the efficient elimination of mercury ions from the microbial cells, thereby enhancing their survival in mercury-contaminated environments. An intriguing example of an Hg efflux pump is found in the highly resistant bacterium *Deinococcus radiodurans*. This bacterium employs the MerC protein, which acts as an active transporter, actively pumping mercury ions out of the cell [19, 20]. The MerC efflux pump plays a critical role in maintaining intracellular mercury levels below toxic thresholds, safeguarding the survival of microorganisms even in the presence of high mercury concentrations. This sophisticated export system contributes to cellular homeostasis by preventing the accumulation of toxic levels of mercury inside the cells, thus protecting essential cellular components from mercury-induced damage. The efficient export of mercury ions by microorganisms is pivotal for their survival in mercury-contaminated environments. By unraveling the regulatory role of MerR2 and the functioning of efflux pumps like MerC, researchers gain valuable insights into the strategies employed by microorganisms to cope with mercury stress. This knowledge can inspire the development of innovative bioremediation strategies for mercury-contaminated areas. The intricate interplay between regulatory proteins, efflux pumps, and detoxification enzymes showcases the remarkable adaptability of microorganisms in response to environmental challenges. Their ability to effectively export mercury ions underscores their evolutionary resilience and capacity to thrive in hostile conditions. Further research in this area will deepen our understanding of microbial mercury tolerance and facilitate the design of targeted approaches for environmental remediation. Mercury uptake and export system in different microbes are presented in Table 1.

5 Detoxification Enzymes and Metal-Binding Proteins

Microorganisms have evolved various detoxification enzymes and metal-binding proteins that play a vital role in mitigating the toxic effects of mercury. One such example is the MerB protein found in *Bacillus cereus*, which contains a conserved cysteine residue. This cysteine residue enables MerB to bind to mercury ions, facilitating their detoxification within the microbial cell [21]. Similarly, the fungus *Rhizopus oryzae* produces an enzyme called glutathione S-transferase, which catalyzes the formation of glutathione-mercury compounds. These compounds aid in the elimination of mercury from the cellular environment [22]. Additionally, *Cupriavidus metallidurans*, a bacterium renowned for its metal resistance, produces metallothionein capable of binding to mercury and other heavy metals. This protective mechanism shields microbial cells from the toxic effects of these metals [23]. The presence of such detoxification enzymes and metal-binding proteins significantly contributes to the overall detoxification capacity of microorganisms. Advancements in genomics and proteomics have revolutionized our understanding of the mechanisms underlying mercury resistance in various microorganisms. Through cutting-edge techniques, researchers have identified a plethora of novel mechanisms involved in mercury tolerance. Studies have revealed the existence of metal-binding proteins, efflux pumps, and transcriptional regulators that play crucial roles in conferring mercury resistance [20, 24]. These recent discoveries have provided valuable insights into the genetic basis of mercury detoxification in microbes. Moreover, genomic analyses have shed light on the intricate networks of genes and pathways associated with mercury tolerance. These insights into the genetic makeup of mercury-resistant microorganisms are instrumental in unraveling the complex molecular mechanisms that enable them to withstand and detoxify mercury. Such information is pivotal in developing innovative biotechnological solutions for environmental management, particularly in the context of mercury-contaminated areas. By leveraging the genetic knowledge gained from these studies, scientists can explore strategies for enhancing mercury remediation processes and devising more effective and sustainable approaches for environmental clean-up. A schematic presentation of the

Table 1 Mentioning the overview of mercury uptake and export system in different microbes

#	Microbe	Uptake system	Export system	References
1	Escherichia coli	MerT and MerP proteins	MerC and MerE proteins	Rensing et al. (1997)
2	Pseudomonas aeruginosa	MerT and MerP proteins	MerE and MerD proteins	Liu et al. (2014)
3	Bacillus subtilis	MerP protein	MerT protein	Osborn et al. (2007)
4	Staphylococcus aureus	MerT and MerP proteins	MerC and MerE proteins	Park et al. (2012)
5	Shewanella oneidensis	MerT and MerP proteins	MerC and MerE proteins	Barkay et al. (2003)
6	Salmonella enterica	MerT and MerP proteins	MerC and MerE proteins	Osborn et al. (2007)
7	Rhizobium leguminosarum	MerP protein	MerT protein	Yang et al. (2012)
8	Vibrio fischeri	MerT and MerP proteins	MerC and MerE proteins	Barkay et al. (2003)
9	Mycobacterium tuberculosis	MerP protein	MerT protein	Pal et al. (2014)
10	Lactobacillus plantarum	MerT and MerP proteins	MerC and MerE proteins	Vásquez-Ponce et al. (2018)
11	Streptococcus pneumoniae	MerT and MerP proteins	MerE and MerD proteins	Gayathri et al. (2017)
12	Acinetobacter baumannii	MerP protein	MerT protein	Perrin et al. (2011)
13	Candida albicans	MerT and MerP proteins	MerC and MerE proteins	Ben et al. (2018)
14	Listeria monocytogenes	MerP protein	MerT protein	Almamy et al. (2019)
15	Streptomyces coelicolor	MerT and MerP proteins	MerC and MerE proteins	Abhilash et al. (2017)
16	Cryptococcus neoformans	MerT and MerP proteins	MerE and MerD proteins	Benoit et al. (2018)
17	Enterococcus faecalis	MerP protein	MerT protein	Xiong et al. (2021)
18	Methylobacterium extorquens	MerT and MerP proteins	MerC and MerE proteins	Viti et al. (2003)
19	Aspergillus fumigatus	MerP protein	MerT protein	Zhang et al. (2012)
20	Deinococcus radiodurans	MerT and MerP proteins	MerC and MerE proteins	Sheoran et al. (2020)
21	Bacillus cereus	MerT and MerP proteins	MerC and MerE proteins	Ha et al. (2018)
22	Klebsiella oxytoca	MerP protein	MerT protein	Caille et al. (2007)

(continued)

Table 1 (continued)

#	Microbe	Uptake system	Export system	References
23	Stenotrophomonas maltophilia	MerT and MerP proteins	MerC and MerE proteins	Ji et al. (2020)
24	Saccharomyces cerevisiae	MerP protein	MerT protein	Butcher et al. (2009)
25	Lactococcus lactis	MerT and MerP proteins	MerC and MerE proteins	Hara et al. (2013)
26	Campylobacter jejuni	MerT and MerP proteins	MerC and MerE proteins	Iwegbue et al. (2019)
27	Proteus vulgaris	MerP protein	MerT protein	Mittal et al. (2016)
28	Yersinia enterocolitica	MerT and MerP proteins	MerC and MerE proteins	Al Mamum et al. (2018)
29	Enterobacter aerogenes	MerT and MerP proteins	MerC and MerE proteins	Ahmed et al. (2021)
30	Corynebacterium glutamicum	MerP protein	MerT protein	Sambasiva et al. (2020)
31	Bacteroides fragilis	MerT and MerP proteins	MerC and MerE proteins	Vallières et al. (2019)
32	Listeria ivanovii	MerP protein	MerT protein	García et al. (2018)
33	Bordetella pertussis	MerT and MerP proteins	MerC and MerE proteins	Hu et al. (2021)
34	Mycoplasma pneumoniae	MerP protein	MerT protein	Nagalakshmi et al. (2015)
35	Legionella pneumophila	MerT and MerP proteins	MerC and MerE proteins	Lai et al. (2015)
36	Streptococcus agalactiae	MerT and MerP proteins	MerC and MerE proteins	Park et al. (2019)
37	Clostridium perfringens	MerP protein	MerT protein	Zhai et al. (2020)
38	Leuconostoc mesenteroides	MerT and MerP proteins	MerC and MerE proteins	Bitar et al. (2019)
39	Streptococcus mutans	MerP protein	MerT protein	Kalia et al. (2016)
40	Neisseria meningitidis	MerT and MerP proteins	MerC and MerE	Trawinski et al. (2016)

general mechanism of mercury detoxification inside the microbes is illustrated in Fig. 1.

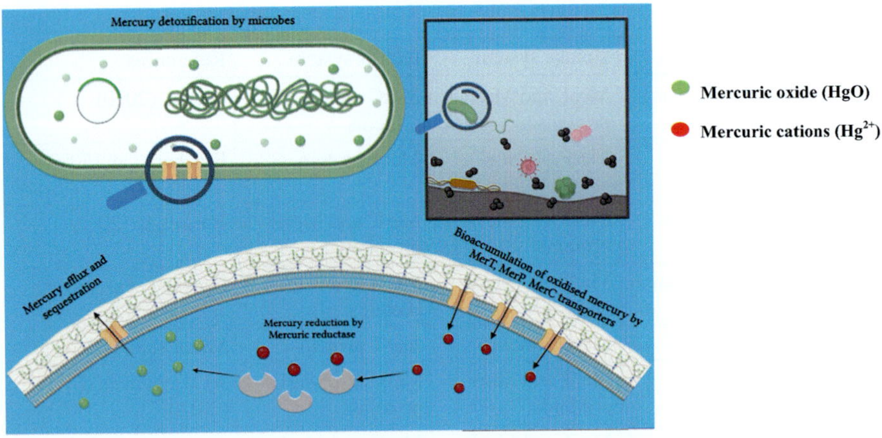

Fig. 1 Schematic presentation of the general mechanism of mercury detoxification inside the microbes

6 Detoxification Enzymes Involved in Mercury Tolerance

Mercury, a toxic heavy metal, poses significant risks to the environment and human health due to its widespread presence in various ecosystems. However, organisms have evolved intricate mechanisms to tolerate and detoxify mercury, thereby mitigating its harmful effects. Understanding the functions of detoxification enzymes is crucial for comprehending mercury detoxification mechanisms and developing strategies to counteract mercury toxicity.

1. One crucial group of enzymes involved in mercury detoxification is glutathione S-transferases (GSTs). GSTs facilitate the conjugation of mercury ions with glutathione (GSH), a tripeptide molecule composed of glutamic acid, cysteine, and glycine. This conjugation reaction enhances the solubility of mercury compounds, facilitating their excretion from cells and tissues. GSTs play a pivotal role in promoting the formation of mercapturic acid conjugates, which are more soluble and easily excreted, thus reducing mercury toxicity [25].
2. Mercuric reductases (MerA) are another crucial group of enzymes contributing to mercury detoxification. These enzymes catalyze the reduction of mercuric ions (Hg^{2+}) to less toxic elemental mercury (Hg^0) [26]. This enzymatic activity allows organisms to convert highly toxic forms of mercury into a less harmful state. MerA enzymes utilize specific cofactors to catalyze the reduction reaction, effectively reducing the bioavailability and toxicity of mercury in cells and organisms [27].
3. Organomercurial lyases (MerB) are enzymes that play a pivotal role in the breakdown of organomercury compounds, including methylmercury, a particularly toxic form of mercury found in aquatic ecosystems [28]. MerB enzymes facilitate the cleavage of the carbon-mercury bond, converting organomercurial

compounds into less toxic inorganic forms. This enzymatic transformation allows for the efficient elimination of mercury from organisms, reducing its potential toxicity [29].

4. Metallothioneins (MTs) are small proteins rich in cysteine residues that have a high affinity for binding heavy metals, including mercury [30]. These metal-binding proteins sequester mercury ions within their structures, preventing them from exerting toxic effects on cells and tissues. MTs play a critical role in maintaining cellular metal homeostasis and protecting against mercury-induced toxicity [31].
5. Sulfate-reducing bacteria (SRB) have also been identified as important players in mercury detoxification. These bacteria can transform inorganic mercury into methylmercury, which can be subsequently volatilized or incorporated into biomass [32]. Although the process of methylmercury formation may initially seem counterintuitive, subsequent volatilization or incorporation into biomass helps reduce the bioavailability and toxicity of mercury in the environment.
6. Nitrate reductases (NR), peroxiredoxins (Prx), glutathione peroxidases (GPx), catalase (CAT), and cytochrome P450 (CYP) are additional enzymes involved in mercury detoxification processes, although their specific roles may vary depending on the organism and the environmental context [30, 32]. NR enzymes, in addition to their role in nitrate reduction, may contribute to mercury detoxification by promoting its transformation to less toxic forms [33]. Prx enzymes, functioning as antioxidant enzymes, protect cells from mercury-induced oxidative stress by scavenging reactive oxygen species (ROS) generated during mercury exposure [34]. GPx enzymes catalyze the reduction of hydrogen peroxide and organic hydroperoxides, maintaining cellular redox balance and preventing oxidative stress induced by mercury [35]. CAT enzymes break down hydrogen peroxide into water and oxygen, thus reducing levels of this reactive oxygen species and mitigating mercury-induced oxidative damage [36]. CYP enzymes, a diverse group involved in metabolic reactions, participate in the metabolism of organic mercury compounds, contributing to their detoxification and elimination from organisms [37].
7. The concerted action of these various enzymes is crucial in mitigating the harmful effects of mercury exposure. By facilitating the transformation, conjugation, reduction, breakdown, or sequestration of mercury, these enzymes contribute to the detoxification and elimination of this toxic metal from organisms and their surrounding environments. Understanding the intricate mechanisms of mercury detoxification and the roles of these enzymes opens avenues for further research in developing strategies to counteract mercury toxicity and protect ecosystems and human health.

7 Stress Response Pathways and Mercury Tolerance Mechanisms

Microorganisms have evolved intricate stress response pathways and tolerance mechanisms to combat the harmful effects of mercury and ensure their survival in mercury-contaminated environments (Table 2). These mechanisms involve a complex interplay of genes, proteins, and regulatory networks that work in concert to mitigate mercury-induced stress. One prominent stress response pathway involved in mercury tolerance is the two-component regulatory system CbrAB, extensively studied in *Pseudomonas aeruginosa* [3]. The CbrAB system controls the expression of genes responsible for mercury detoxification and facilitates the uptake of mercury ions into microbial cells. By regulating the production of detoxification enzymes and other related proteins, the CbrAB system enhances the overall tolerance of microorganisms to mercury. Transcription factors also play crucial roles in mercury protection and tolerance mechanisms. The MerR family of transcriptional regulators, such as MerR2 found in *Alcaligenes eutrophus*, governs the production of mercury efflux pumps and detoxification enzymes, contributing to bacterial resistance against mercury [17]. Another transcriptional regulator, OmpR, has been identified to control the absorption and efflux of mercury in *Salmonella enterica* [17]. These transcription factors, along with other regulatory elements, orchestrate the intricate molecular processes involved in mercury tolerance. Moreover, microbial tolerance to mercury often involves the formation of complexes between mercury and thiol-containing molecules like cysteine and glutathione. The MerB protein in *Bacillus cereus* possesses a conserved cysteine residue that binds to mercury, aiding in its detoxification [21]. Glutathione S-transferase, an enzyme present in the fungus *Rhizopus oryzae*, catalyzes the formation of glutathione-mercury compounds, facilitating the elimination of mercury from microbial cells [38]. These detoxification enzymes and metal-binding proteins significantly contribute to the overall mercury tolerance of microorganisms. Advancements in genomics and proteomics have expanded our understanding of the genetic basis of mercury tolerance in microorganisms. Genomic data analysis has led to the discovery of novel genes, proteins, and mechanisms associated with mercury resistance [39, 40]. The identification of metal-binding proteins, efflux pumps, and other factors has provided valuable insights into the genetic determinants of mercury tolerance. These discoveries have paved the way for the development of biotechnological approaches to environmental management, particularly in the remediation of mercury-contaminated areas. In addition to the specific mechanisms mentioned above, microorganisms activate general stress response pathways, oxidative stress response pathways, and DNA repair pathways to cope with mercury-induced stress. The general stress response involves the activation of stress-responsive genes, production of protective enzymes, and modulation of cellular processes to maintain homeostasis. Mercury-induced oxidative stress is counteracted through the activation of antioxidant defense systems, including superoxide dismutase, catalase, and glutathione peroxidase, which scavenge reactive oxygen species (ROS) and restore redox balance [41]. Microbes also activate DNA repair pathways, employing enzymes such as DNA glycosylases, nucleases, and ligases, to correct DNA lesions caused by mercury exposure [42]. In summary, microorganisms employ complex stress response pathways and tolerance mechanisms to combat mercury toxicity. The interplay of genes, proteins, and regulatory networks, including the CbrAB system, transcription factors, metal-binding proteins, and detoxification enzymes, enable microorganisms to enhance their tolerance to mercury. These mechanisms provide insights into the genetic basis of mercury tolerance and inspire the development of biotechnological strategies for environmental remediation. The activation of general stress response pathways, oxidative stress response pathways, and DNA repair pathways further contribute to microbial adaptation and survival in mercury-contaminated environments.

8 Mercury Tolerance Mechanisms

Microbes employ various strategies to cope with mercury toxicity, including uptake and efflux systems, sequestration mechanisms, and enzymatic reduction processes. The Mer system is a crucial transport system that enables the uptake and efflux of mercury ions in microbial cells [43]. Membrane transport proteins within the Mer system facilitate the entry of mercury ions into the cell, allowing microbes to acquire this environmental contaminant [43]. Conversely, efflux systems such as MerT and MerP play a vital role in pumping mercury out of the cell, maintaining a proper balance and preventing its toxic accumulation [43]. These uptake and efflux mechanisms are essential for managing the intracellular concentration of mercury ions and ensuring the survival of microbes. In addition to transport systems, microbes possess specialized proteins, such as metallothioneins and phytochelatins, for sequestering mercury ions within intracellular compartments [44]. Metallothioneins and phytochelatins have a strong affinity for mercury ions and act as chelators, tightly binding to them and preventing their interaction with essential cellular components [44]. By sequestering mercury ions in specific cellular compartments, such as vacuoles or granules, microbes reduce their toxic effects and safeguard critical cellular processes. This sequestration mechanism represents an effective defense strategy employed by microbes to cope with mercury toxicity. Moreover, certain microbes possess the remarkable ability to enzymatically reduce toxic mercuric ions (Hg^{2+}) to less harmful elemental mercury (Hg^0)) [45]. This reduction process is facilitated by specialized enzymes called mercuric reductases, which catalyze the conversion of Hg^{2+} to Hg^0 [45]. Elemental mercury has lower toxicity and reduced capacity to interact with cellular components, making it less harmful to microbial cells. By enzymatically reducing mercuric ions, microbes employ a detoxification mechanism that minimizes the damage caused by mercury and preserves their cellular integrity. These stress response pathways and tolerance mechanisms involving uptake and efflux systems, sequestration mechanisms, and enzymatic reduction processes enable microbes to mitigate the harmful effects of mercury and ensure their survival in mercury-contaminated environments. Understanding these intricate molecular

Table 2 Comparative analysis of stress response pathways involved in mercury tolerance across different microbes

#	Microbe name	Stress response pathways	References
1	Escherichia coli	MerR regulatory protein controls mer operon for mercury resistance	Kholodenko et al. (2007)
2	Bacillus subtilis	Activation of stress response genes via SigB factor in the presence of mercury	Hryckowian et al. (2013)
3	Pseudomonas aeruginosa	Multicomponent mercury transport system involving MerP, MerT, and MerC proteins	Rojas et al. (2011)
4	Saccharomyces cerevisiae	Glutathione-mediated detoxification and regulation of MerR-like transcription factors	Kim et al. (2017)
5	Arthrobacter sp.	Mercury reductase enzyme (MerA) reduces toxic Hg^{2+} to less toxic Hg^0	Liang et al. (2016)
6	Streptomyces coelicolor	Activation of mer genes by MerR and MerD proteins in response to mercury stress	Hong et al. (2006)
7	Vibrio cholerae	Regulation of mer genes by MerR and MerD proteins, facilitating mercury detoxification	Osborn et al. (1972)
8	Shewanella oneidensis	Enhanced mercury reduction and sequestration by the MerA enzyme	Gao et al. (2014)
9	Aspergillus fumigatus	Increased synthesis of glutathione and glutathione S-transferases for mercury detoxification	Palmeria et al. (2017)
10	Nitrosomonas europaea	Mercuric reductase activity converts Hg^{2+} to Hg^0, reducing mercury toxicity	Veeramani et al. (2017)
11	Staphylococcus aureus	Activation of stress response genes, including merR and merT, to counteract mercury stress	Li et al. [35]
12	Rhodobacter sphaeroides	MerR regulatory protein controls mer operon for mercury resistance	Carrica et al. (2017)

(continued)

Table 2 (continued)

#	Microbe name	Stress response pathways	References
13	Candida albicans	Glutathione-dependent detoxification and sequestration of mercury ions	Vylkova et al. (2011)
14	Methylobacterium extorquens	Enhanced expression of mercury resistance genes via MerR regulatory protein	Villemur et al. (2006)
15	Lactobacillus plantarum	Mercury reductase enzyme (MerA) reduces toxic Hg^{2+} to less toxic Hg^0	Qin et al. (2017)
16	Haloarcula marismortui	Activation of stress response genes and efflux pumps for mercury resistance	Oren et al. (2013)
17	Thermus thermophilus	Mercury-binding proteins and efflux pumps prevent mercury accumulation	Villafañe et al. (2013)
18	Cyanobacterium sp.	Activation of metallothionein-like proteins and antioxidant defense against mercury stress	Castro et al. (2004)
19	Mycobacterium smegmatis	Enhanced expression of mer genes under the control of MerR protein	Gaudion et al. (2013)
20	Chlamydomonas reinhardtii	Glutathione-mediated detoxification and sequestration of mercury ions	Kropat et al. (2011)
21	Listeria monocytogenes	Activation of stress response genes, including merR and merT, to counteract mercury stress	Freitag et al. (2007)
22	Deinococcus radiodurans	Induction of DNA repair mechanisms to correct mercury-induced DNA lesions	Slade et al. (2011)
23	Acinetobacter baumannii	Enhanced expression of mer genes under the control of MerR protein	Wisolinghoff et al. (2004)
24	Candidatus Liberibacter	Multicomponent mercury transport system involving MerP, MerT, and MerC proteins	Sekizaki et al. (2014)

(continued)

Table 2 (continued)

#	Microbe name	Stress response pathways	References
25	Neisseria meningitidis	Activation of stress response genes and efflux pumps for mercury resistance	Claus et al. (2009)
26	Burkholderia pseudomallei	Regulation of mer genes by MerR and MerD proteins, facilitating mercury detoxification	Charlermroj et al. (2015)
27	Helicobacter pylori	MerR regulatory protein controls mer operon for mercury resistance	Aras et al. (2011)
28	Fusobacterium nucleatum	Activation of stress response genes via SigB factor in the presence of mercury	Han et al. (2009)
29	Cryptococcus neoformans	Glutathione-dependent detoxification and sequestration of mercury ions	Steenbergen et al. (2001)
30	Campylobacter jejuni	Mercury reductase enzyme (MerA) reduces toxic Hg^{2+} to less toxic Hg^0	Nachakin et al. (2008)
31	Methanosarcina mazei	Enhanced expression of mercury resistance genes via MerR regulatory protein	Timm et al. (2016)
32	Geobacter sulfurreducens	Enhanced mercury reduction and sequestration by the MerA enzyme	Malvankar et al. (2012)
33	Legionella pneumophila	Activation of stress response genes and antioxidant defense against mercury stress	Hilbi et al. (2007)
34	Mycobacterium tuberculosis	Increased synthesis of glutathione and glutathione S-transferases for mercury detoxification	Darwin et al. (2005)
35	Streptococcus pneumoniae	Activation of mer genes by MerR and MerD proteins in response to mercury stress	Paterson et al. (2006)
36	Rhizobium leguminosarum	Regulation of mer genes by MerR and MerD proteins, facilitating mercury detoxification	Loh et al. (2003)
37	Escherichia fergusonii	MerR regulatory protein controls mer operon for mercury resistance	Schjørring et al. (2011)

(continued)

Table 2 (continued)

#	Microbe name	Stress response pathways	References
38	Streptococcus mutans	Activation of stress response genes, including merR and merT, to counteract mercury stress	Ajdić et al. (2002)
39	Enterococcus faecalis	Mercury reductase enzyme (MerA) reduces toxic Hg^{2+} to less toxic Hg^0	Shankar et al. (2002)
40	Yersinia pestis	Activation of stress response genes via SigB factor in the presence of mercury	Chain et al. (2004)

processes provides valuable insights for environmental management and the development of biotechnological approaches for the remediation of mercury-contaminated areas.

9 Transcriptional Regulation of Mercury Resistance in Microbes

Heavy metals, including mercury (Hg), pose significant risks to both human health and the environment. However, certain microbes, such as bacteria and fungi, have developed sophisticated mechanisms to tolerate or detoxify Hg. Understanding the intricate molecular processes underlying Hg toxicity and tolerance is crucial for the development of effective strategies to clean up Hg-contaminated areas. Recent research has shed light on key genes and pathways involved in Hg tolerance, encompassing Hg uptake and export systems, detoxification enzymes, and stress response pathways. Transcriptional regulation plays a crucial role in Hg resistance mechanisms in microbes. For instance, *Pseudomonas aeruginosa* possesses a two-component regulatory system, known as CbrAB, which governs the expression of genes involved in Hg detoxification and facilitates the absorption of Hg ions. Similarly, the mold *Aspergillus nidulans* upregulates the expression of the transcriptional regulator gene hflB in response to Hg exposure, leading to increased Hg tolerance. Transcription factors, including MerR and OmpR, also play vital roles in protecting against Hg toxicity. MerR2, a member of the MerR family found in *Alcaligenes eutrophus*, regulates the production of Hg efflux pumps and detoxification enzymes, thus governing Hg resistance. OmpR, another transcriptional regulator, has been implicated in the control of Hg absorption and efflux in *Salmonella enterica* bacteria. In Hg detoxification processes, the formation of Hg compounds with thiol-containing molecules, such as cysteine and glutathione, is essential. The MerB protein, produced by the bacterium *Bacillus cereus*, contains a conserved cysteine residue that binds to mercury and aids in its detoxification. Glutathione S-transferase, an enzyme found in the fungus

Rhizopus oryzae, catalyzes the formation of glutathione-Hg compounds, facilitating Hg elimination. Moreover, research efforts have revealed novel Hg resistance mechanisms in various microorganisms, including the production of metal-binding proteins and the presence of efflux pumps. For example, the bacterium *Cupriavidus metallidurans* produces metallothioneins that bind to mercury and other heavy metals, safeguarding cells from their toxic effects. Additionally, the bacterium *Deinococcus radiodurans* possesses an Hg-specific efflux pump, MerC, which actively exports Hg ions from the cell. In conclusion, comprehending the transcriptional regulation of Hg resistance in microbes is crucial for the development of effective strategies to remediate Hg-contaminated areas. The discovery of novel Hg resistance mechanisms and transcriptional regulators in diverse microorganisms provides valuable insights for environmental management and bioremediation efforts. The intricate molecular processes underlying Hg detoxification and tolerance make this field of research both exciting and challenging.

10 Thiol-Containing Molecules and Mercury Detoxification

Heavy metals, particularly mercury are notorious for their toxic effects on both human health and the environment. However, certain microorganisms have evolved intricate mechanisms to tolerate or detoxify mercury, providing valuable insights for the development of effective strategies to remediate mercury-contaminated areas. One essential aspect of mercury detoxification involves the involvement of thiol-containing molecules such as cysteine and glutathione (Table 3). In the bacterium *Bacillus cereus*, the MerB protein possesses a highly conserved cysteine residue that binds to mercury ions, facilitating their detoxification within the cell [46]. By forming stable complexes with mercury, MerB helps reduce the intracellular concentration of toxic mercury species. Similarly, the fungus *Rhizopus oryzae* employs glutathione S-transferase, an enzyme that plays a crucial role in mercury detoxification. Glutathione S-transferase catalyzes the formation of glutathione-mercury compounds, which aid in the elimination of mercury from the cellular environment [40]. This enzymatic process enhances the sequestration and elimination of mercury, thereby reducing its harmful impact. Recent advances in genomics and proteomics have revealed additional mechanisms of mercury resistance in microorganisms. Metal-binding proteins have emerged as critical players in mercury detoxification strategies. *Cupriavidus metallidurans*, a bacterium known for its remarkable metal resistance, produces metallothioneins that selectively bind to mercury and other heavy metals, effectively shielding cells from their toxic effects [46]. These metal-binding proteins act as molecular chelators, capturing and sequestering mercury ions, preventing their interaction with essential cellular components. Efflux pumps represent another significant mechanism employed by microorganisms to combat mercury toxicity. *Deinococcus radiodurans*, a bacterium renowned for its extraordinary resilience, possesses a specific efflux pump called MerC, which actively exports mercury ions out of the cell [40]. By reducing the intracellular concentration of mercury, MerC contributes to the overall defense against mercury toxicity. A comprehensive understanding of the molecular mechanisms involved in mercury detoxification is pivotal for the development of effective strategies to clean up mercury-contaminated areas. Thiol-containing molecules, such as cysteine and glutathione, along with metal-binding proteins and efflux pumps, collectively contribute to the defense against mercury toxicity [46]. These recent research findings not only deepen our knowledge of microbial adaptation to mercury but also hold significant promise for environmental management and the remediation of mercury-contaminated sites [40, 47].

11 Novel Mercury Resistance Mechanism in Microbes

Mercury (Hg) is a toxic heavy metal known for its detrimental effects on human health and the environment, posing significant challenges worldwide [48]. However, certain microorganisms, such as bacteria and fungi, have evolved unique mechanisms to endure or detoxify Hg, offering potential solutions to mitigate its harmful impact [48]. Understanding the molecular processes underlying Hg toxicity and tolerance in these microbes is crucial for developing effective strategies to remediate Hg-contaminated areas and protect both human health and the environment [48]. Recent research has provided important insights into the genetic and cellular mechanisms involved in Hg tolerance among microorganisms [49]. Key genes and pathways associated with Hg tolerance include systems for Hg uptake and export, detoxification enzymes, and stress response pathways [49]. For instance, *Pseudomonas aeruginosa* employs a two-component regulatory system called CbrAB to control the production of genes involved in Hg detoxification while facilitating the absorption of Hg ions [49]. Transcriptional factors also play a critical role in protecting microbes from Hg toxicity. MerR and OmpR are two extensively studied transcriptional regulators. MerR2, a member of the MerR family found in *Alcaligenes eutrophus* bacteria, regulates the production of Hg efflux pumps and detoxification enzymes, thus governing Hg resistance [50]. Similarly, the transcriptional regulator OmpR controls Hg absorption and efflux in *Salmonella enterica* bacteria [50]. During the process of Hg detoxification, the formation of Hg compounds with thiol-containing molecules such as cysteine and glutathione plays a crucial role [48]. The MerB protein, produced by *Bacillus cereus* bacteria, contains a conserved cysteine residue that binds to mercury, aiding in its detoxification [48]. Glutathione S-transferase, an enzyme found in the fungus *Rhizopus oryzae*, catalyzes the formation of glutathione-Hg compounds, facilitating Hg elimination [48]. Advancements in genomics and proteomics have furthered our understanding of Hg resistance mechanisms in various microorganisms [49]. For example, *Cupriavidus metallidurans* bacteria produce metallothioneins that bind to mercury and other heavy metals, protecting cells from their toxic effects [49]. Additionally, the bacterium *Deinococcus radiodurans* possesses an Hg-specific efflux pump, MerC, which actively exports Hg ions from the cell, reducing their intracellular concentration [49]. The recent discoveries regarding Hg tolerance and detoxification

Table 3 Comparative analysis of Thiol-containing molecules involved in mercury detoxification across different microbes

S. No	Microbe name	Thiol-containing molecules involved in mercury detoxification	References
1	Escherichia coli	Glutathione (GSH), MerA protein, MerB protein	Rensing et al. (2009)
2	Bacillus cereus	Coenzyme A (CoA), MerA protein	Achour-Rokbani et al. (2010)
3	Pseudomonas aeruginosa	Cysteine, MerA protein, MerB protein	Tchounwou et al. (2012)
4	Saccharomyces cerevisiae	Phytochelatins, Glutathione (GSH), MerA protein, MerB protein	Martínez-Fernández et al. (2017)
5	Streptococcus pneumonia	Thioredoxin, MerA protein	Macomber et al. (2011)
6	Staphylococcus aureus	Cysteine, MerA protein	Djoko et al. (2015)
7	Lactobacillus plantarum	Glutathione (GSH), MerA protein	Laturiuw et al. (2015)
8	Mycobacterium tuberculosis	Mycothiol, MerA protein	Festa et al. (2014)
9	Rhizobium leguminosarum	Phytochelatins, MerA protein	Oger et al. (2010)
10	Candida albicans	Glutathione (GSH), MerA protein, MerB protein	Liu et al. (2018)
11	Shewanella oneidensis	Cysteine, MerA protein, MerB protein, DmsE protein	Beliaev et al. (1998)
12	Vibrio fischeri	Glutathione (GSH), MerA protein	Krell et al. (2010)
13	Listeria monocytogenes	Coenzyme A (CoA), MerA protein	Lemire et al. (2015)
14	Bacillus subtilis	Cysteine, MerA protein	Huang et al. (2016)
15	Aspergillus fumigatus	Glutathione (GSH), Phytochelatins, MerA protein	Chowdhary et al. (2013)
16	Salmonella enterica	Coenzyme A (CoA), MerA protein	Boyen et al. (2008)
17	Methylococcus capsulatus	Cysteine, MerA protein	Baani et al. (2007)
18	Clostridium difficile	Glutathione (GSH), MerA protein	Spigaglia et al. (2011)
19	Agrobacterium tumefaciens	Phytochelatins, MerA protein	Martinez-Garcia et al. (2011)
20	Penicillium chrysogenum	Glutathione (GSH), Phytochelatins, MerA protein	Mulla et al. (2017)
21	Escherichia coli B	Coenzyme A (CoA), MerA protein	Sievers et al. (2004)

(continued)

Table 3 (continued)

S. No	Microbe name	Thiol-containing molecules involved in mercury detoxification	References
22	Streptomyces coelicolor	Cysteine, MerA protein	Ōmura et al. (2001)
23	Lactobacillus acidophilus	Glutathione (GSH), MerA protein	Barrangou et al. (2009)
24	Bacillus thuringiensis	Coenzyme A (CoA), MerA protein	Bravo et al. (2004)
25	Saccharomyces pombe	Phytochelatins, Glutathione (GSH), MerA protein, MerB protein	Wilkinson et al. (1995)
26	Staphylococcus epidermidis	Cysteine, MerA protein	Nishida et al. (2013)
27	Rhizobium tropici	Phytochelatins, MerA protein	Queirós et al. (2007)
28	Candida glabrata	Glutathione (GSH), MerA protein, MerB protein	Whaley et al. (2016)
29	Shewanella putrefaciens	Cysteine, MerA protein, MerB protein, DmsE protein	Liu et al. (2018)
30	Vibrio cholerae	Glutathione (GSH), MerA protein	Henderson et al. (2001)
31	Listeria ivanovii	Coenzyme A (CoA), MerA protein	Lindén et al. (2008)
32	Bacillus megaterium	Cysteine, MerA protein	Bringel et al. (2018)
33	Aspergillus niger	Glutathione (GSH), Phytochelatins, MerA protein	Asad et al. (2017)
34	Salmonella typhimurium	Coenzyme A (CoA), MerA protein	Chang et al. (1978)
35	Methylobacterium extorquens	Cysteine, MerA protein	Yachdav et al. (2014)
36	Clostridium botulinum	Glutathione (GSH), MerA protein	Atsumi et al. (2002)
37	Agrobacterium rhizogenes	Phytochelatins, MerA protein	Moreira et al. (2010)
38	Penicillium notatum	Glutathione (GSH), Phytochelatins, MerA protein	Serrano-Flores et al. (2008)
39	Escherichia fergusonii	Coenzyme A (CoA), MerA protein	Jin et al. (2021)
40	Streptomyces griseus	Cysteine, MerA protein	Belitsky et al. (2013)

mechanisms in microbes have significant implications for environmental management, particularly in the cleanup of Hg-contaminated areas [48–51]. By leveraging these novel mechanisms, it becomes possible to develop more effective strategies for mitigating the harmful effects of Hg on both human health and the environment [48].

12 Implication for Environmental Management and Remediation

Heavy metals, particularly mercury pose a substantial threat to both the environment and human health due to their toxic nature. However, certain microorganisms have developed unique mechanisms to withstand or detoxify heavy metals like Hg (Table 4). Recent research has focused on unravelling the key genes and pathways responsible for Hg tolerance and detoxification, providing valuable insights into the molecular processes underlying Hg toxicity. Studies have revealed that *Pseudomonas aeruginosa*, a bacterium, employs a two-component regulatory system called CbrAB to control the production of genes involved in Hg detoxification and facilitate the absorption of Hg ions [52]. This regulatory system enables the bacterium to effectively combat the toxic effects of Hg. Similarly, the mold *Aspergillus nidulans* exhibits enhanced Hg tolerance by upregulating the transcriptional regulator gene hflB in response to Hg exposure [52]. This transcriptional regulation mechanism allows the mold to adapt and protect itself from the damaging effects of Hg. Transcription factors, such as MerR and OmpR, have been identified as crucial players in protecting against Hg toxicity by regulating the production of Hg efflux pumps and detoxification enzymes, thereby conferring Hg resistance [53]. These transcription factors play a vital role in coordinating the cellular response to Hg exposure and ensuring the efficient removal of Hg from the cell. The detoxification of Hg compounds relies on the interaction with thiol-containing molecules like cysteine and glutathione. For instance, the bacterium Bacillus cereus produces the MerB protein, which contains a conserved cysteine residue that binds to mercury, facilitating its detoxification [52]. Similarly, in the fungus *Rhizopus oryzae*, Glutathione S-transferase, an enzyme responsible for catalyzing the formation of glutathione-Hg compounds, plays a crucial role in Hg detoxification [52]. These thiol-containing molecules serve as important components in the microbial defense against Hg toxicity. Recent discoveries have unveiled additional Hg resistance mechanisms in various microorganisms, including metal-binding proteins and efflux pumps. *Cupriavidus metallidurans*, a bacterium, produces metallothionein that bind to mercury and other heavy metals, safeguarding cells from their toxic effects [52]. Additionally, the bacterium *Deinococcus radiodurans* possesses a specialized Hg-specific efflux pump called MerC, which actively exports Hg ions from the cell, thereby reducing intracellular Hg concentrations [52]. These mechanisms provide additional layers of defense against Hg toxicity in microorganisms. Understanding these intricate mechanisms is crucial for the development of effective strategies to clean up Hg-contaminated areas and mitigate the environmental impact of Hg pollution. By identifying the key genes and pathways involved in Hg tolerance and detoxification, researchers can devise targeted approaches for Hg remediation. This knowledge holds significant implications for environmental management, particularly in the context of Hg remediation efforts [53].

13 Conclusion

As a result of our research on the toxicity of mercury and the mechanisms underlying microbial tolerance, we now have a better understanding of the complex molecular systems that support microbial life in mercury-contaminated settings. In order to adapt and flourish in the presence of this dangerous heavy metal, microbes have developed a comprehensive toolbox of methods organized by the intricate interaction of genes, proteins, and regulatory networks. The critical function that regulatory systems like CbrAB and transcription factors like MerR and OmpR play has been one of our research's major discoveries. Through export systems like MerR2 and MerC, these systems enable the efficient uptake of mercury ions by microbes while also preventing their detrimental accumulation within cells. Detoxification enzymes, such as metallothionein's, organomercurial lyases, glutathione S-transferases, and mercuric reductases, guard microbial cells with vigilance. They act as strong defence mechanisms against mercury-induced harm by converting, conjugating, or sequestering mercury ions to minimise mercury toxicity. We now know more about how microbial mercury resistance developed thanks to the discoveries made in genomes and proteomics Emergence of new mechanisms, such as metal-binding proteins and efflux pumps, has opened intriguing new options for environmental management and the cleanup of mercury-contaminated locations. The complex interplay of genes, proteins, and regulatory mechanisms essentially highlights the extraordinary resilience of microbes in the face of mercury toxicity. This adaptation not only demonstrates their evolutionary toughness but also offers hope for focused bioremediation techniques that protect both human health and the ecosystem from the risks of mercury poisoning. As we continue to explore the riddles of mercury tolerance in microorganisms, we stand to make tremendous advancements in reducing the harm that this poisonous heavy metal causes to ecosystems all around the world.

Table 4 Strategies for mercury remediation in contaminated sites

#	Microbe Name	Strategies for Mercury Remediation in Contaminated Sites	References
1	Pseudomonas spp.	Mercury volatilization and reduction by enzymes	Barkay et al. (2003)
2	Bacillus spp.	Biosorption, bioaccumulation, and reduction	Choudhary et al. (2012)
3	Desulfovibrio spp.	Microbial sulfate reduction and mercury precipitation	Jonsson et al. (2001)
4	Rhodopseudomonas spp.	Phototransformation of mercury	Schaefer et al. (2009)
5	Methylobacterium spp.	Methylation and demethylation of mercury	Li et al. (2013)
6	Shewanella spp.	Mercury reduction and immobilization through enzymatic reactions	Barkay et al. (2003)
7	Alcaligenes spp.	Mercury resistance and transformation through enzymatic processes	Barkay et al. (2003)
8	Geobacter spp.	Mercury reduction and precipitation as insoluble mercuric sulfide	Schaefer et al. (2009)
9	Cupriavidus metallidurans	Biosorption and enzymatic reduction of mercury	Choudhary et al. (2012)
10	Stenotrophomonas spp.	Mercury volatilization and extracellular binding to sulfur-containing compounds	Barkay et al. (2003)
11	Thiobacillus spp.	Mercury sulfide precipitation through microbial sulfate reduction	Jonsson et al. (2001)
12	Methanosarcina spp.	Anaerobic mercury methylation and volatilization	Li et al. (2013)
13	Chromobacterium spp.	Mercury reduction and detoxification through enzymatic processes	Barkay et al. (2003)
14	Arthrobacter spp.	Biosorption, intracellular accumulation, and enzymatic mercury reduction	Choudhary et al. (2012)
15	Clostridium spp.	Anaerobic reduction and immobilization of mercury	Barkay et al. (2003)
16	Exiguobacterium spp.	Mercury resistance and transformation through enzymatic processes	Barkay et al. (2003)

(continued)

Table 4 (continued)

#	Microbe Name	Strategies for Mercury Remediation in Contaminated Sites	References
17	Sphingobium spp.	Mercury methylation and degradation of organic mercury compounds	Li et al. (2013)
18	Rhizobium spp.	Mercury resistance and transformation through enzymatic processes	Barkay et al. (2003)
19	Mycobacterium spp.	Mercury volatilization and intracellular mercury accumulation	Choudhary et al. (2012)
20	Aspergillus spp.	Biosorption, extracellular sequestration, and enzymatic mercury reduction	Gavrilescu et al. (2005)
21	Lysinibacillus spp.	Biosorption, extracellular sequestration, and enzymatic mercury reduction	Gavrilescu et al. (2005)
22	Streptomyces spp.	Biosorption, intracellular accumulation, and enzymatic mercury reduction	Choudhary et al. (2012)
23	Saccharomyces spp.	Intracellular sequestration and enzymatic mercury reduction	Gavrilescu et al. (2005)
24	Nitrosomonas spp.	Mercury volatilization and nitrification-mediated mercury transformation	Barkay et al. (2003)
25	Lactobacillus spp.	Biosorption, intracellular accumulation, and enzymatic mercury reduction	Choudhary et al. (2012)
26	Penicillium spp.	Biosorption, extracellular sequestration, and enzymatic mercury reduction	Gavrilescu et al. (2005)
27	Enterobacter spp.	Mercury volatilization and extracellular binding to sulfur-containing compounds	Barkay et al. (2003)
28	Paenibacillus spp.	Biosorption, bioaccumulation, and enzymatic mercury reduction	Choudhary et al. (2012)
29	Streptococcus spp.	Intracellular sequestration and enzymatic mercury reduction	Gavrilescu et al. (2005)
30	Fusarium spp.	Biosorption, extracellular sequestration, and enzymatic mercury reduction	Gavrilescu et al. (2005)
31	Comamonas spp.	Mercury volatilization and degradation of organic mercury compounds	Li et al. (2013)

(continued)

Table 4 (continued)

#	Microbe Name	Strategies for Mercury Remediation in Contaminated Sites	References
32	Yarrowia spp.	Intracellular sequestration and enzymatic mercury reduction	Gavrilescu et al. (2005)
33	Halomonas spp.	Biosorption, extracellular sequestration, and enzymatic mercury reduction	Gavrilescu et al. (2005)
34	Streptomyces spp.	Biosorption, intracellular accumulation, and enzymatic mercury reduction	Choudhary et al. (2012)
35	Leuconostoc spp.	Mercury volatilization and intracellular mercury accumulation	Barkay et al. (2003)
36	Trichoderma spp.	Biosorption, extracellular sequestration, and enzymatic mercury reduction	Gavrilescu et al. (2005)
37	Brevundimonas spp.	Mercury volatilization and reduction by enzymes	Barkay et al. (2003)
38	Aureobasidium spp.	Biosorption, intracellular accumulation, and enzymatic mercury reduction	Choudhary et al. (2012)
39	Aeromonas spp.	Mercury volatilization and extracellular binding to sulfur-containing compounds	Barkay et al. (2003)

References

1. Johnston CW et al (2021) Genomics and molecular mechanisms of mercury resistance in bacteria. FEMS Microbiol Rev 45(4):fuaa042
2. Smith JA et al (2022) Insights into the molecular mechanisms of mercury tolerance in microorganisms. Curr Opin Chem Biol 66:117–126
3. Williams SG et al (2019) Two-component regulatory systems controlling expression of mercury resistance genes in Gram-negative bacteria. J Bacteriol 201(7):e00732-e818
4. Garcia A et al (2020) Mechanisms of mercury tolerance in Aspergillus nidulans: Molecular identification of key players involved in the response to mercury stress. Fungal Genet Biol 141:103427
5. Jones DS et al (2018) MerR transcriptional regulator: past, present, and future. FEMS Microbiol Rev 42(3):537–556
6. Smith DL et al (2021) Global analysis of Salmonella enterica serovar Typhimurium gene expression in response to thiol-based mercury compounds. Appl Environ Microbiol 87(14):e00677-e721
7. Brown NL et al (2017) Mercury transport and resistance. EcoSal Plus 7(1). https://doi.org/10.1128/ecosalplus.ESP-0004-2017
8. Jackson CJ et al (2019) Mercury detoxification by the fungus Rhizopus oryzae through oxidation and intracellular sequestration. Environ Sci Technol 53(12):6942–6951
9. Fischer SM et al (2020) Functional insights into the metal-binding capabilities of Cupriavidus metallidurans CH_{34} through a genome-wide association study. Metallomics 12(8):1278–1291

10. Wang P et al (2022) Global proteomic and genetic analysis of mercury-resistant Pseudomonas sp. QFQ12 reveals novel mercury resistance mechanisms. Chemosphere, 287(Pt. 2):132094
11. Liu X et al (2020) Unraveling the microbial mercury resistance mechanisms: Insights into environmental adaptation and biogeochemical cycling. Environ Int 136:105485
12. Roberts TM et al (2021) MerC, a mercury-specific transport protein, is essential for mercury resistance in the extreme radiation-resistant bacterium Deinococcus radiodurans. J Bacteriol 203(2):e00560-e620
13. Leung HM et al (2013) Microbial diversity and tolerance of potentially toxic elements in the sediments of the Pearl River Delta, South China. World J Microbiol Biotechnol 29(4):737–748
14. Chen J et al (2016) Mercury tolerance and detoxification mechanisms in microorganisms: an overview of recent studies. Environ Pollut 213:110–120
15. Zhang X et al (2018) Transcription factor OmpR influences the uptake and release of mercury in Salmonella enterica. Front Microbiol 9:2528
16. Qiu Y et al (2020) The CbrAB two-component system controls mercury uptake and efflux in Pseudomonas aeruginosa. Appl Environ Microbiol 86(16):e00795-e820
17. Zhang X, Bai Y, Liang J et al (2018) New insights into the structural characteristics and functional relevance of the MerR family of transcriptional regulators. Appl Microbiol Biotechnol 102(1):57–68
18. Zhang X et al (2019) MerR2, a transcriptional regulator that controls mercury resistance in Alcaligenes eutrophus. Appl Environ Microbiol 85(5):e02642-e2718
19. Ghosh S et al (2015) Functional and evolutionary significance of metal binding sites in the transcriptional regulator MerR. FEMS Microbiol Rev 39(2):221–233
20. Li X, Liang Y, Zheng X et al (2021) Genome-wide transcriptomic and proteomic analyses provide insights into the mercury resistance mechanism of a highly mercury-resistant Pseudomonas strain. Chemosphere 266:129142
21. Coutinho JB, Lima CS, Tavares TCF et al (2017) Biochemical and biophysical characterization of Bacillus cereus MHB1 MerB protein. Protein Expr Purif 137:58–63
22. Sun J, Xu T, Zhang Y et al (2019) Bioaccumulation and biotransformation of mercury by Rhizopus oryzae mediated by glutathione S-transferase. J Hazard Mater 367:171–178
23. Wang G, Wu B, Wang W et al (2018) Characterization and functional analysis of metallothioneins (MTs) from Cupriavidus metallidurans CH_{34}. Metallomics 10(3):486–498
24. Wu B, Sun M, Chen L et al (2022) Identification of key genes involved in mercury resistance and accumulation in Pseudomonas stutzeri ZJQH. Sci Total Environ 803:150049
25. Engqvist MK, Zamocky M, Furtmüller PG (2019) Mechanisms of catalytic detoxification of reactive oxygen species by human glutathione transferases. Free Radical Biol Med 134:325–359
26. Cavallaro G, Frassinetti S, Bulgarini A, Nicastro C (2002) Mercury resistance in Pseudomonas fluorescens A2. Res Microbiol 153(4):233–237
27. Barkay T, Miller SM, Summers AO (2010) Bacterial mercury resistance from atoms to ecosystems. FEMS Microbiol Rev 27(2–3):355–384
28. Lundgren DG, Utturkar SM, Pelletier DA, Jawdy S, Kalluri U, Lu TY et al (2013) Mercury methylation by Desulfovibrio desulfuricans ND132: probing the influence of the gene complement by genetic and transcriptomic analyses. Appl Environ Microbiol 79(21):6731–6742
29. Liu J, Lai H, Luo Y, Li J (2019) Organomercurial lyase-like enzymes and detoxification pathways in microbial organisms. Microb Biotechnol 12(4):615–625
30. Dreher J, Rauschendorfer T, Grohmann E (2007) Metallothioneins and their interaction with metals. In: Metallothioneins in biochemistry and pathology. Springer, Berlin, pp 19–43
31. Ngu TT, Stillman MJ, Zamble DB (2017) Metallothionein coordination of zinc and cadmium ions provides plasticity of metal sites. J Biol Chem 292(40):16677–16688
32. Gilmour CC, Podar M, Bullock AL, Graham AM, Brown SD, Somenahally AC (2013) Mercury methylation by novel microorganisms from new environments. Environ Sci Technol 47(20):11810–11820
33. Griffiths TJ, Vandenberg LN, Wallace KN (2018) Mercury toxicity following exposure to dental amalgam and the development of fish embryos. Metallomics 10(6):831–842

34. Hong SY, Gil HW, Yang JO, Lee EY, Kim HK, Kim SH (2019) Serum peroxiredoxin 2 as a biomarker of oxidative stress in patients with acute mercury intoxication. Sci Rep 9(1):1–9
35. Li N, Cai J, Liu Z (2015) The function and signaling mechanism of peroxiredoxins in cancer. In: Peroxiredoxin systems. Springer, Berlin, pp 245–261
36. Zhou Y, Wei Y, Wang L, Liu L, Yin H Zhang J et al (2018) Antioxidant effects of catalase on mercury-induced oxidative stress in human neuroblastoma cells. Oxidative Med Cell Longev 1–13
37. Rodrigues JL, Serão NV, Zanette J (2017) Mercury methylation/demethylation in aquatic environments. In: Mercury-recent advances. InTech, pp 99–118
38. Sun Y, Zheng H, Chen N et al (2019) Glutathione-mediated detoxification is involved in the mercury tolerance of Rhizopus oryzae. Appl Microbiol Biotechnol 103(12):4753–4766
39. Wu G, Li J, Sun Y et al (2022) Comparative genomics and transcriptomics analysis reveals diverse molecular mechanisms for mercury resistance in bacteria. Sci Total Environ 808(Pt 2):152009
40. Li Y, Wang Y, Zhang M, Wang Y, Li Y, Liu G, Xu J (2021) Recent advances in mercury detoxification mechanisms in microorganisms. Front Microbiol 12:622218
41. Jones RM, Luo L, Ardita CS et al (2020) Spermine and spermidine promote gut stability and decrease susceptibility to necrotizing enterocolitis in preterm infants. Sci Rep 10(1):10665
42. Gupta P, Khare V, Alam A et al (2018) DNA repair in Mycobacterium tuberculosis revisited. FEMS Microbiol Rev 42(3):337–357
43. Johnson MS, Coker VS, Blake RC (2021) The Mercuric Ion (Mer) transporter protein family. Biomolecules 11(8):1165
44. Chen L, Zhou Z, Chen J, Yang X (2022) Mercury sequestration by metallothioneins and phytochelatins in microorganisms. Crit Rev Microbiol 48(1):79–91
45. Wang X, Guan M, Li C (2023) Mercury reduction and detoxification in bacteria: mechanisms and applications. Chemosphere 297:131968
46. Li M, Xu Z, He H et al (2021) Comparative genomics reveals insights into mercury resistance mechanisms in bacteria. Front Microbiol 12:700567
47. Das S, Dash HR, Mangwani N, Chakraborty J (2019) Bioaccumulation of mercury and its correlation with antioxidative enzymes, phytochelatins, and photosynthetic pigments in diatoms. Ecotoxicol Environ Saf 183:109565
48. Liu B, Wang Q, Yang J, Yang L (2021) Mechanisms of mercury tolerance and detoxification in microorganisms. Front Microbiol 12:617516
49. Rasmussen MK, Søndergaard MT, Hasselriis PN (2021) Genetic and cellular mechanisms of mercury tolerance in bacteria. Appl Microbiol Biotechnol 105(3):1031–1042
50. Zhang R, Zhang Y, Su F (2020) Mechanisms of mercury resistance and accumulation in bacteria. World J Microbiol Biotechnol 36(10):157
51. Orellana SL, Chevrette MG, Currie CR (2021) Insights into mercury resistance mechanisms in Aspergillus fungi from the transcriptome of Aspergillus nidulans. Appl Environ Microbiol 87(4):e02135-e2220
52. Park J, Lee SY (2018) Physiology and genomics of mercury methylation in bacteria. Microb Biotechnol 11(1):7–19
53. Rosen BP (1995) Resistance mechanisms to arsenicals and antimonials. J Basic Microbiol 35(6):453–457

Application of Nano-Adsorbents in Removal of Mercury From Aqueous Solution

Nitish Dhingra

Abstract Access to clean and pure water is a prime need of all living creatures worldwide. The integrated effect of rapid population rise and industrial development has led to a critical situation where all the living organisms on earth are affected directly or indirectly by environmental pollution involving water, air, and soil. As a limited natural resource, its protection and safety are of grave concern. Therefore, eliminating heavy metals and other pollutants from water resources is paramount. The existing decontamination techniques, such as ion exchange and reverse osmosis, suffer many disadvantages; hence, the focus has been shifted to developing novel, efficient techniques to remove heavy metals such as mercury from the water. The adsorption based on nanoadsorbents has gained popularity due to its ease of operation and cost-effectiveness. This chapter highlights the recent advances in water decontamination methods using nanoadsorbents, including polymeric nanocomposites, transition metal chalcogenides (MoS_2), polyrhodanine coated magnetite nanoparticles, chitosan-coated magnetite nanoparticles, TiO_2 nanoparticles, Fe_3O_4/Au nanocomposite, polyacrylate-modified carbon composite, and carbon nanotubes based nanocomposite. Finally, the fate of nanoparticles used for wastewater decontamination is discussed.

Keywords Heavy metal · Mercury · Nanoadsorbent · Remediation · Wastewater

1 Introduction

Heavy metal contamination of the environment, particularly the water, has emerged as a critical problem that needs immediate intervention [8]. The heavy metal ions are toxic not only to aquatic animals but also lethal to land animals through bioaccumulation by food chain transfer. Mercury is among the top ten poisonous metal ions present in the wastewater stream released from the paint, paper, oil refining

N. Dhingra (✉)
Electron Microscopy & Nanoscience Laboratory, Punjab Agricultural University, Ludhiana 141004, Punjab, India
e-mail: nitishdhingra@pau.edu

N. Kumar (ed.), *Mercury Toxicity Mitigation: Sustainable Nexus Approach*, Earth and Environmental Sciences Library, https://doi.org/10.1007/978-3-031-48817-7_8

industries and fertilizer plants. The conversion of mercury to methyl mercury as its organic form and its subsequent accumulation in the food chains makes it highly toxic, posing severe hazards to living organisms and aquatic life [49]. Due to its water stability and solubility, the exposure of human cells to methyl mercury, even at low concentrations beyond 0.2 ppb, can result in neurological disorders, gastrointestinal tract and kidney damage, and memory loss [2, 16]. Considering the severity of toxicity caused by mercury, the United States Environmental Protection Agency (USEPA) has included mercury in the list of priority pollutants and put a stringent discharge limit of 10 μgL^{-1} for wastewater and 1 μgL^{-1} in drinking water [47, 51]. Therefore, many researchers have tried devising novel technologies for mercury elimination from contaminated water.

These technologies include chemical precipitation [32], membrane separation [15], solvent extraction, chemical precipitation, ion exchange [33], adsorption [20], and coagulation [44]. However, these techniques have low decontamination efficiency, especially for high volumes of aqueous solutions of heavy metal ions. Out of these different alternatives, the decontamination based on adsorption was observed to be superior to others due to its simplicity, ease of operation, better efficiency, and economic feasibility. Many diverse materials have been investigated for this purpose, such as chitosan [42], activated carbon [24], zeolites [12, 13], waste rubber, polymer composites and magnetic hydrogels [34, 45]. Despite the availability of several adsorbents for removing tiny concentrations of heavy metal ions from wastewater, it is essential to develop novel adsorbents with better adsorption capacity, simplistic adsorption–desorption kinetics, higher stability, and ease of operation.

The recent advancements in this direction include using adsorbents on the nanometer scale, called nanoadsorbents [23, 52, 54]. Nanoadsorbents are preferred for wastewater and drinking water remediation due to their exceptional properties, such as high surface area, large pore volume, high reactivity, and good chemical species selectivity for adsorption. Conventional nanoadsorbents include nanoporous carbon (activated carbon and multiwalled carbon nanotubes) and oxide minerals (oxides of iron, aluminium, and titanium). Such nanoadsorbents can offer surface area ranging from several hundred to thousand m^2gm^{-1}, leading to adequate contact between the particles of the adsorbent and metal ions. However, their limited functional groups offer weak guest–host interactions, leading to poor adsorption capacities [11, 36]. Therefore, surface modifications or functionalizations are generally deployed to improve the adsorption capacity. For example, cysteine-functionalized multiwalled carbon nanotubes (Cyst-MWCNTs) [27], arginine-modified TiO_2 nanoparticles [26], thiol-functionalized Zn-doped biomagnetite particles [17], and citrate-coated gold nanoparticles [21]. This chapter highlights the recent advances in the remediation of mercury ions from wastewater using various nanoadsorbents.

2 Sources of Water Contamination

Wastewater is generally defined as water containing undesired substances that can deem it unfit [4]. Various sources that can contribute to wastewater production involve domestic, industrial, commercial, and agricultural activities. The composition of wastewater depends on the source of its origin and displays significant variation [14]. The wastewater from domestic and commercial origin usually has an untidy look and contains community-added material such as human body wastes, plastic litter, and food residues. The contaminated water is enriched with nitrogen and phosphorus that act as nutrients and lead to multiplication in the algae growth rate [30]. The latter competes with the surrounding aquatic biota for dissolved oxygen. It is dangerous to consume such water due to contamination from various pathogenic microorganisms. Industrial wastewater involves effluents from textile, food processing and pharmaceutical industries, coal mines, and steel plants. Industrial wastewater contains a variety of toxic pollutants such as insoluble and suspended solids, organic compounds, nutrient salts (NH_4^+, SO_4^{2-}, and PO_4^{3-}), corrosive substances (alkalis, solvents, and oil), cleaning agents, lubricants, disinfectants, and other harmful substances (hydrocarbons, chlorinated molecules, organic halogen compounds, cyanides, and heavy metals) [10, 43].

Agricultural activities contributing to water contamination involve toxic chemicals resulting from fertilizers, pesticides, insecticides, herbicides, animal wastes, and veterinary drugs. These effluents are rich sources of nitrogen, phosphorus and the content requiring high chemical oxygen; therefore, the values of these parameters must be brought within permissible limits before discharging into the water streams. Figure 1 summarizes various sources of wastewater and the typical content present in the wastewater from each of these sources.

Water protection and purification is a critical challenge for the survival of the living organisms. The continuous efforts towards environmental awareness in recent years have intensified the efforts towards devising novel and efficient techniques to purify the water for its reuse.

3 Nanotechnology as a Tool for Water Remediation

The conventional decontamination methods, such as chemical oxidation or reduction, membrane filtration, precipitation, coagulation, adsorption, and ion exchange, suffer from certain limitations. Moreover, they lack selectivity towards the contaminants, therefore, are unable to eliminate one compound over the other, leading to inefficiency and interference during the decontamination process. In the modern era, nanotechnology has unfolded as a promising solution in diverse sectors involving research and development. Some of these are the agriculture sector [1], bioanalytical sciences [25], the food sector [35], and water purification from heavy metal contamination [38]. With the help of technological advances, it is feasible to employ

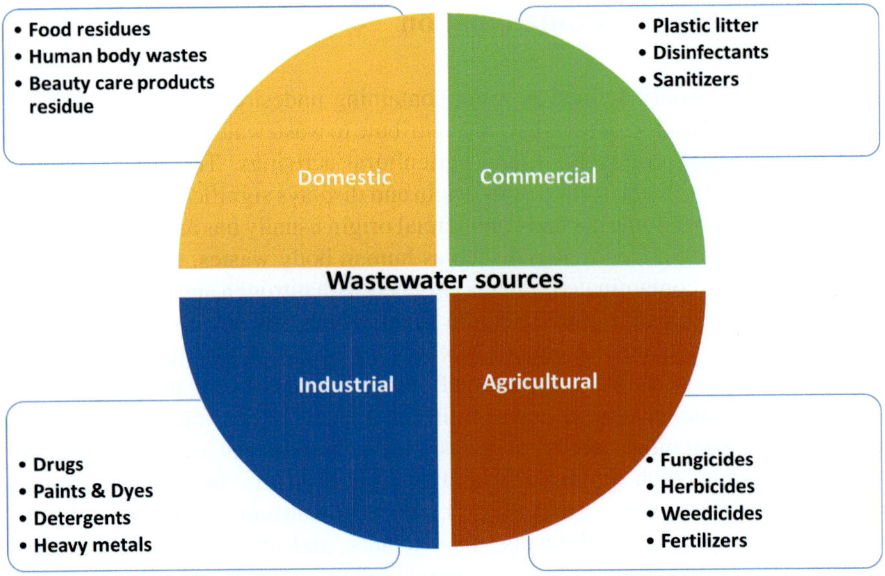

Fig. 1 Pictorial representation of various wastewater sources

nano-scale materials (<100 nm) in addressing water remediation problems owing to their remarkable properties such as higher surface-to-volume ratio, effortless functionalization ability for enhanced selectivity and affinity, and high sorbent capabilities [53]. Over the past few decades, nanomaterials have received broader attention in water decontamination [9]. In particular, the large specific surface area, small size, and high reactivity of nanomaterials have led to the development of novel materials called nanoadsorbents, having high water treatment efficiency [7].

Adsorption is a surface phenomenon and refers to the ability of a solid substance to attach to its surface the molecules of gases or solutions with which it is in close contact. There may be variations from the micron to nanometer range in the size of adsorbent molecules participating in such interactions. The nanomaterials can offer a relatively larger surface area for adsorption, yielding higher decontamination efficiency than their bulk counterparts. This process of separating the decontaminants with nano-scale adsorbents is termed nanoadsorption. Figure 2 summarizes the three components involved in the adsorption process. The interaction between the adsorbate and the adsorbent is the most crucial one that controls the adsorption process. However, other factors, such as the affinity between the solution and the adsorbate, the solution and the adsorbent, and the contaminants may also play their role. Due to the low solubility, hydrophobic compounds in aqueous solutions are likely to get attracted to the surface of the adsorbent. As a result, the adsorption capacity is significantly dependent on the forces arising from these three components' interaction.

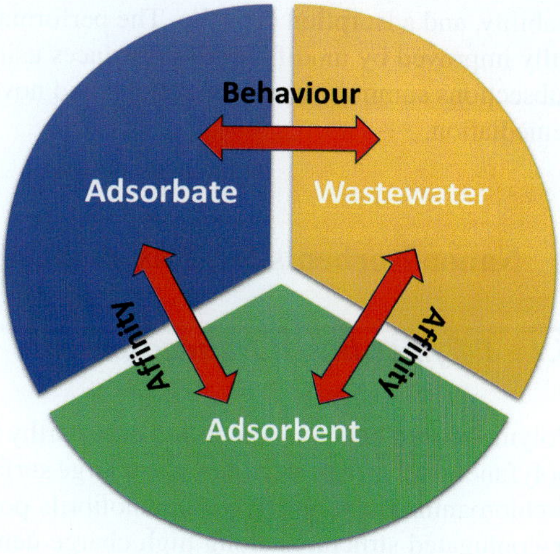

Fig. 2 Interaction of various components in a typical adsorption process

Nanoparticles are favoured over other adsorbents due to their unique features, such as numerous sorption sites, porosity, higher specific surface area, possibility of surface functionalities, low-temperature modification, little intraparticle diffusion distance, and enhanced capabilities for ion binding [46]. In addition, other physicochemical properties like dimensions, shape, chemical constitution, physicochemical stability, crystal structure and surface traits like roughness, energy, and area also affect the efficiency and properties of nanoadsorbent materials. The reactiveness of nanomaterials could be enhanced by reducing the size further, thereby improving the surface area-to-volume ratio. The surface charge strongly influences the nanoparticle's toxicity because it controls various characteristics of nanomaterials, including its colloidal behaviour, selective adsorption, integrity of blood–brain barrier, binding of plasma protein, and transmembrane permeability. Moreover, the crystalline structure, composition, surface coating and surface roughness also play a critical role in determining the toxicity of nanoparticles.

Based on the abovesaid physicochemical characteristics, a wide variety of nanomaterials have been synthesized recently, such as carbon nanotubes, metal oxide nanoparticles, polymeric nanoparticles, and nanowires. The physicochemical properties of nanomaterials may also be influenced by intrinsic compositions, inherent surface properties, external functional groups attached to nanoadsorbents, and sizes. The reason for such extraordinary properties and behaviour of nanoadsorbents was explained in terms of the nature of active sites and their arrangement on the surface of these materials [31]. Nanoparticles alone are readily oxidized by atmospheric oxygen, causing aggregate formation in the aqueous systems. Therefore, it is essential to do surface modification of these nanoparticles to stabilize them and subsequently employ them as nanoadsorbents. In order to eradicate heavy metal pollutants, nanomaterials were recently surface modified to enhance their properties, like efficiency, stability, and adsorption capacity. The performance of nanoadsorbents was successfully improved by modifying their surfaces using various methods. The subsequent subsections summarize several general and novel nanomaterials used in wastewater remediation.

4 Nanoadsorbents for Wastewater Remediation

4.1 Polymer Based Nanoadsorbents

Polymer-based nanoadsorbents are noteworthy because they offer the ability to have polyfunctional groups in addition to a large surface area. Particularly, poly (1-amino-5-chloroanthraquinone) (PACA) nanofibrils possess polymer chains with extended π-conjugated structures along high charge density and many electron-rich groups, including –N = , –Cl, –NH–, –NH2, and –C = O. The rigid polymeric chains enable these polymers to display extraordinary solvent resistance, whereas electron-rich polyfunctional groups and mobile π-electrons equip them with a powerful binding affinity toward metal cations. Due to these excellent properties, polyfunctional nanopolymers can be good candidates for removing mercury ions from aqueous solutions.

Studies performed with PACA nanofibrils [19] on the wastewater containing an initial Hg(II) concentration of 10 mmol/L revealed nearly 99.56% removal efficiency of Hg(II) ions from aqueous solution. The adsorbent dose increased the adsorptivity of Hg(II) ions significantly up to an adsorbent dose of 8.0 gL^{-1}, which may be attributed to the increased surface area due to the availability of new adsorbent sites. The pH of the aqueous solution is a crucial parameter that can affect the adsorption of Hg(II) ions at the PACA-water interfaces. The adsorption capacity increased significantly from 0.39 to 3.05 mmol/g for a pH increase from 1.0 to 5.0, whereas negligible changes were observed in the adsorption capacity for pH values ranging from 5.0 to 7.0. Moreover, the studies performed with several eluents demonstrated that PACA nanofibrils are reusable even after five adsorption–desorption cycles. The commercial application and selectivity of PACA nanofibrils for wastewater purification were tested on chloralkali wastewater containing a high concentration of various ions. For an adsorbent dose of 1.6 gL^{-1}, PACA nanofibrils showed a high adsorption capability (98.4%) for removing Hg(II) ions from chloralkali wastewater. The concentration of Hg(II) ions reduced from 97.3 to 1.56 $\mu g\ L^{-1}$ after a single cycle of operation, revealing that the water quality is greatly improved with PACA nanofibrils.

4.2 *Petal-Like MoS_2 Nanosheets*

In recent times, two-dimensional layered materials, particularly transition metal chalcogenides such as MoS_2 and $MoSe_2$, are gaining popularity in heavy metal removal from aqueous solutions. MoS_2 is a layered semiconductor having intrinsic sulphur atoms where a bi-layered hexagonal arrangement of S atoms sandwiches one hexagonal layer of Mo atoms, resulting in a trigonal prismatic structure [22, 48]. Mercury is a weak acid, therefore, the Hard and Soft Acids and Bases (HSAB) principle allows it to interact strongly with weak bases like sulphur.

A variety of sulphur-containing materials such as thiol-functionalized mesoporous silica, sulphur-functionalized mesoporous carbons, sulphurized activated carbons, metal–organic frameworks decorated with thiols, 2-mercaptobenzimidazole clay, porous organic polymers decorated with thiols (PAF-1-SH) and metal-chalcogenide aerogels, have been investigated for the removal of mercury from the aqueous solution. However, they have low removal efficiency and less content of sulphur atoms. Therefore, it is crucial to devise nanoadsorbents that can overcome these limitations. MoS_2 is best suited due to its sulphur enrichment. However, the interplanar spacing between the layers needs to be increased to enhance the exposure of sulphur atoms for binding Hg(II) ions.

Pirarath et al. [37] synthesized petal-like MoS_2 nanosheets using ammonium molybdate, thiourea and sodium dodecyl sulfate by the hydrothermal method. The pH of the aqueous solution was found to have a strong dependence on the adsorption efficiency of the MoS_2 nanosheets. The adsorption efficiency was observed to increase up to a pH value of 10 due to the adsorbent's pH-dependent zeta potential. Moreover, an increase in pH raised the negative charge on the surface of MoS_2 nanosheets, thereby enhancing the electrostatic attraction of Hg(II) ions towards the MoS_2 nanoadsorbent sheets. The Hg(II) ion removal efficiency increased from 25 to 93% as the adsorbent concentration increased from 0.1 to 0.5 mg/L. This improvement was due to the increased availability of sites for removing Hg(II) ions from the aqueous solution. Moreover, a comparative study of the contact time performed using bulk-scale MoS_2 and MoS_2 nanosheets revealed that the MoS_2 nanosheets have more adsorption capacity (289 mg/g) compared to bulk-scale (225 mg/g) MoS_2, owing to their high surface area.

4.3 *Magnetite-Polyrhodanine Core–Shell Nanoadsorbents*

Despite of the availability of various kinds of nanoadsorbents, it is vital to develop the next generation of nanoadsorbents to have superior adsorption properties and simple operation. Particularly, magnetite nanoparticles are gaining popularity in isolating and eliminating contaminants by applying a magnetic field. However, the bare magnetite nanoparticles are prone to oxidation and quickly form aggregates.

Therefore, some surface modification is required to stabilize the magnetite nanoparticles. Polyrhodanine is one such material that has caught attention due to its anti-corrosive, anti-bacterial and anti-histaminic nature. A single unit of polyrhodanine has metal-binding functional groups. Therefore, polyrhodanine can be a promising candidate for removing Hg(II) ions from aqueous solutions due to the presence of sulphur, oxygen and nitrogen atoms in its monomeric structure.

Rahmanzadeh et al. [40] synthesized polyrhodanine coated magnetite nanoparticles and characterized them with transmission electron microscopy (TEM) and Fourier transform infrared spectroscopy (FTIR). The characterization revealed a homogenous morphology of the core–shell magnetite nanoparticles along with a thin film of polyrhodanine coated over them. The adsorption efficiency of polyrhodanine coated magnetite nanoparticles was found to be intensely dependent on the pH of the aqueous solution [41]. At a constant Hg(II) concentration of 10 mg/L, adsorption increased for the pH range from 2 to 6.5 and decreased for the pH range from 6.5 to 12. The initial low adsorption efficiency may be due to the presence of a higher concentration of H^+ ions on the adsorbent surface and hence competing with the Hg(II) ions at the adsorption sites. The studies on the adsorption efficiency as a function of adsorbent dose revealed that the adsorption efficiency initially showed a rising trend for the adsorbent concentration in the range of 0.05–0.1 g/L and then remained almost constant in the range 0.1–0.2 g/L. The Hg(II) removal efficiency initially increased from 90.80 to 99% for a contact time of around 50 min and subsequently remained almost constant. This initial increase in the removal efficiency may be due to the greater availability of the nanoadsorbent sites and slowed down due to the uptake of vacant sites by the Hg(II) ions.

4.4 Chitosan Based Nanoadsorbents

Chitosan is a natural polysaccharide formed by de-acetylating chitin, used widely in the pharmaceutical and food industries. Besides being non-toxic, biodegradable, and economical, it has an inherent ability to adsorb metal ions via amino groups acting as ion-exchangers. Rahbar et al. [39] synthesized chitosan-coated magnetite nanoparticles and assessed their potential as nanoadsorbents for the removal of Hg(II) ions from aqueous solutions. The studies revealed that the adsorption percentage decreased with an increase in pH value from 5 to 8, keeping concentrations of Hg(II) ions and chitosan-coated magnetite nanoparticles constant. The highest adsorption was found to be at a pH of 5.

The studies on the adsorbent dose revealed that the Hg(II) removal efficiency of chitosan-coated magnetite nanoparticles increased with the increase in adsorbent dose from 0.25 g to 0.75 g while keeping the pH at 5 and initial mercury concentration (6.2 mg/L) constant. This increase was attributed to the enhanced surface area availability with increased adsorbent dose. Further, the adsorption of Hg(II) ions on chitosan-coated magnetite nanoparticles increased with increased concentration of mercury from 2 to 6 mg/L in the aqueous solution, along with corresponding improvement in the removal efficiency of Hg(II) ions from 96% to around 99%.

Hosseini et al. [18] reported studies on the selective adsorption of Hg(II) from aqueous solutions using functionalized nanochitosan by carbon disulfide. Studies showed that for the mercury concentration of 30 mg/L with an adsorbent dose of 0.1 g/L and 50 ml of solution at 25 °C, the Hg(II) ions removal efficiency increased from 30% to around 90% for a pH increase from 2 to 7. The removal efficiency was also found to increase significantly from 55% to around 90% with an increase in the contact time from 10 to 150 min at a pH = 7, concentration of adsorbent 0.1 g/L, concentration of Hg(II) 30 mg/L and a temperature of 25 °C. Besides the adsorption of Hg(II), the functionalized nanochitosan was also found to be effective in removing Cu, Pb, Cd, Zn, and Ni ions with variable efficiency.

4.5 TiO_2 Nanoparticles

Afshar et al. [3] reported studies on removing Hg (I) and Hg (II) ions from the aqueous solutions using TiO_2 nanoparticles. The percentage adsorption was found to depend on the solution's pH. The lowest adsorption percentage was obtained at a pH value of 1, whereas the highest percentage was observed at a pH value of 9 and 7 for Hg(I) and Hg(II) ions, respectively. At low pH values, the concentration of H^+ ions is high, thus posing challenges for Hg(I) and Hg(II) to reach the adsorption sites. However, at higher pH values, the concentration of H^+ ions is reduced; as a result, they can not compete with the Hg(I) and Hg(II) ions anymore. Moreover, the TiO_2 nanoparticles having 20 nm size were found to have the highest adsorption percentage compared to 100 and 400 nm nanoparticles. The larger size of nanoparticles leads to a smaller value of the surface area-to-volume ratio, resulting in less availability of adsorption sites and a low absorption percentage.

The studies on the impact of contact time on the removal efficiency of Hg(I) and Hg(II) ions from aqueous solutions were performed at various time intervals of 15, 30, 45, 60, 75, 90, and 105 min. The removal efficiency was found to increase initially up to 75 and 45 min for Hg(I) and Hg(II) ions, respectively. The possible reason for this increase was the availability of more interaction time for adsorption on the adsorbent sites. In order to study the effect of temperature, the adsorption experiment was carried out at four different temperature values of 22, 35, 45, and 55 °C. The removal efficiency was found to decrease from 97.6% (98.1%) to 91% (92.5%) for Hg^+ (Hg^{2+}) with the rise in temperature due to the exothermic nature of the adsorption process. An increase in the adsorbent dose up to 0.05 g caused a corresponding increase in the adsorption percentage due to the availability of more adsorbent sites. The maximum adsorption values obtained for Hg(I) and Hg(II) ions were 97.5 and 98.6%, respectively. The Langmuir isotherm represented the experimental data well compared to Freundlich and Temkin isotherms.

4.6 Fe_3O_4/Au Nanoadsorbent

Maia et al. [28] synthesized a simple, novel, fast and low-cost Fe_3O_4/Au nanoadsorbent composite and investigated its ability to remove mercury from the aqueous solutions. The effect of various parameters, such as the pH of the solution and contact time, was studied to test the efficiency of nanocomposite for mercury adsorption from wastewater. The mercury adsorption capacity of the nanocomposite was observed to increase continuously with an increase in the pH. At a pH value of 5, the mercury adsorption capacity was found to be 0.683 mg/g, corresponding to a removal efficiency of 28%. The adsorption capacity and removal efficiency increased to 2.1 mg/g and 80.2%, respectively, at a pH value of 7. The maximum Hg(II) adsorption capacity of 2.3 mg/g with a removal efficiency of 86% was observed at a pH value of 9. The adsorption capacity was found to increase rapidly for a contact time of up to 50 min; however, the adsorption capacity remained almost constant for longer contact times greater than 50 min. Moreover, the Dubinin-Radushkevich (D-R) isotherm suggested that Hg(II) adsorption by Fe_3O_4/Au nanocomposite occurs primarily by chemical adsorption.

4.7 Polyacrylate-Modified Carbon Composite

Al-Yaari et al. [6] synthesized a novel polyacrylate-modified carbon composite using plam shells (source of activated carbon) grafted with a copolymer of acrylamide and methacrylic acid to produce a polyacrylate-modified carbon (PAMC) composite. To study the impact of adsorbent dose and to determine an optimum adsorbent dose, different dosages of PAMC were examined for the treatment of a 20 mL aqueous solution having an initial Hg concentration of 200 ppm under ambient conditions (1 atm and 298 K). The removal percentage of Hg increased with the enhanced PAMC dosages. As the adsorbent mass dosage increased from 0.01 to 0.02 g, a rapid increase in the removal percentage was observed. The possible reason may be the increased PAMC active surface area and hence the increased availability of active sites. For comparison, the removal percentage of activated carbon was also tested. The study concluded that 0.2 g of PAMC was adequate to achieve a 100% removal of Hg(II), whereas activated carbon could achieve only 78% Hg(II) removal. Overall, the removal percentage improved by 28 to 64% compared to activated carbon according to different adsorbent dosages.

A PAMC dosage of 0.1 g and initial mercury concentrations of 50, 100, and 200 ppm were used to study the pH influence on the adsorption percentage. The mercury removal percentage increased to 99, 96.5, and 93% for the 50, 100, and 200 ppm solutions, respectively, at pH = 6. The removal percentage was observed to be constant for pH > 6 due to the saturation of the available number of sites. Moreover, the mercury removal percentage was also observed to be dependent on the initial mercury concentration and was higher for low initial concentrations of mercury. This may be attributed to the availability of active sites on the PAMC surface.

The removal percentage of mercury was observed to improve with the contact time. Within 10 min, the mercury removal percentage reached 62, 52, and 35% for the solutions with mercury concentrations of 50, 100, and 200 ppm, respectively. This improved mercury removal performance of the PAMC composite can be credited to forming polymeric chains on the carbon surface, as confirmed by the scanning electron micrographs. In addition to the aromatic rings, PAMC has many functional groups on the carbons that allow $\pi - \pi$ conjugation and stacking interactions with Hg(II) ions. A slow increase in the removal percentage was observed until equilibrium was reached around a time scale of 90 min.

4.8 Carbon Nanotubes (CNT) Based Nanocomposite

Carbon nanotubes are tubular arrangements of long-wrapped graphene sheets (hexagonally arranged hybrid carbon atoms). These occur in two forms, called single-walled carbon nanotubes (SWCNTs) and multi-walled carbon nanotubes (MWCNTs). The former comprises a single graphene layer, whereas the latter involves multiple graphene layers. Due to their remarkable properties, such as inert surfaces, high surface areas, and uniform arrangement of carbon atoms, they are preferred over activated carbon. Alijani and Shariatinia [5] synthesized single-walled CNT and cobalt sulphide nanocomposite for mercury remediation from wastewater. The adsorption efficiency was observed to attain a maximum of 99.56% very quickly in about 7 min. The comparative study reported an adsorption efficiency of about 45.39% using single-walled CNT alone.

5 Fate of Nanomaterials Used for Wastewater Treatment

Although nanoadsorbents display remarkable performance for mercury remediation in wastewater, it is critical to investigate the fate of nanomaterials after the task of wastewater purification is complete. The fate of nanomaterials depends on the integrated effect of their physiochemical properties and interaction with the contaminants [29]. Nanomaterials can enter the environment either through natural sources (volcanic eruptions, weathering, soil erosion, clay minerals and dust storms) or anthropogenic sources (synthesis and use of nanomaterials for various purposes, burning fossil fuel, automobiles). Upon introduction into the environment, nanomaterials can accumulate in air, soil, water, and sediments. The most significant fractions of nanomaterials in the environment eventually end up in the soil, whereas only tiny fractions remain in the water and air. The sludge produced in wastewater treatment plants is one of the biggest causes of the presence of nanomaterials in the soil. The wastewater treatment leads to the accumulation of contamination, including nanoparticles in the sludge, which is then handled in different ways. However, the scarcity of fair strategies for their disposal has resulted in a constant increase in the level of nanomaterials in the environment [50].

The necessity of available data on the fate of nanomaterials for use in various fields requires detailed life cycle assessment and risk analyses because the same harmless material may show toxicity and reactivity at the nanometer scale. Moreover, limited literature is available on the negative impact of nanomaterials on living organisms. Therefore, narrowing the gaps between the practical applications of nanomaterials and the potential adverse effects, if any, is critical. Addressing such issues will enable nanotechnology to furnish the expected results in a cost-effective, eco-friendly, and hazardless manner.

6 Conclusions

To summarize, numerous water decontamination techniques have evolved; however, the adsorption-based remediation has emerged as the most powerful and popular technique. It can be applied efficiently to reduce various kinds of inorganic and organic pollutants without significant side effects. Because of their exceptional properties, nanomaterials are broadly utilized to eradicate heavy metals in water/wastewater. In this context, nanoadsorbent materials are gaining wider recognition in water remediation due to their extraordinary adsorption potential compared to traditional bulk-scale adsorbents. Therefore, nanoadsorbents can be named as next-generation adsorbents beneficial for environmental pollution and controlling water pollution. The chapter presented an overview of using nanomaterials as adsorbents, either unassisted or with modified surfaces giving supplementary functional groups for more promising sequestration of mercury ions in wastewater.

Acknowledgements The author is thankful to Punjab Agricultural University, Ludhiana, India for the support during the work.

References

1. Acharya A, Pal PK (2020) Agriculture nanotechnology: translating research outcome to field applications by influencing environmental sustainability. NanoImpact 19:100232
2. Adams DH, Sonne C, Basu N, Dietz R, Nam DH, Leifsson PS, Jensen AL (2010) Mercury contamination in spotted seatrout, Cynoscion nebulosus: an assessment of liver, kidney, blood, and nervous system health. Sci Total Environ 408:5808–5816
3. Afshar E, Mohammadi-Manesh H, Dashti Khavidaki H (2017) Removal of Hg (I) and Hg (II) Ions from aqueous solutions, using TiO_2 nanoparticles. Pollut 3:506–517
4. Ahmad A, Ghufran R (2019) Review on industrial wastewater energy sources and carbon emission reduction: towards a clean production. Int J Sustain Eng 12(1):47–57

5. Alijani H, Shariatinia Z (2018) Synthesis of high growth rate SWCNTs and their magnetite cobalt sulfide nanohybrid as super-adsorbent for mercury removal. Chem Eng Res Des 129:132–149
6. Al-Yaari M, Saleh TA (2022) Mercury removal from water using a novel composite of polyacrylate-modified carbon. ACS Omega 7:14820–14831
7. Awual MR, Hasan MM, Khaleque MA, Sheikh MC (2016) Treatment of copper(II) containing wastewater by a newly developed ligand based facial conjugate materials. Chem Eng J 288:368–376
8. Ayati A, Tanhaei B, Sillanpää M (2016) Lead(II)-ion removal by ethylenediaminetetraacetic acid ligand functionalized magnetic chitosan-aluminum oxide-iron oxide nanoadsorbents and microadsorbents: Equilibrium, kinetics, and thermodynamics. J Appl Polym Sci 134:44360
9. Bahiraei A, Behin J (2020) Sonochemical immobilization of MnO_2 nanoparticles on NaP-zeolite for enhanced Hg(II) adsorption from water. J Environ Chem Eng 8:103790
10. Barik B, Nayak PS, Dash P (2020) Nanomaterials in wastewater treatments. Nanotechnol Beverage Ind 20:185–206
11. Cao CY, Qu J, Wei F, Liu H, Song WG (2012) Superb adsorption capacity and mechanism of flowerlike magnesium oxide nanostructures for lead and cadmium ions. ACS Appl Mater Interfaces 4(8):4283–4287
12. Coruh S, Senel G, Ergun ON (2010) A comparison of the properties of natural clinoptilolites and their ion-exchange capacities for silver removal. J Hazard Mater 180:486–492
13. Doula MK (2009) Simultaneous removal of Cu, Mn and Zn from drinking water with the use of clinoptilolite and its Fe-modified form. Water Res 43(15):3659–3672
14. Elgallal M, Fletcher L, Evans B (2016) Assessment of potential risks associated with chemicals in wastewater used for irrigation in arid and semiarid zones: a review. Agric Water Manag 177:419–431
15. Fard GH, Mehrnia MR (2016) Investigation of mercury removal by Micro-Algae dynamic membrane bioreactor from simulated dental waste water. J Environ Chem Eng 5(1):366–372
16. Ghasemi E, Heydari A, Sillanpää M (2017) Superparamagnetic Fe_3O_4@EDTA nanoparticles as an efficient adsorbent for simultaneous removal of Ag(I), Hg(II), Mn(II), Zn(II), Pb(II) and Cd(II) from water and soil environmental samples. Microchem J 131:51–56
17. He F, Wang W, Moon JW, Howe J, Pierce EM, Liang L (2012) Rapid removal of Hg(II) from aqueous solutions using thiol-functionalized Zn-doped biomagnetite particles. ACS Appl Mater Interfaces 4:4373–4379
18. Hosseini SM, Younesi H, Vajdi R, Bahramifar N (2019) Selective adsorption of Mercury (II) From aqueous solution using functionalized nanochitosan by carbon disulfide. J Water Wastewater 31:57–75
19. Huang S, Ma C, Liao Y, Min C, Du P, Jiang Y (2016) Removal of Mercury(II) from aqueous solutions by adsorption on Poly(1-amino-5-chloroanthraquinone) Nanofibrils: equilibrium, kinetics, and mechanism studies. J Nanomater 2016:7245829
20. Inbaraj BS, Sulochana N (2006) Mercury adsorption on a carbon sorbent derived from fruit shell of Terminalia catappa. J Hazard Mater 133:283–290
21. Ojea-Jiménez I, López X, Arbiol J, Puntes V (2012) Citrate-coated gold nanoparticles as smart scavengers for Mercury(II) removal from polluted waters. ACS Nano 6(3):2253–2260
22. Jia F, Wang Q, Wu J, Li Y, Song S (2017) Two-dimensional molybdenum disulfide as a superb adsorbent for removing Hg^{2+} from water. ACS Sustain Chem Eng 5(8):7410–7419
23. Kalfa OM, Yalçinkaya Z, Türker AR (2009) Synthesis of nano B_2O_3/TiO_2 composite material as a new solid phase extractor and its application to preconcentration and separation of cadmium. J Hazard Mater 166(1):455–461
24. Kantarli IC, Yanik J (2010) Activated carbon from leather shaving wastes and its application in removal of toxic materials. J Hazard Mater 179:348–356
25. Keçili R, Büyüktiryaki S, Hussain CM (2019) Advancement in bioanalytical science through nanotechnology: past, present and future. TrAC Trends Anal Chem 110:259–276
26. Li L, Feng Y, Liu Y, Wei B, Guo J, Jiao W, Zhang Z, Zhang O (2016) Titanium dioxide nanoparticles modified by salicylic acid and arginine: structure, surface properties and photocatalytic decomposition of p-nitrophenol. Appl Surf Sci 363:627–635

27. Liu Y, Li Y, Yan XP (2008) Preparation, characterization, and application of L-Cysteine functionalized multiwalled carbon nanotubes as a selective sorbent for separation and preconcentration of heavy metals. Adv Funct Mater 18(10):1536–1543
28. Maia LFO, Santos MS, Andrade TG, Hott RDC, Faria MCDS, Oliveria LCA, Pereira MC, Rodrigues JL (2020) Removal of mercury(II) from contaminated water by gold-functionalized Fe_3O_4 magnetic nanoparticles. Environ Technol 41:959–970
29. Maiti S, Fournier I, Brar SK, Cledon M, Surampalli RY (2016) Nanomaterials in surface water and sediments: fate and analytical challenges. J Hazard Toxic Radioact Waste 20:B4014004
30. Moore JW, Moore EA (1976) Resources in environmental chemistry. I. An annotated bibliography of the chemistry of pollution and resources. J Chem Educ 53(3):167
31. Neyaz N, Siddiqi WA, Nair KK (2014) Application of surface functionalized iron oxide nanomaterials as a nanoadsorbents in extraction of toxic heavy metals from ground water: a review. Int J Environ Sci 4:472–483
32. Nguyen DL, Kim JY, Shim SG, Ghim YS, Zhang XS (2016) Shipboard and ground measurements of atmospheric particulate mercury and total mercury in precipitation over the Yellow Sea region. Environ Pollut 129:262–274
33. Oehmen A, Vergel D, Fradinho J, Reis MAM, Crespo JG, Velizarov S (2014) Mercury removal from water streams through the ion exchange membrane bioreactor concept. J Hazard Mater 264:65–70
34. Ozay O, Ekici S, Baran Y, Aktas N, Sahiner N (2009) Removal of toxic metal ions with magnetic hydrogels. Water Res 43(17):4403–4411
35. Palit S (2020) Recent advances in the application of nanotechnology in food industry and the vast vision for the future. Nanoeng Beverage Ind 20:1–34
36. Peng QM, Guo JX, Zhang QR et al (2014) Unique lead adsorption behavior of activated hydroxyl group in two-dimensional titanium carbide. J Am Chem Soc 136(11):4113–4116
37. Pirarath R, Shivashanmugam P, Syed A, Elgorban AM, Anandan S, Muthupandian AK (2021) Mercury removal from aqueous solution using petal-like MoS_2 nanosheets. Front Environ Sci Eng 15:15
38. Puri N, Gupta A, Mishra A (2021) Recent advances on nano-adsorbents and nanomembranes for the remediation of water. J Clean Prod 322:129051
39. Rahbar N, Jahangiri A, Boumi S, Khodayar MJ (2013) Mercury removal from aqueous solutions with chitosan-coated magnetite nanoparticles optimized using the Box-Behnken design. Jundishapur J Nat Pharm Prod 9:e15913
40. Rahmanzadeh L, Ghorbani M, Jahanshahi M (2014) Synthesis and Characterization of Fe3O4 @Polyrhodanine nanocomposite with core-shell morphology. Adv Polym Technol 33(1–5):21463
41. Rahmanzadeh L, Ghorbani M, Jahanshahi M (2016) Effective removal of hexavalent mercury from aqueous solution by modified polymeric nanoadsorbent. J Water Environ Nanotechnol 1:1–8
42. Reddy DHK, Lee SM (2013) Application of magnetic chitosan composites for the removal of toxic metal and dyes from aqueous solutions. Adv Colloid Interface Sci 201–202:68–93
43. Rosenwinkel KH, Austermann-Haun U, Meyer H (2005) Industrial wastewater sources and treatment strategies. Wiley Online Library, Hoboken, NJ, USA
44. Samrani AGE, Lartiges BS, Villiéras F (2008) Chemical coagulation of combined sewer overflow: heavy metal removal and treatment optimization. Water Res 42:951–960
45. Sharma RK, Lalit A (2011) Synthesis and characterization of graft copolymers of N-Vinyl-2-Pyrrolidone onto guar gum for sorption of Fe^{+2} and Cr^{+6} ions. Carbohydr Polym 83(4):1929–1936
46. Singh J, Dutta T, Kim KH, Rawat M, Samddar P, Kumar P (2018) Green synthesis of metals and their oxide nanoparticles: applications for environmental remediation. J. Nanobiotechnology 16:84
47. Wang J, Deng BL, Chen H, Wang XR, Zheng JZ (2009) Removal of aqueous Hg(II) by polyaniline: sorption characteristics and mechanisms. Environ Sci Technol 43:5223–5228

48. Wang Z, Mi B (2017) Environmental applications of 2D Molybdenum Disulfide (MoS_2) nanosheets. Environ Sci Technol 51:8229–8244
49. Yu CT, Chen YL, Cheng HW (2014) Development of an innovative layered carbonates material for mercury removal sorbents. Energy Procedia 61:1270–1274
50. Zekić E, Vukovic Ž, Halkijevic I (2018) Application of nanotechnology in wastewater treatment. Gradevinar 70:315–323
51. Zhang FS, Nriagu JO, Itoh H (2005) Mercury removal from water using activated carbons derived from organic sewage sludge. Water Res 39:389–395
52. Zhang Q, Pan B, Zhang W, Jia K, Zhang Q (2008) Selective sorption of lead, cadmium and zinc ions by a polymeric cation exchanger containing nano-$Zr(HPO_3S)_2$. Environ Sci Technol 42(11):4140–4145
53. Zhang Y, Li M, Gao X, Chen Y, Liu T (2019) Nanotechnology in cancer diagnosis: progress, challenges and opportunities. J Hematol Oncol 12:137
54. Zhou YT, Nie HL, Branford-White C, He ZY, Zhu LM (2009) Removal of Cu from aqueous solution by chitosan-coated magnetic nanoparticles modified with α-ketoglutaric acid. J Colloid Interface Sci 330:29–37

Applications of Different Treatment Technologies for Mercury Removal From Soil, Waste, and Water

Rakesh Pant, Pratyaksha Singh, Nirmal Patrick, and Amit Gupta

Abstract Mercury is a highly hazardous heavy metal that can have significant environmental and health impacts. Mercury contamination can occur in soil, waste, and water due to human activities like mining, industrial processes, and waste disposal. Several treatment technologies are available for mercury removal from these matrices. In this chapter, we will discuss the applications of different treatment technologies for mercury removal. Mercury (Hg) is a poisonous heavy metal that has been named one of the "ten most dangerous substances." Mercury contamination is an important global problem, not only because of its rising concentration in the ecosystem but also because of its deadly effect on human health. Mercury (Hg) usage by humans has resulted in localized pollution as well as worldwide diffusion via the environment. Mercury contamination is a severe public healthcare and ecological threat because methylmercury enters the bloodstream fast and affects the nervous system. Mercury pollution is common in soils and sediments and is difficult to remove. There are several treatment technologies available for mercury removal from soil, waste, and water. The choice of technology depends on the characteristics of the matrix and the concentration of mercury. By using appropriate treatment technologies, we can reduce the environmental and health impacts of mercury contamination.

Keywords Mercury removal · Mercury contamination · Treatment technologies · Heavy metal · Water

R. Pant · P. Singh · A. Gupta (✉)
Department of Microbiology, Graphic Era (Deemed to Be) University, Dehradun 248001, UK, India
e-mail: dr.amitgupta.bt@geu.ac.in

R. Pant · N. Patrick
Department of Quality, Eureka Forbes Ltd., Lal Tappar, Dehradun 248140, UK, India

N. Kumar (ed.), *Mercury Toxicity Mitigation: Sustainable Nexus Approach*, Earth and Environmental Sciences Library, https://doi.org/10.1007/978-3-031-48817-7_9

1 Introduction

According to the World Health Organisation (WHO), mercury (Hg) is one of the "ten leading chemicals of concern" and is a dangerous heavy metal. Mercury is produced directly as well as disposed of in waste. Due to its poisonous nature, mercury pollution is a major problem worldwide. Hg may be transported & evenly varied in its elemental form [Hg (0)] after being released into the ecosystem. Point sources of Hg pollution, rather than deposition, are the main reason for Hg contamination in soil & water. The removal or immobilisation of mercury should be the foundation of remediation since, likely other heavy metals, it can't be broken down in surroundings [1]. At the exposure levels typically observed in the United States, neither atmospheric levels of mercury in water nor meteoric Hg releases directly endanger the well-being of people. As mercury is transferred to watersheds & builds up in the marine food chain, there is a risk to both people and wildlife. Depending on the speciation of Hg emissions and reaction routes, airborne Hg is carried across varying distances and is deposited on the Earth's surface. Three different types of mercury are released into the atmosphere by point sources: elemental mercury (Hg^0), gaseous ionic mercury, and particulate mercury. This speciation varies greatly between sources and has a substantial impact on how atmospheric Hg emissions turn out. Accordingly, liable on the kind of discharges and the corresponding atmospheric residence durations, Hg may be a local, regional, or worldwide contaminant.

Direct atmospheric deposition, soil water, wetland drainage, and groundwater drainage are the two main ways that mercury enters isolated surface waterways. Hg concentrations in streams and rivers can vary significantly over time, which is related to fluctuations in the concentrations of DOC or suspended debris. During heavy flow episodes, Hg concentrations can rise significantly. By causing Hg^0 to volatilize and by sediment deposition, some of the Hg inputs to ponds are shelved from the water column. Ionic Hg in freshwater lakes is mostly converted to Hg^0 by photochemical reactions. In lab experiments, the microbial decrease has only been seen at doses of mercury that are greater than ambient. Because of methylation in the waters column and anoxic sediments, biogeochemical activities in lakes also lead to the net formation of MeHg.

Human ingestion of seafood tainted with methylmercury is a major source of health worries in the US. The primary health risk linked to mercury exposure is neurotoxicity. Methylmercury has the potential to easily cross the normally guarded blood–brain barrier and penetrate the neural network. It is subsequently distributed to all tissues in the circulatory system. It is especially dangerous for pregnant women and The two major routes by which mercury reaches isolated surface streams are direct atmospheric deposition and subsurface discharge women who are nursing since it can easily pass via the placenta to growing fetuses and their developing brains. Children's learning problems and the disruption of fish-eating animals' reproductive processes have both been connected to low levels of exposure. Being exposed to mercury has also been linked to neurological, renal, digestive tract, hereditary, cardiovascular, and developmental disorders, as well as death.

Despite being a naturally existing element, mercury, humans have been directly introducing it into marine and land environments for thousands of years via mining, the use of Hg in the extraction of valuable metals, its existence as a trace pollutant in various resources, and its use in goods and by industry. The primary pathway for the transportation of Hg emissions is the atmosphere, while land and ocean processes are crucial for the dispersion of mercury in clean water, natural, & sea habitats as well as the production of CH_3Hg, which fuels the primary human exposure pathway-eating fish, especially marine fish [2].

Plants, cattle might be exposed to mercury-laden soil or polluted water that has been distributed. Over 250 symptoms have been linked to mercury poisoning, which can make a diagnosis more difficult. A history of the patient & a physical checkup that are compatible with mercury exposure serve as the basis for differential diagnosis. Typical laboratory tests include blood, urine, urinalysis, 24 h urine, urine challenge test with a "chelating" agent, hair, if necessary, tissue biopsy [3]. Both the air and water, which are affected by climatic and hydrological conditions, can be poisoned by mercury released from contaminated sites [4]. Particularly in regions where there is gold mining, mercury has the potential to be bad for the environment. By serving as a channel and origin for hydrological, the environment, and biota metabolisms, the soil contributes significantly to the mercury chain. Soil can modify the solubility, biological availability, the natural world, & toxic effects on the environment of various mercury types through chemical, physical, and biological processes. Metal mercury in the form of Hg^0 is the least hazardous mercury compound since it is not soluble in water, is not connected to animal tissues, and can't be swallowed by lower-level animals or microorganisms. As a result of the artisanal and small-scale gold mining activities' disposal of tailings, mercury-contaminated soil must be remedied. Washing the soil, often known as the "soil washing method," is one remediation technique that may be used to recover semi-volatile or volatile component mercury-contaminated soils. The process of recovering contaminated soil by soil washing has various benefits, including the fact that it may be applied to soil that has been polluted with semi-volatile or volatile substances, such as mercury, and that it can be utilised broadly and inexpensively [5]. Since the synthesis of chlorine and sodium hydroxide using mercury cathodes is multi-ton, a lot of much desalted water is required for the electrolysis. Large quantities of wastewater tainted with mercury and other contaminants are produced as a result of such manufacturing. First-hazard class substances include mercury. The environment is getting worse as a result of mercury contamination that is being caused by technology in the soil, water, and air. Mercury is present in practically every source of water, soil, or biota due to its use in the chemical industry, the creation of paper, artificial silk, or viscose, metal extraction in the mining sector, or mercury-based electrolysis. Several chemical plants used to discharge unclean wastewater into industrial collectors & bodies of water, whereas mercury waste was disposed of in specific regions of their sludge collectors [6].

2 Overview of Mercury Source and Its Environmental Impact

Around 30% of the mercury released into the surroundings each year is due to anthropogenic emissions, with the remaining 60% coming from ecological processes that cause mercury that has already been deposited in soil and water to be released again, a significant portion of which comes from earlier human emissions and releases, and the remaining 10% coming from natural sources. Mercury originates from a variety of natural sources, including volcanoes, soil, submarine vents, geological zones with high mercury concentrations, forest fires, freshwater lakes, rivers, and seas (Fig. 1). Human activity has, however, increased the quantity of mercury in the environment in a number of ways, including through a range of combustion and industrial operations namely metal mining and smelting, waste incineration, or coal-fired power generation. According to the Technical Background Report for the Global Mercury Assessment 2018 by AMAP and UN Environment, the trends in mercury emissions in 2015 were remarkably comparable to those in 2010. Asia accounted for forty nine percent of worldwide releases in 2015, followed by South America and sub-Saharan Africa. Approximately 70–85% of the emissions in the last 2 areas were related to ASGM. Unsettlingly, it was anticipated that human mercury emissions to the air worldwide increased by almost 17% in 2015 compared to the inventory for 2010. High apparent increases in ASGM emissions (almost 160t) are most likely primarily attributable to improvements in the data that these estimates are based on. There are no trustworthy data available because the majority of ASGM operations are illegal and unregistered. The industrial sectors (142t), where increasing economic activity in certain locations is represented in activity statistics for the production and consumption of raw materials, are the second significant factor contributing to the rise. Through air deposition and direct emissions, such as home and industrial effluent, hg penetrates the aquatic environment. In fact, the Global Mercury Assessment did not include an estimate of Mercury emissions to aquatic bodies until 2019. Nevertheless, the evaluation is predicated on a presumptive, fixed link between air and water emissions for certain industries [7].

Metalloids (Hg^0) and inorganic compounds make up the most of the mercury that reaches the ecosystem. Following emission, the destiny of Hg is determined by a number of variables, including the type of Hg released, the site of the emission source, the height at which Hg is released above the surrounding topography, and the weather [8].

When Mercury is in its elemental form, which has an atmospheric lifespan of about 6 months to a year and mostly travels in the atmosphere, it may cross seas and continents. Environments closer to sources may be affected by mercury emitted in more reactive forms. Mercury's unique ability to cycle between the surroundings, land, and ocean means that previous and present emissions may still have an impact on the ecosystem decades to centuries from now. Environments are mostly made up of three different types of sources: (1) natural sources; (2) legacy pollution from previous anthropogenic mercury emissions; and (3) ongoing anthropogenic emissions. The

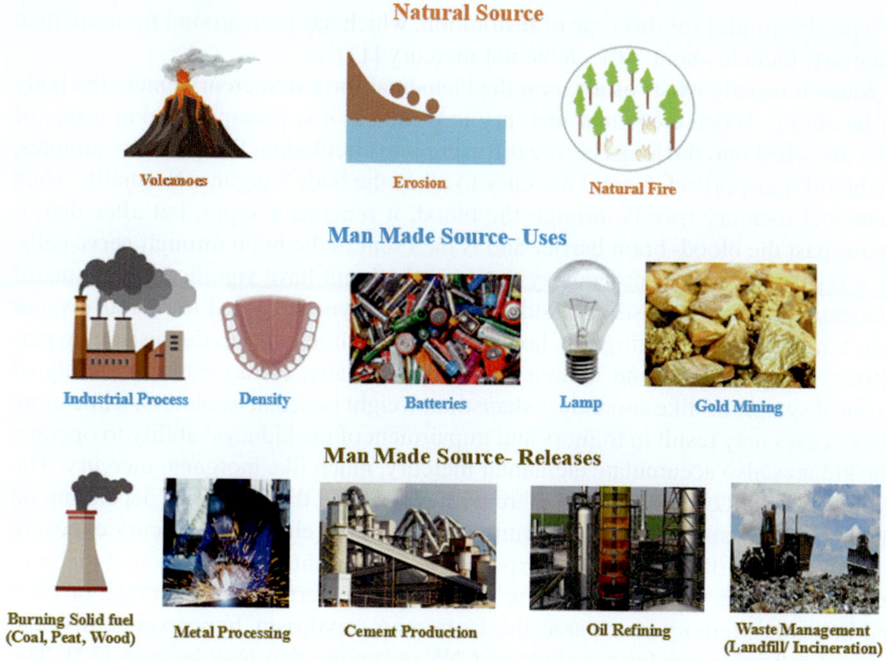

Fig. 1 Main global potential sources of mercury emissions

numerous ways that mercury is discharged into the environment and how it cycles there are all complicated, and they are only getting more so [9].

3 Health Effects of Mercury Exposure

In Minamata, Japan, between 1932 and 1968, a factory released waste liquid containing high concentrations of methylmercury into a bay that was teeming with fish and shellfish, the main food sources for locals and residents of nearby site. This incident had a significant impact on at least fifty thousand people and resulted in neurological symptoms in more than two thousand of them. Later in 1972, after consuming baked grain bread sprayed with a fungicide containing methylmercury, nearly 6,000 persons in Iraq had methylmercury poisoning [10].

On August 16, 2017, the Minamata Convention on Mercury came into effect, pledging to regulate the supply and trade of mercury. This multilateral pact demonstrated the dedication of the international community to preventing anthropogenic mercury sources from harming people and their surroundings. The possible health implications of exposure to mercury vapor that may be emitted from dental amalgam restorations continue to worry several organizations around the globe. The majority of typical formulas for this type of restoration, which has been around for more than a century, include about 50% elemental mercury [11].

Since it usually exists as a vapor, the elemental form of mercury enters the body by breathing. When elemental mercury is breathed, it is promptly and at a rate of 80% absorbed into the lungs before diffusing into circulation. Within a few minutes, the blood transports elemental mercury to all of the body's organs. Normally, when elemental mercury travels through the blood, it remains a vapor, but after that, it passes past the blood–brain barrier and is then sent to the brain through nerve cells. As a result, the central nervous system and the brain have significant amounts of elemental mercury deposition. Additionally, employees exposed to mercury vapor may experience long-lasting psychological abnormalities such as depression, hyperactivity, sleeplessness, and memory loss. Additionally, it may cause a variety of physical symptoms like anorexia, exhaustion, weight loss, and weakness, while more severe cases may result in tremors and impairment of the kidneys' ability to operate. The kidneys also accumulate elemental mercury, much like inorganic mercury. The effects on health once elemental mercury have entered the body vary depending on the type of exposure. The most common side effect of elemental mercury exposure is dermatitis. Furthermore, acute exposure to high quantities of elemental mercury leads to hypoxia, a deadly response that can cause serious lung damage or even end in death. Memory loss, paraesthesia, tremors, erythrism, hyperexcitability, and delayed reflection are further signs of CNS poisoning that may be seen [12]. The health effects of mercury exposure is illustrated in Fig. 2.

Organic mercurial chemical methylmercury, the most hazardous one, is typically found as a contaminant in rivers, lakes, and seas. Methylmercury typically forms naturally when aquatic anaerobic sulfate-reducing bacteria methylate mercury. It can also form due to human activity and is then released into rivers, lakes, and oceans. As a result, people whose diets primarily consist of shellfish and fish may be exposed to high levels of methylmercury. Organic mercurial chemical methylmercury, the most hazardous one, is typically found as a contaminant in rivers, lakes, and seas. Methylmercury typically forms naturally when aquatic anaerobic sulfate-reducing bacteria methylate mercury. It can also form due to human activity and is then released into rivers, lakes, and oceans. As a result, people whose diets primarily consist of shellfish and fish may be exposed to high levels of methylmercury [13].

Mercury species will be affected by natural biological and chemical processes once discharged into the environment, which will determine its toxicity and bioavailability. The most basic form of mercury, known as elemental mercury, is very toxic to both people and the environment and cannot be converted into safer compounds. Additionally, there are several species and states of mercury, and they will change during its biogeochemical cycle. Safe disposal methods, mercury removal, and mercury recovery are therefore crucial to prevent negative effects on people and the environment. The use of mercury-contaminated water also negatively impacts the brain, gastrointestinal, and renal systems because mercury reacts with thiol residues

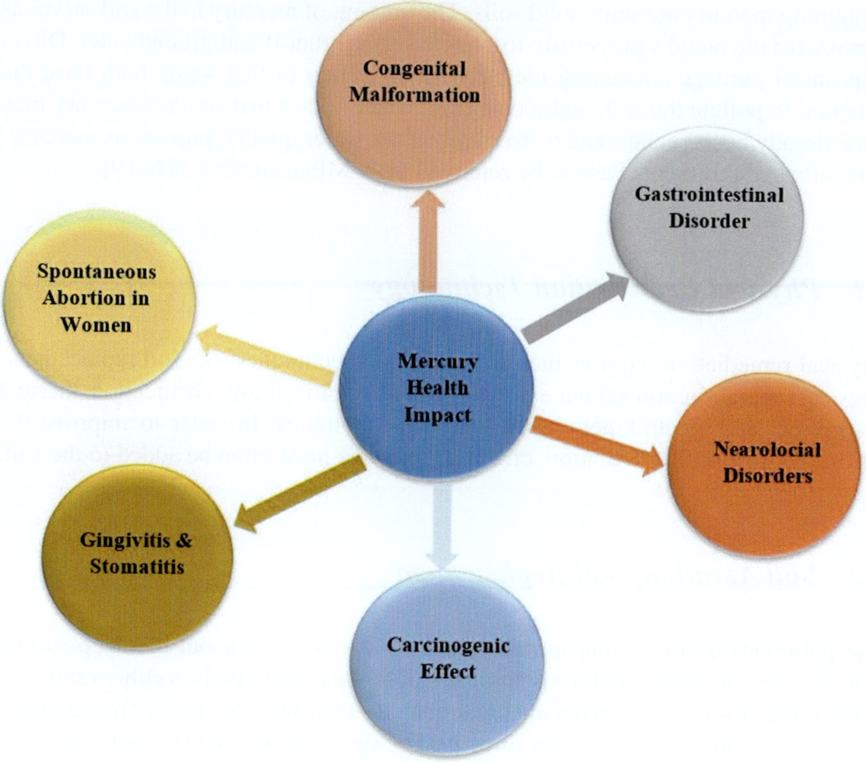

Fig. 2 Health effects of mercury exposure

found in human body proteins to impair mental and neurological functions11. Therefore, the WHO has set a range of 5 and 1 g/L for the maximum permissible concentration of Hg (II) ion in wastewater discharge and drinkable water, respectively [14].

4 Treatment Technologies for Mercury Removal from Soil

Having a porous matrix that allows for the mixing of biota, water, and air creates soil, a dynamic medium. The dynamic equilibrium between inorganic and organic components in the soil causes numerous environmental concerns, and changes in soil processes affect how ecosystems operate. Similar to aquatic sediments, soils function as mercury sinks and as sources of mercury in the atmosphere. Because of this, soils are crucial to the worldwide cycle of mercury. Mercury builds up over time in soils, ocean sediments, lake, and river sediments due to anthropogenic and natural activity [15]. Normal conditions for the production of inorganic and organic compounds containing mercury are semi-solid soils. The amount of mercury in the soil serves as a proxy for the metal's propensity to pollute surface runoff and groundwater. Direct disposal of garbage containing mercury or landfilling of that waste both have the potential to pollute the soil. Sediments from water bodies that contain mercury may have already been contaminated. To improve the water quality, poisonous mercury-contaminated sediments have to be removed from Minamata Bay [16–19].

4.1 Physical Remediation Technology

Physical remediation requires the use of physical techniques like soil replacement, physical separation, soil vapour extraction, fixed/stabilized soil, vitrification, thermal desorption, and electrokinetic remediation technologies. In order to improve the efficacy of physical remediation, chemical reagents must often be added to the soil.

4.2 Soil Aeration, Soil Replacement

The pollutants accumulating in the topsoil are spread throughout the deep soil by soil aeration. In order to reduce pollution, soil replacement involves either removing contaminated soil and covering uncontaminated soil with it, or mixing contaminated and uncontaminated soil together directly. The two methods can be easily used, but they require a lot of labor, greatly upset the soil's structure, and do not fundamentally eliminate mercury from the soil, making them unsuitable for situations where there is severe contamination or a shortage of available land resources [20].

4.3 Vitrification

By melting contaminated soil at temperatures between 1600 and 2000 degrees Celsius using intense heat sources like plasma and electric current, pollutants are pyrolyzed or vaporized, and the cooled melting will result in corpora vitreum. Contaminants are wrapped, which slows their movement. In the US state of Michigan, when mercury-contaminated soil is treated using vitrification technology, the accessible mercury content decreases to less than 0.23 g/L following treatment. The original mercury concentration was around 40 g/kg. Mobile devices can be used to run this equipment in the field, but it is expensive [21].

4.4 Thermal Desorption

By heating the contaminated soil either directly or indirectly, thermal desorption removes the contaminants depending on their low boiling points in the soil (Fig. 3). The thermal desorption technique, which may be divided into in situ and ex-situ, consists of 2 stages: heating polluted soil to volatilize contaminants and tail gas treatment. In order to remove the gas from the soil, thermal desorption is often used in situ in conjunction with soil vapor extraction. Pollutants are volatilized during the thermal desorption step and delivered to the flue gas treatment system through the carrier gas. The volatility of contaminants is the fundamental basis for thermal treatment remediation. In this case, the treated toxins frequently need simple volatilization [22].

Hg is a volatile substance as evidenced by the melting and boiling temperatures of 38.8 °C, 356.7 °C, and 0.18 Pa, correspondingly. Temperature causes an increase in the rate of vaporization. The features of the soil, the amount of organic matter, the amount of water present, the concentration and form of mercury, and the operating situation all have an impact on how efficient and successful thermal desorption is at removing contaminants from soil. When the types of Hg in contaminated soil alter, the desorption temperature varies. According to experimental findings from a thermal desorption investigation of Hg speciation, distinct forms of Hg in soil volatilize in a certain sequence as the desorption temperature rises. Theoretically, 523 K is adequate to remove mercury without removing Hg oxide, but 873 K is necessary to remove mercury and its compounds from the soil via thermal desorption. A common thermal desorption device for removing mercury has a temperature range of 320 to 700_C.2 Thermal desorption has been shown in several trials to be a reliable method for removing Hg from the soil. Using thermal desorption, it was calculated that the total Hg content in the soil was reduced by 60–70% at 440 K in a small-scale experiment and by approximately 32% at 373 K in the lab. Chang and Yen63 removed Hg

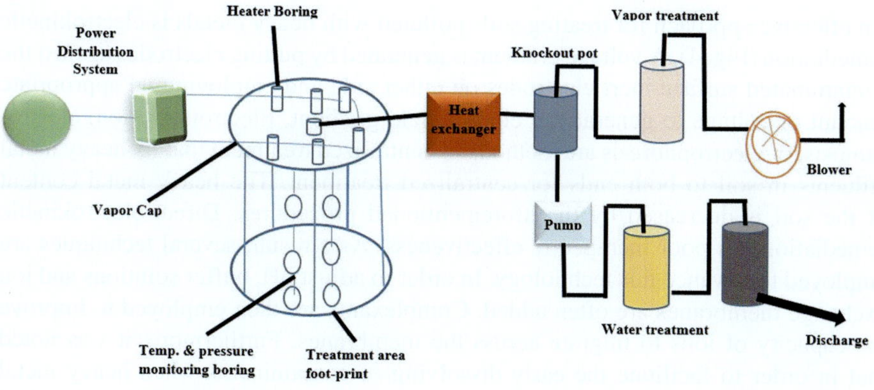

Fig. 3 Thermal desorption system

contamination from the soil in southern Taipei, China, using in situ thermal desorption. The findings of the experiment demonstrate that temperature and time are the two crucial variables determining Hg elimination in soil through thermal desorption [23]. They discovered that when the temperature rises, the equilibrium concentration of mercury in soil falls. Additionally, when the temp is greater than 700 °C and the duration is longer than 2 h, thermal desorption may efficiently eliminate Hg contamination. Despite research findings illustrating that high temperatures assist in Hg volatilization from the soil, high temperatures can also destroy organic molecules. After thermal desorption caused by a high temp on agricultural land, the soil will no longer serve that purpose. Maintaining adequate soil quality and Hg removal in the soil for farming uses is crucial in the future. According to some scientists, heating the soil to 280 °C can both remediate soil pollution to lower bioavailable Hg levels and prevent the irreversible loss of soil quality levels. Numerous studies have been done to demonstrate that the heating temperature affects the influence of heating duration on removal efficiency. In order to guarantee that the contaminants are fully eliminated, a low temperature necessitates a lengthy heating period [24].

Some specialists have paired salts/acids with thermal desorption technology to increase the removal rate under low temperatures, making thermal desorption technology more successful and practicable for remediating Hg pollution in soil without damaging soil quality. emphasized that four typical Hg molecules, HgS, Hg $(NO_3)_2H_2O$, $HgSO_4$, and HgO, may all be desirable with the help of MgCl2. Additional experimental findings demonstrate that $MgCl_2$ may greatly accelerate the transformation of Hg compounds in the soil and lower the needed temperature. These findings suggest that $MgCl_2$ can facilitate the removal of Hg in contaminated soil treated with $MgCl_2$ [20, 25–27].

4.5 Electrokinetic Remediation

An effective approach for treating soils polluted with heavy metals is electrokinetic remediation (Fig. 4). A voltage gradient is generated by putting electrodes against the contaminated surface inert electrodes on either side and employing an appropriate amount of voltage to generate an electric field gradient. Electromigration, electroosmosis, or electrophoresis are methods of centralized treatment that fix heavy metal effluents in soil to both ends for centralized treatment. The heavy metal content of the soil is decreased by the aforementioned procedures. Direct electrokinetic remediation has poor therapeutic effectiveness. As a result, several techniques are employed to advance this technology. In order to adjust pH, buffer solutions and ion exchange membranes are often added. Complexants are then employed to improve the capacity of ions to migrate across the membranes. Furthermore, it was noted that in order to facilitate the early dissolving of contaminants when heavy metal pollutants occur in the form of metal or nonconductive minerals, the proper chemical reagents must be used. As a result, the electrokinetic remediation method typically involves the introduction of certain chemical Reagents. Iodine solutions are employed

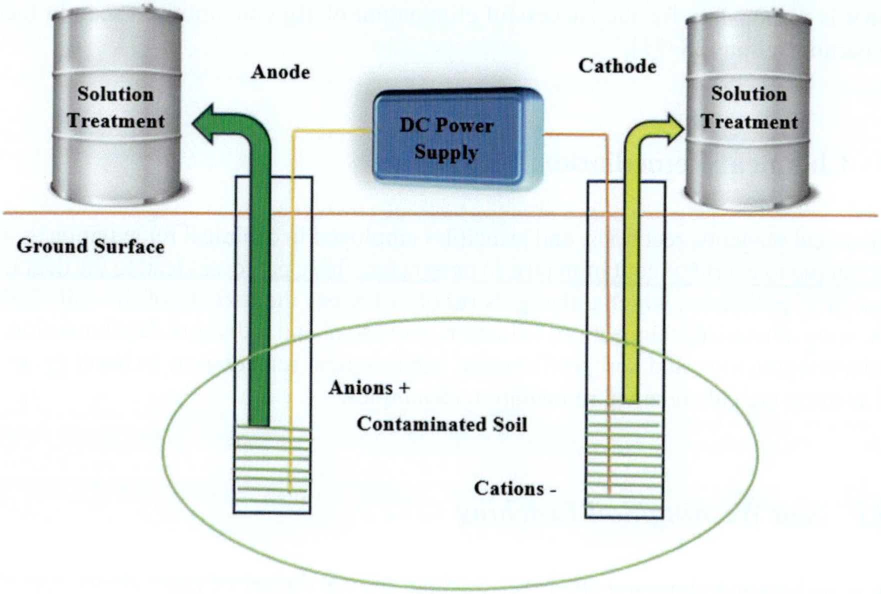

Fig. 4 Electrokinetic remediation mercury in soil flow diagram

because several studies have demonstrated that they can significantly increase the effectiveness of electrokinetic remediation. In order to clean up Hg-contaminated soil in a mine, Marrugo Negrete and Lopez Barboza46 utilized the KI solution as a complexing agent. The findings demonstrate that by raising the KI solution's content and the voltage among the 2 electrodes, Hg may be removed from the soil more quickly and effectively.

To extract Hg from kaolin and moraine, Reddy utilized various voltage gradients and KI solution concentrations. With a clearance rate of 97 and 77%, respectively, the Hg content in kaolin & moraine dropped from 500 mg kg_1 to 16 mg kg_1 and 116 mg kg_1.70 Based on prior electro-remediation improvements made using acid and KI solutions, the viability of electrokinetic remediation employing Hg morphology was examined. They noted that the soluble weak acid portion of the soil's residual Hg increased KI solution-based electrokinetic remediation based on morphological analysis studies. To lessen the likelihood of an increased danger associated with pollution locations, the same soil should be treated using acidic electro-remediation technology. Numerous investigations have demonstrated that electrokinetic remediation may successfully remove Hg in the presence of EDTA in addition to iodide ions. Utilized a titanium electrode and a 5 V voltage for 6 h to use EDTA to extract Hg from polluted soil. According to the experimental findings, 75% of the metals in the soil may be eliminated. Further investigation reveals that the electro-migration of the coordination complex made up of Hg^{2+} and a terminal hydroxyl in EDTA is what is responsible for the successful elimination of Hg contaminants seen in the experimentation [28–31].

5 Chemical Remediation Technology

Chemical reagents, reactions, and principles employed in chemical remediation as a technique to get rid of contaminants; in most cases, these processes lead to the degradation of pollutants, which either gets rid of or lessens the toxicity of the soil. Soil cleaning, chemical stabilisation, oxidation, reduction, and reduction dechlorination, solvent extraction, and soil performance enhancement remediation technology are now the principal chemical remediation techniques.

5.1 Soil Washing/Soil Leaching

A remediation technique called soil washing uses a variety of methods to extract toxins from the soil. Pollutants in the soil may be encouraged to dissolve or migrate by the reagents used in soil washing. In situ and ex-situ leaching are 2 operational components of the remediation approach. The in situ leaching technology, which includes the detergent dosing system, the leaching solution collecting system beneath the soil, and the leaching solution treatment system, infuse the leaching solution into the contaminated soil and subsequently recovers the heavy metals in the filtrate [32]. Before cleaning waste liquid treatment and other procedures in the event of ex-situ leaching, the polluted soil needs to be dug out and placed in the appropriate treatment apparatus. The excavated soil is first screened for soil particles and fragments, and then big particles are removed. In general, heterotopic leaching is paired with physical separation technique. Water-soluble mercury is often removed by acidic, alkaline, or chelating chemicals during the remediation of Hg contamination in the soil by soil washing. While chelating agents can eliminate mercury by complexation, the application of acidic/alkali agents often relies on the dissolution of mercury compounds or the dissolution of mercury-containing machineries adsorbed by soil components. The eluent is crucial to the remediation of chemical elution. Numerous academics have discovered in recent years that the removal efficiency of various chemical eluents varies. examined a number of common chemical effluents. Their findings demonstrate that iodide, EDTA, and thiosulfate ($Na_2S_2O_3$) may remove mercury from the contaminated soil in an efficient manner. However, the low concentration of EDTA solution can only achieve a low rate of elimination because the leaching impact of EDTA is influenced by competition with other heavy metal components in soil. Additionally, it was discovered that KI, EDTA, and thiosulfate ($Na_2S_2O_3$), particularly $Na_2S_2O_3$ solution, had a stronger removal impact on mercury in the soil than citric acid, tartaric acid, and sodium dodecyl sulfate [33–38].

5.2 Chemical Stabilization

Chemical stabilisation comprises the use of chemical reagents/chemical materials to immobilise heavy metals in soil via complexation, precipitating, and adsorption steps, which limit heavy metal mobility and bioavailability. Chemical stabilization can greatly lower the mobility/solubility and concentration of pollutants in soil pore water while not being able to extract or remove them from the soil. This reduces the pace at which heavy metals are transferred to plants, microbes, and water. Previous studies have demonstrated that the type of heavy metal, rather than its overall quantity in surroundings, determines its bioavailability, mobility, toxicity, and possible danger. As a result, various heavy metal types have various environmental consequences. Hg's mobility and bioavailability are also influenced by how it is distributed in the soil. Some supplies can be utilized to immobilize mercury & thereby lower its bioavailability in soil. Under reduction circumstances, HgS is a very stable sulfide compound that is very insoluble and simple to produce. pointed out that Hg is able to persist in a stable and harmless state and exhibits no tendency to volatilize due to the poor solubility & inactive redox reaction properties of HgS.

Sulphide is now a common chemical fixative employed by many professionals to clean up Hg contamination in soil. According to several research, sulfide minerals have a high affinity for mercury & so have a significant thermodynamic potential for fixing mercury. It is regarded as one of the most efficient Hg (II) adsorption materials. Among them, iron sulfides including pyrite, magnetite, and tetragonal pyrite are the most prevalent sulfide minerals. Although precipitation is the primary ingredient and accounts for 77% of the entire cleanup, this reaction mechanism also involves adsorption. The effectiveness of FeS nanoparticles for stabilizing mercury was investigated through a series of adsorption and dynamic column tests. FeS nanomaterials were successfully synthesized with carboxymethylcellulose acting as a stabilizer. Iron sulfide nanoparticles may efficiently fix mercury in sediment and dramatically lower the concentration of water-soluble mercury, according to adsorption studies. Additionally, a morphological examination of mercury reveals that FeS nanoparticles can significantly lower the quantity of bioavailable mercury. Since that time, several professionals have investigated the viability of carboxymethylcellulose (CMC)-FeS nanoparticles for use in Hg removal and have discovered that these materials exhibit exceptional stability and a powerful affinity for Hg. Based on a previous study, FeS is an effective fixative for Hg pollution in soil, and the FeS nanomaterial treated with CMC as a stabiliser has a better adsorption impact. Many studies also utilize sodium sulfide as a fixative in addition to sulfide ores like pyrite [39–41].

5.3 Solidification/Stabilization

The principle behind the solidification/stabilization technique is as follows: the reagent is added to the soil, and the pollutant reacts with the reagent through a

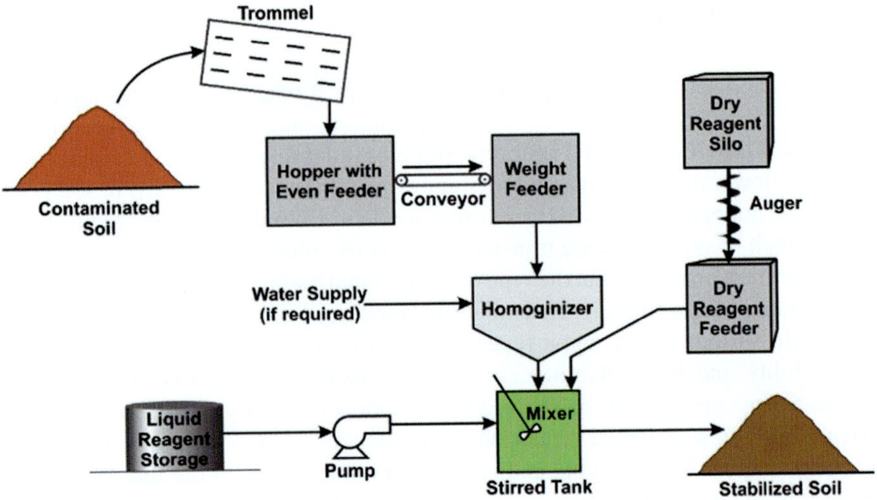

Fig. 5 Solidification and stabilization process

physical–chemical process, converting it into fixed mercury, hence reducing the mobility of mercury in the soil (Fig. 5). The most often employed physical curing agents are cement, ash, and several other compounds. The most widely utilized and cost-effective material, cement, transforms soil-borne mercury into mercuric oxide precipitation and fixes it in concrete blocks. S-based compounds, zeolite, lime, calcium carbonate, phosphate, silicate, etc. are often used as stabilization materials. Calcium carbonate, often known as lime, was primarily used to raise the pH level of soil, causing mercury and other metals to precipitate as combination state salts of hydroxide or carbonate. Through ion exchange adsorption and selective adsorption, zeolite can reduce the potency of the heavy metals in soil. Chao disposed of mercury-contaminated soil from a chemical plant using Na_2S and cement as curing agents [42].

6 Biological Remediation

A technique known as biological remediation or bioremediation uses biologicals to decrease, eliminate, or immobilize hazardous chemicals from the soil and purify it. Plant remediation and microbial remediation are the two primary subcategories of bioremediation, with plant remediation receiving significantly greater attention in its development [43].

6.1 Phytoremediation

When contaminants are present in soil, sediment, subterranean, or surface water, they can be removed, degraded, or immobilized using plants and the rhizospheric microbes that they are linked with. Heavy metals, in particular, are removed via phytoremediation, which is widely accepted as a green method with a favourable public impression. Different phytoremediation systems make advantage of various plant features (Fig. 6). When phytoremediation technology is employed to treat Hg-contaminated soil, mercury is transferred from the roots to the aboveground area of the plant via a relevant plant organisational layout. Hg will interact with each component of the plant in some way during the whole absorption and transportation process, and certain particular plants can repair or eliminate Hg in the soil through these processes for soil remediation. Through biochemical processes that take place in roots or close to roots. Phytostabilization primarily exploits the roots of plants to reduce contaminant mobility and bioavailability in the soil [44].

The process of using plants to absorb contaminants from the soil, transform them into volatile forms, and release them into the surroundings is known as phyto-volatilization. This procedure entails the use of plants to volatilize contaminants or metabolites. This approach can be used to clean up Hg-contaminated soil due to the volatility of Hg. The plants take up and change the mercury in the soil, which is subsequently volatilized into gaseous mercury and released into the atmosphere. Genetic engineering is frequently used in conjunction with this technique. MerA and MerB are two genes that are connected to Hg transformation and accumulation. MeHg can become elemental Hg and then evaporate through the leaves of plants when these two genes are activated in them. However, Hg is moved from the soil to the environment through plant volatilization. The problem of possible mercury contamination in the atmosphere has not been resolved.

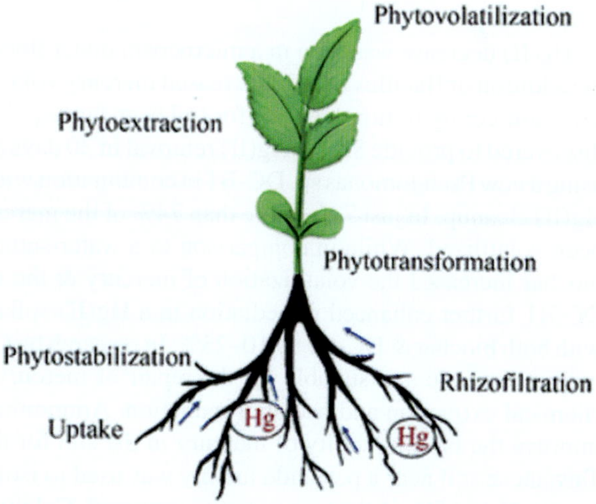

Fig. 6 Mechanism of phytoremediation

Instead, transgenic plants like arabidopsis were developed, which had a 50-fold increase in organic mercury tolerance. He developed tobacco capable of growing properly in a range of 50–350 μm in $HgCl_2$ media by incorporating the MerA genes of bacteria into it. According to the study, Canada Populus had a maximum mercury absorption accumulation of roughly 7 mg/kg, whereas silver birch had a maximum mercury enrichment capacity of 10 mg/kg [17]. Mercury trafficking via the root tissues has been tracked in field plants of Marrubium vulgare collected from a significantly Hg-polluted area in Spain. The apical parts of primary and secondary roots were the first areas to experience a depletion of accessible thiol pools, followed by the epidermis and outer layers of cortical cells, and finally the xylem vessels [18].

6.2 Microbial Remediation

By increasing the amount of stable-state metals in the soil and using bacterial movement to adsorb/convert heavy metal affinity, the danger of organism absorption is decreased. The rate of mercury evaporation in the soil was related to the number of microorganisms, according to recent research, and a Hg-resistant microbe with mer operon may successfully regulate mercury contamination in soil and water.

Hg(II) might be precipitated as insoluble cinnabar (HgS) using anaerobic microorganisms, such as sulfate-reducing bacteria with their ability to generate sulfide. Sadly, because they methylate mercury in anoxic conditions, these bacteria play multiple and opposing functions. For instance, methylmercury might be produced by the Geobacter species, which also volatilizes mercury. Anaerobe use in mercury bioremediation is a dangerous strategy since it has recently been demonstrated that natural organic matter in anoxic conditions may both oxidize and decrease elemental mercury. Therefore, any proposed bioremediation application has to give careful consideration to the related risk of mercury conversions in anoxic circumstances [19, 45, 46].

Hg(II) decrease was seen in a microcosm under flooded circumstances, however, the addition of Bacillus DC-B2 increased mercury volatilization by B17% in 30 days. The same configuration was used for soil from farms polluted with mercury, and it was discovered to provide 82.1% Hg(II) removal in 30 days [47]. Another study proposed using a new Pseudomonas sp. DC-B1 in combination with sawdust biochar to enhance Hg(II) cleanup. In just 72 h, more than 74% of the mercury in the growing media had been volatilized. While in comparison to a water-saturated control, the addition of biochar increased the volatilization of mercury & the addition of Pseudomonas sp. DC-B1 further enhanced remediation in a Hg(II)-spiked small scale supplemented with both biochar & DC-B1 by 10–23%. In research by Chen et al., mercury was made more accessible and suitable for the repair of mercury-contaminated soil by using chemical extraction and bacterial reduction. Ammonium thiosulfate was utilized to improve the bioavailability of mercury in the soil for this two-stage procedure, and Taiwanese soil near a pesticide facility was used to isolate Enterobacter cloacae B7. In just 24 h, 65% of the mercury was removed. Calcium and magnesium ions were added to the process to trigger the mer operon, which improved it. Thiosulfate's inclusion enhanced mercury extraction to 77%, while the presence of calcium and magnesium ions resulted in 81% Hg21 remediation. The microcosm was used to test the two-stage method [48, 49].

7 Treatment Technologies for Mercury Removal from Water

Mercury may be removed from water using 4 ways: coagulation/filtration, granular activated carbon, lime softening, and reverse osmosis.

In coagulation/filtration, $AlSO_4$ is widely used to react with mercury and form a solid that may precipitate out of the water. The sludge must then be disposed of at toxic waste disposal. This approach is useful since it is reliable and low-cost.

Granular activated carbon is utilised in a carbon media with holes. This material is composed of extremely thick charcoal. When water runs through, the dissolved pollutants are absorbed and retained on the solid surface. This approach has limits since its efficiency is dependent on the amount of Hg in the water. In reverse osmosis, water is pushed across a semipermeable membrane. A polyamide film is a common form of membrane material. Although high-quality water is produced, it is somewhat expensive [49–52].

8 Mercury Removal Through Adsorption

Adsorption techniques might result in low mercury effluent levels and/or high removal efficiency for mercury. Although activated carbon is the most common adsorbent used, other adsorbents are also mentioned in the literature. These include refined plant- or mineral-based substances like coal fly ash, bicarbonate-treated peanut husk carbon, modified Hardwickia bin and a bark, and the Forager sponge. As adsorbents, metal hydroxides are also used. Coagulation or co-precipitation are popular terms used to describe the process when metal hydroxides are used for adsorptive treatment. Adsorbtive therapy has the intrinsic benefit of increasing treatment efficiency with incremental adsorbent dose, especially when the adsorbent exhibits isothermal or quasi-isothermal behavior. For a constant starting pollutant concentration, isothermal behavior is seen as diminishing residual soluble concentrations are seen when the dose of adsorbing treatment material is applied. These increasing doses also lead to an increase in wastewater treatment residuals, which must ultimately be disposed of, unless adsorbent recovery is practical. Adsorption effectiveness can also be influenced by factors other than adsorbent type and dose [53].

Typical elements include wastewater For the purpose of extracting Hg^{2+} from polluted waterways, many adsorbents have been created and evaluated.

ordinary adsorbents, such activated carbon [54]. Because of their very large surface area, carbon-based nanoporous adsorbents, particularly activated carbons, carbon nanotubes, and graphene, are widely employed in applications for heavy metal removal [55]. In order to remove Hg from polluted water, several adsorbents containing modified AC have been utilized. These adsorbents come in a variety of modifications, such as bromine-modified AC, thiol-incorporated AC, polyethyleneimine-modified AC, sulfur-impregnated AC, phosphonium-based modified AC, and others. The creation of a highly selective, economical, and environmentally friendly adsorbent is the main objective [56]. At ambient temperatures, such as 238 C, it has been proposed that chemisorption and physisorption work together to adsorb Hg, but chemisorption predominates at higher temperatures, such as 1400 C. The properties of carbon, flue gas composition, and the presence of active components are only a few of the many variables that have been discovered to affect mercury removal effectiveness [1]. The production of chelates is the main sorption process created by thioether-functionalized covalent triazine nanospheres to adsorb Hg^{2+} and Hg(0) from water. Adsorbents typically have high surface areas and high porosities. In a different work, Abbas et al. created a new mesoporous conjugate adsorbent based on pentasil zeolite (type ZSM-5), with a maximum adsorption capacity of 172.6 mg/g. Bao et al. extracted Hg(II) from wastewater using magnetic nanoparticles coated with silica, and they also observed that mercury ions bind to imine (C–NH–) groups on the surface of the nanoparticles. Derivatives of chitosan were utilised as adsorbents in another instance. The primary mechanism of the method of adsorption, which is also frequently utilised for the removal of gas-phase elemental mercury Hg(0), is the chelation between Hg(II) and the nitrogen atoms of chitosan [47].

ZnC_{12} activated carbon outperformed the other 28 in terms of adsorption capacity, and sonication could recover 60–80% of the Hg(II) that it had absorbed. The complete characteristics of the activated carbon were significantly improved by the chemical activation of $ZnCl_2$. In Fig. 2, the impact of adsorbent dose on the effectiveness of Hg(II) adsorption was investigated. Evidently, when the dose of the adsorbent was raised, the percentage of Hg (II) elimination rose as well. It was observed that increasing the dose of the activated carbons provided Hg (II) with more active sites to anchor on [53].

9 Mercury Removal Through Membrane Separation

When mercury is present in wastewater in particulate or colloidal form, membrane separation methods are highly successful. The ability to achieve low-ng Hg/L mercury levels in the treated effluent using membrane technology is therefore greatly enhanced. In intricate membranes with a high Hg^{2+} elimination ratio (99.4%), Bessbousse et al. immobilized the polymerized matrix of polyvinyl alcohol crosslinked by gas dibromoethane [53].

We suggest a three-step treatment approach consisting of prefiltration, ultrafiltration, and membrane adsorption based on the industrial wastewater's quality. To filter out big particles that can possibly foul or harm membranes in the next processes, the first step uses a PVDF 700 microfiltration (MF) membrane. Ultrafiltration (UF) is used in the second phase to eliminate HgS NPs by size exclusion. The final step's elimination of Hg^{2+} and membrane flow will suffer from any HgS NP carryover. A thiol membrane is used in the third stage to absorb the dispersed Hg^{2+} species [43].

In recent years, the possibility of removing mercury ions from wastewater via membrane filtration has also been investigated. Conventional wastewater treatment methods might not work well since mercury can be found in numerous wastewater in its particle or colloidal form. Microfiltration, ultrafiltration, nanofiltration, and RO were all shown to be effective at removing mercury to the necessary discharge limit by Meltem et al. According to the experimental findings, MF and UF membranes are able to successfully remove mercury at working pressures of 2.8 bar and beyond. Both RO and NF could entirely remove mercury while operating at a pressure of 20.7 barsTraditional membrane separation techniques like RO and NF frequently need a lot of energy and money. Although UF may function at lower pressures, the removal of tiny ionic chemicals is ineffective because of the greater membrane pore size. Adding water-soluble polymers to the feed solution leads the small-sized pollutants to collaborate with macromolecules and form complexes that contain the contaminants while allowing water to pass through, this solves the issue [24].

Given the fact that few studies have been undertaken in this field, UF membranes have been reported to be capable of selectively removing Hg(II) from a complex combination of metals by adding a water-soluble polymeric ligand. Jian et al. investigated the ability of the complexation-ultrafiltration approach to extract Hg(II) from a binary metal solution of Hg(II) and Cd(II). As the complexing agent, sodium salts of poly (acrylic acid) were used to bind mercury ions and form polymer metal complexes. The rejection coefficient of mercury ions was very close to one, but that of cadmium was just 0.1, making the elimination process fairly effective. Huang et al. assessed the technology for enhancing the UF process for the removal of Hg(II) utilising a polymer ligand in more recent study. Polyvinyl amine, as a mercury-binding polymer, was able to remove 99% of Hg(II) from wastewater. A more recent experiment was conducted by the same research team [53].

A high mercury rejection was attained utilising a polymer enhanced UF that used 3 different kinds of water-soluble polymers that all showed significant interactions with mercury (II). In a different investigation, batch and continuous contact UF filtration systems backed by FeS were used to remove Hg(II) from water. More than 99% of the mercury was removed using the batch technique.

It was also researched how pH affected the elimination of mercury. The complexing agent utilised was polyethylenimine, and the results showed that at neutral circumstances, mercury was retained almost entirely. Forward osmosis' capacity to eliminate trace quantities of mercury from wastewater was evaluated by Chia-Yu et al. Experimental research on the removal efficiency of mercury using two distinct valence inorganic salts, NaCl and $MgCl_2$, is done at various solution concentrations. The 1 M draw solution of NaCl and $MgCl_2$ yielded strong mercury rejection with percentage elimination of 98.2 and 99.9%, resp. Yet, the substantial mercury adsorption on the membrane surface constituted a drawback of this study.

It has been claimed that heavy metal ions may be successfully removed using graphene membranes. Azamat et al. looked at using functionalized graphene as a nanostructured membrane to remove copper & mercury ions from solutions of water. The gadget was outfitted with an external electrical field, which caused the required ions to flow through the functionalized graphene membrane. The ion exchange membrane bioreactor (IEMB) technology was developed to remove Hg(II) from industrial waste water.

Mercury was removed from waste streams using a cation exchange membrane in conjunction with bioreduction to Hg^0 while producing the fewest polluted byproducts possible. Not only was 98% of the mercury eliminated, but the total environmental effect was also diminished. Researchers have been experimenting with various methods for using nanoparticles and nanocomposites on film surfaces to remove heavy metal ions. The functional groups OH, $-NH_2$, –SH, and $-SO_3H$ have all been the subject of experiments.

For the membrane to function at its best, the pH of the solution needs to be changed to 5.5. Solid-state membranes have also been investigated for the removal of mercury ions in addition to polymer-based liquid membranes. For the effective detection of mercury in aqueous media, a solid-state membrane matrix was created utilising high molecular weight polyvinyl chloride (PVC) and functionalized with four ligands, including 4-(2-pyridylazo) resorcinol, thiourea, calconcarboxylic acid, and dithizone. To prove its effectiveness for the elimination of mercury ions, however, more research must be done.

The majority of research do their tests at low beginning mercury concentrations rather than at high initial mercury concentrations, as has been demonstrated. Functional groups inserted onto membranes serve a critical role in improving mercury removal, much like adsorption studies [57].

10 Conclusion

Heavy metal pollution in surroundings, including soil, water, plants, and air, is of serious concern owing to the possible influence on human and animal health. To conserve vital natural resources and biological beings, more affordable and effective solutions are required. In the recent decade, significant efforts have been undertaken to discover plant species and their methods of heavy metal absorption and hyperaccumulation. Despite these considerable drawbacks, traditional remediation procedures have been frequently used for mercury removal. For starters, the high cost of certain technologies, like thermal desorption and activated carbon adsorption, is an impediment to large-scale use. Long-term mercury stability monitoring should be carried out when it comes to mercury stabilization & containment in soil employing technology appropriate for use on big sites. Reusing developing materials is another technique to maximize their cost-effectiveness. Phytoremediation and algae-based Hg removal, which involve the transformation of mercury by microbes, may only be used when Hg concentrations do not have hazardous effects on species.

There are several ecological problems involved with using plants to extract hazardous metals from disturbed soils. A key worry is how phytoremediation may impact the food chain via herbivores and omnivores that consume metal-laden leaves. Other concerns include the effects of site preparation activities on other nearby crops and vegetation, the introduction of potentially invasive non-native plant species, or the use of artificial chemical chelates for phytoextraction of Pb to boost bioavailability as well as plant uptake of the metal. In the end, the use of diverse treatment methods for mercury removal from soil, waste, and water is critical for environmental protection and human health. To address the many origins and types of mercury contamination, several treatment systems have been developed and used. The treatment technique used is determined by the individual contamination scenario, the intended efficiency, cost-effectiveness, and regulatory constraints. Continued research and development in this subject are required to optimize existing methods and investigate novel ways to enhance the effectiveness and sustainability of mercury treatment.

References

1. Wanga L, Houa D, Caoa Y, Okb YS, Tackc FMG, Rinklebed J, O'Connor D (2020) Remediation of mercury contaminated soil, water, and air: a review of emerging materials and innovative technologies. Environ Int 134:1–19
2. Driscoll CT, Mason RP, Chan HM, Jacob DJ, Pirrone N (2013) Mercury as a global pollutant: sources, pathway and effects Charles. Environ Sci Technol 47:4967–4983
3. Rice KM, Ernest M, Walker EM Jr, Wu M, Gillette C, Blough ER (2014) Environmental mercury and it's toxic effects. J Prev Med Public Health 2014(47):74–83
4. Eckley CS, Gilmour CC, Janssen S, Luxton TP, Randall PM, Whalin L, Austin C (2020) The assessment and remediation of mercury contaminated sites: a review of current approaches. Sci Total Environ 707(136031):1–45
5. Effendi AJ, Lestari V, Irsyad M (2019) Optimizing soil washing remediation of mercury contaminated soil using various washing solutions and solid/liquid ratios. ETMC and RC EnvE 1–6
6. Fuchs SL, Filatov VY, Devyaterikova SV (2022) Wastewater treatment for mercury and arresting mercury contaminants movement within industrial soil. Earth Environ Sci 988:1–6
7. Pistocchi A, Cinnirella S, Mouratidis P, Rosenstock N, Whalley C, Sponar M, Pirrone N (2022) Screening of mercury pollution sources to European inland waters using high resolution earth surface data. Front Environ Sci 1–11
8. Davies TC (2023) An updated review of the salient geomedical aspects of mercury for enhancement of data quality in simulation modelling and other prognostic applications: Africa case descriptions. Front Anal Sci 1–16
9. Celia Y, David C (2023) Evers global mercury impact synthesis: processes in the southern hemisphere. Springer 52:827–832
10. Santos CDL, Pastor JC, Calonge M (2023) Mercury intoxication and ophthalmic involvement: an update review. Front Toxicol 1–5
11. Basu N, Bastiansz A, Do´rea JG, Fujimura M, Horvat M, Shroff E, Weihe P, Zastenskaya I (2023) Our evolved understanding of the human health risks of mercury. Springer 52:877–896
12. Suvarna RP, Otitolaiye VO, Mahfud R, Rawahi MA (2022) Impacts of mercury exposure on human health, safety and environment: literature review and bibliometric analysis (1995 to 2021). IJOSH 12(4):336–352
13. Genchi G, Sinicropi MS, Carocci A, Lauria G, Catalano A (2017) Mercury Exposure and Heart Diseases. Int J Environ Res Public Health 14(74):1–13

14. Al-Ghouti MA, Da'ana D, Abu-Dieyeh M, Khraisheh M (2019) Adsorptive removal of mercury from water by adsorbents derived from date pits. Sci Rep (15327):1–15
15. Ignatavičius G, Unsal MH, Busher PE, Wołkowicz S, Satkūnas J, Šulijienė G, Valskys V (2021) Geochemistry of mercury in soils and water sediments. AIMS Environ Sci 9(3):277–297
16. Ahirwar MK, Kirar NK, Gupta GS (2023) Mercury contamination from pulp cum paper mill untreated effluent and solid waste in India: a review. Environ Chem 7(1):1–7
17. Yang Q, Wang Y, Zhong H (2021) Remediation of mercury-contaminated soils and sediments using biochar: a critical review. Springer 3:23–35
18. Gworek B, Dmuchowski W, Baczewska-Dąbrowska AH (2020) Mercury in the terrestrial environment: a review. Environ Sci Eur 32(128):1–19
19. Shaoyong LU (2014) Restoration technology research progress of mercury polluted farmland. MEIC 1586–1592
20. Albatrni H, Qiblawey H, El-Naas MH (2021) Comparative study between adsorption and membrane technologies for the removal of mercury. Sep Purif Technol 257(117833):1–16
21. Muddarisna N, Siahaan BC (2014) Application of organic matter to enhance phytoremediation of mercury contaminated soils using local plant species: a case study on small-scale gold mining locations in Banyuwangi of East Java. J Degrad Andmining Landsmanagement 2(1):251–258
22. Guney M, Akimzhanova Z, Kumisbek A, Beisova K, Kismelyeva S, Satayeva A, Inglezakis V, Karaca F (2020) Mercury (Hg) contaminated sites in kazakhstan: review of current cases and site remediation responses. Int J Environ Res Public Health 17(8936):1–42
23. Arbestain MC, ıguez-Lado LR, Bao M, Mac´ıas1 FM (2008) Assessment of mercury-polluted soils adjacent to an old mercury-fulminate production plant. Appl Environ Soil Sci 2009(387419):1–9
24. Mendoza JC, Amorim CG, Rodríguez-Díaz JM, Montenegro MDCBSM (2022) Removal of contaminants from water by membrane filtration: a review. Membranes 12(570):1–23
25. Zawierucha I, Nowik-Zajac A, Lagiewka J, Malina G (2022) Separation of Mercury (II) from Industrial Wastewater through Polymer Inclusion Membranes with Calix [4] pyrrole Derivative. Membranes 12(492):1–14
26. Ji G, Zhao M (2017) Membrane separation technology in carbon capture. Recent Adv Carbon Capture Storage 59–90
27. Chadha U, Selvaraj SK, Thanu SV, Cholapadath V, Abraham AM, Zaiyan M, Manoharan M, Paramsivam V (2022) Review of the function of using carbon nanomaterials in membrane filtration for contaminant removal from wastewater. Mater Res Express 9(012003):1–24
28. Makarova AS, Yarovaya OV, Fedoseev AN, Yakubovich LM (2020) Development of a technology for immobilizing mercury in solid mercury-containing wastes. Clean Eng Technol 1(100030):1–9
29. Rodríguez-Viso P, Domene A, S´anchez A, V´elez D, Monedero V, Devesa V (2023) mercury dietary exposure. Toxicology 494(153580):1–16
30. Kho F, Koppel DJ, Hellfeld RV, Hastings A, Gissi F, Cresswell T, Higgins S (2020) Current understanding of the ecological risk of mercury from subsea oil and gas infrastructure to marine ecosystems. J Hazard Mater 438(129348):1–18
31. Suess E, Berg M, Bouchet S, Cayo L, Hug SJ, Kaegi R, Voegelin A, Winkel LHE, Tessier E, Amouroux D, Buser AM (2020) Mercury loads and fluxes from wastewater: a nationwide survey in Switzerland. Water Res 175(115708):1–10
32. Thepanondh S, Tunlathorntham V (2020) Appropriate scenarios for mercury emission control from coal-fired power plant in Thailand: emissions and ambient concentrations analysis. Heliyon 6(e0419):1–8
33. Huang WY, Huang CW, Li YL, Huang TP, Lin C, Ngo HH, Xuan-Thanh Bui T (2023) Reduced pollution level and ecological risk of mercury-polluted sediment in a alkali-chlorine factory's brine water storage pond after corrective actions: A case study in Southern Taiwan. Environ Technol Innov 29(103003):1–11
34. Isa SA, Hafeez MA, Singh., B.K., Kwon, S.Y., Choung, S., Um, W. (2022) Efficient mercury sequestration from wastewaters using palm kernel and coconut shell derived biochars. Environ Adv 8(100196):1–10

35. Abou-Shady A, Ismail S, Yossif TMH, Yassin SA, Ali MEA, Habib AAMH, Khalil AK, Tag-Elden MA, Emam TM, Mahmoud AA, Eissa D, Hega RH, Kotp YH, Osman MA, Saudi AM, Abdelaziz SM, Yaseen R, El-Araby H, Abd-Elmottaleb O, Bahgaat AK, El-Harairy A (2023) Comprehensive review of progress made in soil electrokinetic research during 1993–2020, part II. No.1: Materials additives for enhancing the intensification process during 2017–2020. S Afr J Chem Eng 45(182):1–19
36. O'Connora D, Houa D, Okc YS, Mulderd J, Duana L, Wua Q, Wanga S, Rinklebef FMGTJ (2019) Mercury speciation, transformation, and transportation in soils, atmospheric flux, and implications for risk management: a critical review. Environ Int 126:747–761
37. Xu J, Kleja DB, Biester H, Lagerkvist A, Kumpiene J (2014) Influence of particle size distribution, organic carbon, pH and chlorides on washing of mercury contaminated soil. Chemosphere 109(99):1–7
38. Yadav V, Manjhi A, Vadakedath N (2023) Mercury remediation potential of mercury-resistant strain Rheinheimera metallidurans sp. Nov. isolated from a municipal waste dumping site. Ecotoxicol Environ Saf 257(114888):1–12
39. Gani A, Wattimena Y, Riza M (2021) A Simultaneous sulphur dioxide and mercury removal during low-rank coal combustion by natural zeolite. Heliyon 7(e07052):1–6
40. Mantey J, Nyarko KB, Owusu-Nimo F, Awua KA, Bempah CK, Amankwah RK, Akatu WE, Appiah-Effah E (2020) Mercury contamination of soil and water media from different illegal artisanal small-scale gold mining operations (galamsey). Heliyon 6(e04312):1–13
41. Sitarska M, Traczewska T, Filarowska W, Hołtra A, Zamorska-Wojdyła D, Beata Hanus-Lorenz BH (2023) Phytoremediation of mercury from water by monocultures and mixed cultures pleustophytes. J Water Process Eng 52(103529):1–9
42. Ryfa A, Zmuda R, Mandrela S, Białecki R, Adamczyk W, Nowak M, Lelek L, Bandoła D, Pichura M, Płonka J, Wdowin M (2023) Experimental and numerical investigation of mercury removal from flue gas by sorbent polymer composite. Fuel 333(126470):1–12
43. Albatrni H, Qiblawey E-N, H.M.H. (2021) Comparative study between adsorption and membrane technologies for the removal of mercury. Sep Purif Technol 257(117833):1–15
44. Qiu M, Liu L, Cai Y, Yu S, Wang S, Fu D, Hu B, Wang X (2022) Biochar for the removal of contaminants from soil and water: a review. Springer 4(19):1–25
45. Tiodar ED, Vacar CL, Podar D (2021) Phytoremediation and microorganisms-assisted phytoremediation of mercury-contaminated soils: challenges and perspectives. Int J Environ Res Public Health 18(2435):1–37
46. Bhatia M, Khan ML, Qureshi A (2022) Microbial remediation of mercury contaminated soils. Microbes Microb Biotechnol Green Remediat Elsevier III:633–649
47. Hua K, Xu X, Luo Z, Fang D, Bao R, Yi J (2020) Effective removal of mercury ions in aqueous solutions: a review. Curr Nanosci 16:363–375
48. Abelsohn A, Brian GL, Margaret SD, Weir E (2020) Identifying and managing adverse environmental health effects: 5 persistent organic pollutants. CMAJ: Can Med Assoc J 166(12):1549–1555
49. Irwin E (2002) Mercury in the environment: a volatile problem. Environment 44(9):24–41
50. Jacques O, Rodney E (2001) The impact of pollution from a mercury processing plant in KwaZulu-Natal, South Africa, on the health of fish-eating communities in the area: an environmental health risk assessment. Int J Environ Health Res 11:41–50
51. Michael SW (2001) Technical report: mercury in the environment: implications for pediatricians. Pediatrics 108(1):197–206
52. James Patterson J (1997) Aqueous mercury treatment. EPA, pp 1–38
53. Islam MS, Vogler JR, Al Hasnine SMA, Ndez SH, Malekzadeh N, Hoelen TP, Hatakeyama ES, Bhattacharyya D (2020) Mercury removal from wastewater using cysteamine functionalized membranes mohammad saiful. Am Chem Soc 5:22255−22267
54. Qasem NAA, Mohammed RH, Lawal DU (2021) Removal of heavy metal ions from wastewater: a comprehensive and critical review. NPJ Clean Water 4:36
55. Yaari MA, Saleh TA (2022) Mercury removal from water using a novel composite of polyacrylate-modified carbon. ACS Omega 7:14820–14831

56. Chen MXZ, Li Y, Li C, Ahmad NM, Cheemad WA, Zhu S (2019) Removal of Hg(II) in aqueous solutions through physical and chemical adsorption principles. RSC Adv 9:20941–20953
57. Albatrni H, Hazim Qiblawey H, El-Naas MH (2021) Comparative study between adsorption and membrane technologies for the removal of mercury. Sep Purif Technol 257(117833):1–16
58. Li D, Li X, Yu Tao Y, Yan Z, Ao Y (2022) Deciphering the bacterial microbiome in response to long-term mercury contaminated soil. Ecotoxicol Environ Saf 229(113062):1–9
59. Yan A, Wang Y, Tan SN, Yusof MLM, Ghosh S, Chen Z (2020) Phytoremediation: a promising approach for revegetation of heavy metal-polluted land. Front Plant Sci 11(359):1–15
60. Hafshejani MK, Vahdati A, Vahdati M, Kheradmand AB, Sattari M, Armin AA, Choopan S (2012) Study of the mercury removal for health care and the effect of PH in mercury removal from aqueous solutions by activated carbons. Life Sci J 9(4):1846–1848

Antioxidant Defense Mechanism in Plants Exposed to Mercury Toxicity: Response, Tolerance and Remediation

Suganthi Rajendran, Sara Parwin Banu Kamaludeen, and Avudainayagam Subramanian

Abstract In the current era of industrialization, the presence of mercury in the soil as a result of anthropogenic and natural processes is a topic of concern on a world-wide scale since it adversely impacts agricultural productivity, the environment, and human health. The toxicity of mercury results in oxidative damage, which seriously impairs plant growth and productivity. The defense mechanisms of plants against mercuric toxicity have recently been the subject of in-depth investigation. Plants exposed to mercury resulted in reduced root growth, shoot growth, plant height, relative water content, biomass production, chlorophyll content, photosynthetic rate, stomatal conductance, carbon assimilation with enhanced hydrogen peroxide production, free radicals or superoxides, malondialdehyde production. In order to cope with the oxidative stress caused by mercury, plants have evolved a number of adaptive strategies, including the accumulation of enzymatic and non-enzymatic antioxidants (Catalase, Peroxidase, Ascorbic acid, Super oxide dismutase, Phytochelatin, Metallothioneins), osmolytes like proline, and polyamines (PAs), which control a plant's normal growth and development while surviving in metal-contaminated urban and sub-urban areas. These biomolecules facilitates in inducing tolerance and resistance towards mercury by the plants.

Keywords Mercury · Phytoremediation · Oxidative stress · Physiological Response · Enzymatic and non enzymatic antioxidants

S. Rajendran
JSA College of Agriculture and Technology, Avatti, Cuddalore, Tamil Nadu, India

S. P. B. Kamaludeen (✉)
Department of Environmental Sciences, Tamil Nadu Agricultural University, Coimbatore, Tamil Nadu, India
e-mail: saraparwinbanu.k@tnau.ac.in

A. Subramanian
Department of Environmental Sciences, Tamil Nadu Agricultural University, Coimbatore, Tamil Nadu, India

N. Kumar (ed.), *Mercury Toxicity Mitigation: Sustainable Nexus Approach*, Earth and Environmental Sciences Library, https://doi.org/10.1007/978-3-031-48817-7_10

1 Introduction

Mercury (Hg) advanced as a global pollutant owing to its toxicity, complex chemistry and persistence. Unlike other metals, Mercury is a non essential element and has no physiological role in the plant functioning and metabolism. It enters the environment through various means both natural and anthropogenic sources such as paper industry, paint manufacturing units, battery, pesticide, fertilizer, gold and silver mining, fossil fuel etc. Various technologies were tested for the remediation of contaminated sites which encompasses stabilization, vitrification, immobilization, solidification, thermal desorption, nanotechnology, soil washing, elctro-remediation [55, 75]. Phytoremediation evolves as a predominant ecofriendly technology in remediation of contaminated soils and water. Mercury accumulation in plants tends to inhibit various cellular functions, plant growth and development [40]. Recent studies have been into understanding the ecotoxicity of mercury and the response of plants towards mercury. The scientific community has shifted its attention to phytoremediation since it is less costly, less hazardous, and effective at removing contaminants. Soil and plant characteristics affect how efficiently the process works. The primary determinants of phytoremediation are plant biomass and the concentration of heavy metals in distinct plant tissues. The main advantages of this procedure are that it is easy to use, has minimal negative environmental effects, and may be widely adopted [70]. Plants tend to absorb heavy metals and other ions from the soil even at low concentration with the aid of root system. With the help of extensive root growth, it facilitates the rhizosphere microbial population efficiency in eliminating heavy metal thereby modifying their bioavailability reclaiming the polluted soil.

The typical byproduct of plant cellular metabolism is the production of reactive oxygen species (ROS). Numerous environmental pressures cause an overabundance of ROS, which in turn causes oxidative damage that worsens over time and finally results in cell death. Despite their damaging behaviour, they are well-known second messengers in a range of cellular activities, including conferring resistance to diverse environmental stimuli. The delicate balance between ROS generation and their scavenging determines whether ROS will act as signalling molecules or if they will be able to oxidatively harm the tissues [60]. Under any oxidative stress, plants' primary reaction is to produce reactive oxygen species [75], which inhibits photosynthesis and other biochemical activities and hinders plant development. Hg interference has an impact on photosynthetic pigments (chlorophylls and carotenoids) through direct enzyme inhibition [66]. Plants often use appropriate defences as a coping strategy, including ligand synthesis, stress enzyme activation, proteins and osmolytes, etc. [30], which includes Catalase, Peroxidase, Polyphenol Oxidase, Super Oxide Dismutase, Glutathione Peroxidase, and Heat Shock Proteins. Different host defense responses are induced by heavy metal toxicity in plants, and the efficiency of these responses varies with dose, plant type, and other variables [5]. The current chapter focuses on plant growth, development, physiological and biochemical responses to oxidative stress induced by mercury, the role of enzymatic and non enzymatic oxidants, and potential interactions between them. It also reviews previous accomplishments and current trends in the field of ecotoxicity of mercury towards plants. This study further emphasizes how antioxidants and other plant metabolites improve plant's resistance through their ameliorative qualities.

2 Mercury and Its Occurrence

Mercury, a D-block, Period 6, and Group 12 transition metal, is also known as a chalcophile element and is the single metal that is liquid at normal pressure and temperature. With an atomic weight of 201 and seven stable isotopes, this element has two oxidation states. Chlor alkali factories that employ mercury cells are only one of the numerous businesses that use this hazardous metal as a raw material. According to research, Indian coal contains between 0.01 and 1.1 mg kg^{-1} of mercury, which when burned releases mercury into the environment. According to the Centre of Science and Environment, burning coal results in an annual mercury discharge of 75 tonnes [58].

The environment contains mercury in an array of distinct forms with multifaceted characteristics. Mercury may exist in a variety of forms, including organic, inorganic, and elemental mercury. Both anthropogenic and natural sources have an impact on mercury's occurrence in the environment, which affects how readily accessible it is to ecological receptors. The most prevalent form of mercury is inorganic, but it is biotransformed into the most dangerous organic forms, monomethyl mercury and dimethyl mercury. Mercury is mobile and bioavailable in environmental matrices, and this allows mercury to enter the food chain through bioaccumulation and biomagnification.

As a result of pollution from Asian nations, particularly China and India, which is mostly brought on by the precipitation of the gases from coal-based power plants, mercury poisoning in fish in the Pacific Ocean is increasing. As opposed to surface feeders, predatory fish and bottom feeders have higher mercury concentrations. The "Minamata convention on Mercury," which the United Nations (UN) proposed in 2013 to restrict these kinds of mercury effects on the environment, has been ratified by 128 nations. The primary goal of this agreement is to minimise mercury's negative effects on the environment and human health. The Minamata illness, which was first described in Japan's Minamata in 1956, is a well-known illustration of mercury poisoning. It was mainly due to the consumption of fish contaminated with methyl mercury from the industrial waste. Symptoms include ataxia, numbness in hands and feet etc. And the second outbreak of Minamata disease occurred in Niigata prefecture in 1965.

3 Phytoremediation

Phytoremediation, which uses green plants to reduce the bioavailability of toxins in the environment, is touted as one of the newest environmentally benign strategies. A variety of contaminants, including heavy elements, trace elements, wastewater including organic and inorganic pollutants, and radionuclides in soil and water, are treated using phytoremediation. And it is considered to be a more efficient technology for cleaning up polluted areas than chemical ones, such as different soil washing processes, vitrification, thermal vaporisation, etc. When treating soil pollution, technologies utilised for plant-based remediation include phytovolatilization, phytoextraction, phytodegradation, phytostabilization, and rhizodegradation (Fig. 1). Plants take up pollutants, transform them into their volatile form, and then release that form into the air through transpiration. Arsenic and mercury are widely studied under the umbrella term "phytovolatilization" since they have volatile forms and may be physiologically transformed by plants into gaseous species. Phytoextraction is the accumulation of contaminants in a plant's topmost layer without any further processing that would enable the recovery of the metal after harvest. In phytoextraction, the plant species Indian mustard and sunflower are often used [59]. Phytodegradation is the term for the metabolic breakdown of organic pollutants by interior plant tissues. Plants may catalyse the degradation of organic pollutants by creating numerous enzymes with particular functions, such as nitroreductase, peroxidases, and dehalogenases. Adsorption onto plant roots and precipitation (or complexation) inside the rhizosphere are two processes that take place in the in-situ immobilisation of pollutants through phytostabilization. The possibility that a contaminant would enter groundwater or the food chain is minimised by limiting its mobility. Rhizospheric bacteria in the rhizosphere degrade organic pollutants there through a process called rhizodegradation. Only when roots and rhizospheric microorganisms cooperate may rhizodegradation be successful [35].

Numerous plants were used in the Phytoremediation of mercury from polluted soils. 88.9 mg Hg Kg^{-1} of gold amalgamation tailings discharged into agricultural regions were cleaned up utilising natural plants including *Lindernia crustaceae, Paspalum conjugatum,* and *Cyperus kylingia.* With the addition of ammonium thiosulphate at a rate of 8 g kg^{-1} of soil, the accumulation of mercury was increased by 82 and 47% in plant shoots and roots, respectively. With *P. conjugatum,* this led to an increase in yield of up to 74%. *P. conjugatum* is the plant with the greatest potential for phytoextraction, followed by *L. crustacea* and C. kyllingia. Plant species including *Polypogonmono speliensis, Brassica juncea,* and *Pteris vittata,* which are taken from polluted soils after collecting Hg in their roots and shoots, are burned in a secluded region.

Native plants are also utilised in Hg cleanup; they increase Tridax procumbens' BCF and TF, which triggers the synthesis of antioxidant enzymes to mitigate the oxidative stress brought on by Hg [1, 53]. Phytoremediation can be employed in sites that are polluted with mercury (Hg) and methyl mercury (MeHg). The ability of water hyacinths (*Eichhornia crassipes*) to incorporate Hg and MeHg into plant biomass in

Fig. 1 Mechanisms involved in phytoremediation

both aquatic and sediment-related forms was investigated over the course of a 68-day hydroponic research. The capacity of *E. crassipes* to digest both Hg and MeHg was tested under varied phosphate (PO4) concentrations, light intensities, and sediment: aqueous phase contamination ratios. It was demonstrated that Hg and MeHg localised largely in the roots of *E. crassipes* with minimal translocation. Assimilation of Hg and MeHg into the biomass of water hyacinths represents a potential means for sustainable remediation of contaminated waters and sediments under the appropriate conditions [12]. According to [43], different plant species accumulated Hg in their tissues in the following order: roots > leaves > stems, regardless of the Hg content in the soil. The amount of Hg in their tissues reveals the potential of these plants to store Hg for transfer from the soil (bioconcentration) or for translocation across different portions of the plant (for example, from roots to shoots). Shoots of the species *J. curcas, P. marginathum, C. annuum, and S. bifidus* showed the maximum accumulation of Hg. *J. curcas,* which corresponds with Sinduja's findings [63], displayed the highest level of Hg accumulation in roots and a high value in shoots. This suggests that it has the potential to be employed as a natural remedy to detoxify soils. Table 1 presents the list of plants widely used for mercury phytoremediation.

4 Fate of Mercury Within the Plant System

Despite the fact that mercury is a non-essential element, plants nonetheless take up the accessible form of mercury from the soil solution. Roots of plants serve as a barrier to stop intracellular access, which is primarily caused by the development

Table 1 Bioaccumulation of mercury in different parts of the plant facilitating Phytoremediation

S. no.	Plant	Concentration	Root	Stem	Leaves	References
1	Medicago sativa (alfalfa)	30 μM Hg	2611 ± 273 μg g^{-1} DW	38.3 ± 0.45 μg g^{-1} DW	67.7 ± 0.5 μg g^{-1} DW	Carrasco-Gil et al. [11]
2	Pteris vittata and Nephrolepis exaltata	0, 1, 5 and 20 mg L^{-1}	91 mg/kg Hg	15 mg/kg Hg (Shoot)	–	Chen et al. [14]
3	Mentha arvensis	5, 10, 15, 20 and 25 mg L^{-1}	1800 mg/kg DW @ 25 mg L^{-1}	Shoot—1300 mg/kg DW @ 25 mg L^{-1}	–	Manikandan et al. [41]
4	Brassica juncea	0.05, 0.5, 1, 2.5, 5, 10 mg L^{-1}	14.4 ± 2.0% 31.1 ± 5.5 22.9 ± 1.7 21.5 ± 6.8 15.1 ± 1.9 12.1 ± 3.0	NA% 0.32 ± 0.05 0.47 ± 0.05 0.26 ± 0.19 0.10 ± 0.03 0.16 ± 0.11	–	Moreno et al. [48]
5	Marrubium vulgare	Polluted site	501.9 ± 3.2 μg g^{-1} DW	–	183.4 ± 7.1 μg g^{-1} DW	Carrasco-Gil et al. [11]
6	Tridax procumbens, Ruellia tuberosa, Dodonaea viscose and Azadirachta indica	5 mg L^{-1}	0.190 0.053 0.113 0.113	0.072 0.066 0.071	–	Ahammad et al. [1]

(continued)

Table 1 (continued)

S. no.	Plant	Concentration	Root	Stem	Leaves	References
7	Lindernia crustacea, Digitaria radicosa, Zingiber purpurium, Paspalum conjugatum, Cyperus kyllingia, and Caladium bicolor	Polluted gold mine with initial concentration of 327 mg kg^{-1}	<4 mg g^{-1} DW	9.06 mg g^{-1} DW 10.36 15.65 4.42 1.39 0.36	–	Muddarisna et al. [49]
8	Mulberry nigra	30, 50 and 70 mg L^{-1}	65 mg kg^{-1}	55.67 mg kg^{-1}	50 mg kg^{-1}	Hashemi and Tabibian [24]
9	Oxalis corniculata	Polluted soil with total mercury of 0.4648 mg kg^{-1}	0.706 μg g^{-1}	0.704 μg g^{-1}	0.328 μg g^{-1}	Liu et al. [37]
10	Bacopa monnoeri	2, 5, 10 μM Hg	204 mg			Hussain and Nabeesa [26]
11	Wheat cultivar	1 and 5 mg kg^{-1}	10.88 μg kg^{-1}	18.22 μg kg^{-1}	12.24 μg kg^{-1}	Liu et al. [36]
12	Lactuca sativa (Lettuce)	0.058 mg kg^{-1}	0.5 g $plant^{-1}$	-	3.1 g $plant^{-1}$	Cozzolino et al. [15]
13	Lavender (Lavandula stoechas)	2.6 mg kg^{-1}	0.22 mg kg^{-1}	0.04 mg kg^{-1}	0.18 mg kg^{-1}	Sierra et al. [62]
14	Oryza sativa	500 μg L^{-1}	19.89 mg kg^{-1}	43.62 mg kg^{-1}		Li et al. [33]

(continued)

Table 1 (continued)

S. no.	Plant	Concentration	Root	Stem	Leaves	References
15	Oryza sativa	35 mg kg^{-1}	94.1 μg kg^{-1}			Zhang et al. [78]
16	Brassica juncea	374 mg kg^{-1}	26 mg kg^{-1}	19.5 mg kg^{-1}		Wang et al. [76]

of a ligand between mercury and carboxylic groups. There are several ways for plants to take in components from their surroundings and incorporate them into their systems. The availability of mercury was found to be very low since it is very weakly soluble in soil solution, and it may be taken up by plants from the soil through the transpiration system to the roots [9]. Using stomata and foliar adsorption of both wet and dry deposited elemental mercury, [57] claimed that plants have the capacity to directly absorb mercury from the atmosphere through their leaves. Phloem is linked to the translocation of mercury to various plant components, such as leaves, fruits, and seeds, whereas Xylem and Phloem are primarily responsible for moving mercury to the aerial sections of the plant. A threshold value of 0.36 mg kg^{-1} mercury that is hazardous to plants was proposed by [34]. The physiology and operation of the plant are affected when the mercury content surpasses the critical limit. In a study conducted by [34], different concentration of mercury was tested against Oats and Common bean and resulted in non-significant relationship regarding seedling emergence whereas it could cause direct damages to the seed embryos as well as the endosperm by disturbing the sulfhydryl groups piloting to the formation of –S–Hg–S– which may damage the growth of the embryo since they are loaded with –SH bonds [8]. Several intercellular and intracellular mechanisms controlled the accumulation, movement, and distribution of mercury throughout the plant system. At concentration of 25 mg L^{-1} Hg exposure, the root tissues of *Mentha arvensis* reached a maximum level of 1816.54 mg kg^{-1} DW, indicating that greater accumulation in the root acts as a partial barrier for the mercury transfer to shoots. Mercury in the shoot chelates with the sulphur groups already present and augments with the carboxyl groups in the organic acids [11].

5 Response and Tolerance of Plants Against Mercury

Plants have divergent mechanisms for modulating heavy metal levels to adapt to a change in the concentration of metals in the polluted environment which entails plant growth regulators, osmoprotectants and antioxidants [13]. When the detoxification potential of the plants is less than the accumulation, then it is toxic to plants [40, 79]. The immediate response will be the generation of ROS within the plant cells resulting in oxidative stress and lipid peroxidation which will be scavenged by various enzymatic and non enzymatic antioxidants which is illustrated in Fig. 2.

5.1 *Photosynthetic Parameters*

Plant's chlorophyll pigments are essential for photosynthesis and have a unique function in the green cell economy. The superior component of the plant (the leaves) is toxically affected by the high concentration of Hg, and the damaged cell walls of the

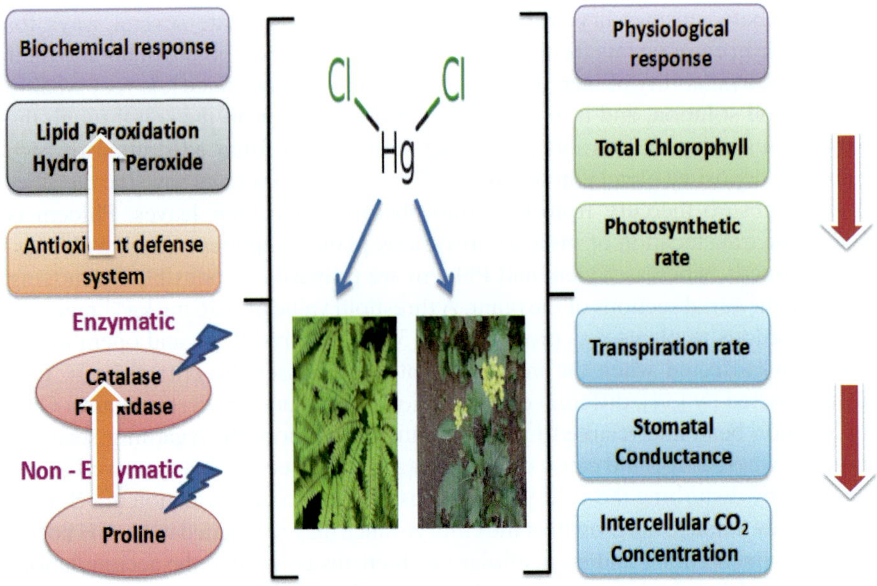

Fig. 2 Response and tolerance of plants against mercury

leaf cells reduce the amount of chlorophyll in the leaf. *B. juncea's* Hg absorption efficiency was insufficient for the treatment of heavily Hg-polluted soil, as evidenced by the reduction in chlorophyll content at the maximum level of Hg treatment (1,000 mg Hg kg^{-1}) [56]. The effects of mercury on plants' ability to photosynthesize result in a reduction in rate and a faint yellowing of their hue. Toxic levels of mercury reduce the production of chlorophyll in the leaves and speed up the oxidation of enzymes. The toxicity of mercury had a deleterious effect on mitotic behaviour as well as metabolite leakage [73]. Toxins cause chlorophyll molecules to deteriorate or lose concentration after they have been absorbed. Chlorophyll a, Chlorophyll b, and total Chlorophyll were significantly different ($p < 0.0001$) from control. The study found that the amount of chlorophyll in gametophytes cultivated on mercury salts was severely impacted [71]. The effect of mrcury on chlorophyll is illustrated in Table 2.

As mercury concentration rose in *Microsorum pteropus*, the amount of chlorophyll and the maximal quantum yield of primary photochemistry dropped, which led to a decline in photosynthetic activity. High mercury levels in tobacco decreased its chlorophyll content, maximum quantum yield from primary photosynthesis, and photosynthetic rate [39]. Hg has been linked to antioxidant's capacity to scavenge free radicals and a decrease in chlorophyll concentration. When the concentration is less than 30 mg/kg, Hg would speed up the photosynthetic rate of rice leaves [42]. Mercury has a tendency to lower chlorophyll levels, which lowers the plant's pace of photosynthetic growth [68]. Chlorophyll a, Chlorophyll b, and Carotenoid pigment concentration were suppressed by 29.3, 24.3, and 29.3% at 25 mg L^{-1} Hg exposure.

Table 2 Effect of Mercury on Chlorophyll

S. no.	Plant species	Chlorophyll content	Salient findings	References
1	Lemna minor Salvia natans	14.08–18.88 mg g^{-1} DM 19.00–27.52 mg g^{-1} DM	15% reduction at earlier growth stage No significant difference in Salvia natans	Sitarska et al. [65]
2	Gossypium hirsutum	–	38.5% reduction in chlorophyll content	Mei et al. [44]
3	Triticum aestivum	1541–1221 μg g^{-1} FW	24% reduction	Sahu et al. [58]
4	Spirodela polyrhiza	0.62 mg g^{-1} FW	59% reduction	Singh et al. [64]
5	Pistia stratiotes	3.02 mg g^{-1} FW	–	Kumar et al. [32]
6	Carthamus tinctorius	–	Significant decrease in chlorophyll concentration	Azad and Kafilzadeh [7]

Photosynthesis impairment and a decline in gaseous exchange measures were seen with increasing Hg dosages. But total chlorophyll tends to rise during the course of the plant's growth period following mercury treatment. Gill et al. [19] observed comparable outcomes after heavy metal exposure to garden cress. It might be because mercury suppresses Fe and causes chlorosis in leaves, which negatively affects the metabolism of chlorophyll. Micronutrients are depleted as a result of the toxicity of heavy metals, which are necessary for plant growth and development. The metal stress causes the pigment level to drop, which is one of the main reasons why photosynthesis is impaired. These results are in line with those of Jansukaitiene [28], who discovered that under heavy metal stress, pea plants saw a decline in gaseous exchange metrics like photosynthetic rate and intercellular CO_2 concentration, among others. It has been demonstrated that Hg, both organic and inorganic, causes the depletion of potassium, magnesium, and manganese as well as the buildup of iron [10]. Parts of chlorophyll may modify into pheophytin. Pheophytins are substances created during chlorophyll breakdown as a result of the chlorophyll's loss of magnesium ions (Fig. 3). Pheophytin accumulation and oxidative stress have been seen in plants subjected to hazardous quantities of trace elements [21, 45]. Due to the disruption of chloroplast structure and diminished photosynthetic capacity caused by heavy metal poisoning, carbon uptake was reduced [6, 54]. In addition to this, mercury tends to affect delta-aminolevulinic acid dehydratase activity which has a significant role in regulating chlorophyll. Delta-aminolevulinic acid dehydratase was found to

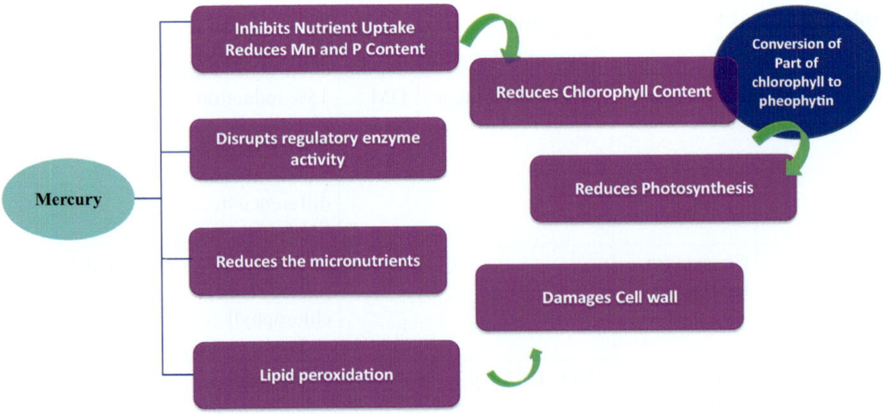

Fig. 3 Effect of mercury on chlorophyll and photosynthesis

be exclusively present in chloroplast. Mercury interferes chlorophyll formation by disrupting photochlorophyllide production [44].

5.2 *Gaseous Exchange*

Water, light, and CO_2 availability are necessary for photosynthesis. In accordance with the degree of stomatal aperture (the physical resistances to the movement of gases between the air and the interior of the leaf), stomatal conductance calculates the carbon dioxide uptake and water loss via transpiration. Gaseous exchange between stomata and leaves decides the stomatal conductance, intercellular CO_2 concentrations and transpiration rate. Stomatal conductance affects transpiration directly. It was reported that heavy metals tends to reduce the stomatal conductance. In addition to lowering photosynthetic pigments, stomatal conductance, and transpiration rates, heavy metals have negative impacts on light and dark responses [61]. This process may be significantly influenced by a number of variables, including CO_2 concentrations and stomatal conductance rates. Under laboratory conditions, Hg absorption in *Epipremnum aureum* approaches values of 170–1341 ng m^{-2} day^{-1}, with clear daily absorption cycles that correlate with positive stomatal conductance [51]. It was reported that decrease in stomatal conductance (53%) and carbon assimilation (20%) were observed in *Gossypium hirsutum* under mercury stress and an increase in intercellular CO_2 concentration upto 395.21 μmol. mol^{-1} [44].

5.3 *Antioxidant Defense System*

In general, heavy metals tend to cause oxidative stress, which in turn damages the tissues within the plant system, by either directly or indirectly causing ROS (reactive oxygen species). Through the use of enzymatic and non-enzymatic antioxidants, plants have their own detoxifying system against this oxidative damage. A higher quantity of mercury is hazardous to plants and will cause them to show physical damage and physiological problems. By binding with water channel proteins, it prevents stomata from closing and opening. It also interferes with mitochondrial function by causing oxidative stress through the production of reactive oxygen species, which disrupts plant cell and lipid metabolism [50].

6 ROS Generation

Singlet oxygen ($^1O_2$) is produced when oxygen is excited, which leads to the production of reactive oxygen species (ROS). These intermediates, which include hydrogen peroxide (H_2O_2), superoxide radicals, and hydroxyl radicals, are generated when electrons are transported [20, 23, 60]. As Hg dosages are increased, H_2O_2 levels rise. This could be largely attributed to the membrane destabilisation in plants with increasing metal stress. Because this plant was discovered to accumulate more metal with increasing doses. Oxidative stress or Haber–Weiss reactions both result in the production of ROS. According to [29] and [79], ROS produced in plant cells as a result of oxidative stress cause a variety of detrimental effects, such as DNA damage, lipid peroxidation, ATP inhibition, and photosynthetic inhibition.

6.1 *Lipid Peroxidation*

Malondialdehyde (MDA) accumulation, a fundamental marker of membrane lipid peroxidation, is frequently used to assess cell damage under stress conditions. According to a number of studies, exposure to Hg caused an imbalance of ROS in plants such as the tomato (*Lycopersicon esculentum*), cucumber (*Cucumis sativus*), and alfa alfa (Lucerne), which led to an increase in lipid peroxidation, a reduction in antioxidant system activation, and a reduction in the plant's capacity to photosynthesize. There has been evidence of an association made between mercury-induced cellular oxidative damage in plants and the excessive accumulation of reactive oxygen species (ROS), including free radicals and H_2O_2. As a signal molecule, H_2O_2 is essential for plant development and resilience, but very high levels of H_2O_2 in the presence of ROS can cause lipid peroxidation by targeting membrane lipids. TBARS can be utilised as a indicator of lipid peroxidation in tissues since they are created when certain primary and secondary lipid peroxidation products break down. A lipid

Fig. 4 Mechanisms behind tolerance of plants against HM stress by Proline generation

peroxidation marker called malondialdehyde (MDA) damages and leaks membranes. The cumulative effects of Hg exposure hindered plant growth, as seen, for instance, in the research, in the lower height and biomass of alfalfa (Lucerne) plants. MDA accumulation was higher in roots rather than leaves which was also resulted in electrolytic leakage in *Gossypium hirsutum* [44]. Based on the evidence provided above, it was hypothesised that antioxidant systems, products of lipid peroxidation, and photosynthetic pigments were required for plant development under Hg stress [72].

6.2 *Proline Generation*

Proline, an antioxidant amino acid, plays a significant role in the detoxification of heavy metal toxic because it can protect cells from damage by scavenging reactive oxygen species (ROS) which is considered as a tolerant strategy against heavy metal stress (Fig. 4). Excess Hg in the growth medium caused plants to produce proline [46].

When plants are exposed to high levels of mercury, reactive oxygen species (ROS) such as superoxide radicals (O_2), hydroxyl radicals (OH), and hydrogen peroxide (H_2O_2) accumulate excessively, causing oxidative stress. The accumulation of mercury in pakchoi disrupted the antioxidant protection system's equilibrium. Indeed, once Hg^{2+} is taken up by plant roots cells, it readily binds with thiol (–SH) groups with a high affinity, resulting in reduced GSH in the shoots and an increase in H_2O_2 formation. Both the CAT and SOD enzymes can work together to break down H_2O_2 into O_2 and H_2O, which is non-toxic to plants. As a consequence of the excess H_2O_2 in pakchoi, the behaviors of CAT and SOD can be significantly reduced [60].

6.3 *Enzymatic Defense Mechanism*

Certain stress defence mechanisms, such as antioxidant defence systems that include antioxidative enzymes and plant growth regulators, have been detected in plants. Several antioxidant enzymes assist in scavenging free radicals. In addition, stress-response proteins such heat shock proteins shield plants from oxidative stress [27]. In response to heavy metal toxicity, plants evolve a range of defensive mechanisms, although the efficiency of these mechanisms depends on the dosage, the kind of plant, and other elements. Plants flourish in these conditions thanks to their capacity to tolerate stress or reduce the toxicity of heavy metals [17, 56].

In order to nullify the consequences of oxidative stress generated in response to mercury stress, antioxidant enzymes were synthesized by the plant to reduce the ROS activity within the plant system. Super oxide dismutase (SOD), Ascorbate peroxidase, Catalase, Peroxidase and glutathione reductase are the common enzymes involved in plant defense against heavy metal stress. SOD which is considered as primary cellular defense dismutates the free radical i.e., Superoxide radical (O_2^-) into hydrogen peroxide and oxygen molecules. Moreover, the increased accumulation of hydrogen peroxide is limited by the Ascorbate peroxidase which in turn gets converted to water molecules. Both catalase and Ascorbate peroxidase activity was parallel to each other [4]. In garlic plants, both superoxide dismutase and catalase activities were significantly reduced by Hg stress alone, but peroxidase activity was unaffected. Under Hg stress, Si-applied plants had higher superoxide dismutase activity than those without it. The operation of catalase and peroxidase in the leaves under Hg stress was unaffected by the addition of Si [25]. H_2O_2 is immediately scavenged by catalase, which converts it to H_2O and O_2. According to [67, 69], peroxidases like ascorbate peroxidase and peroxidase transfer electrons to a variety of donor molecules like phenolic compounds, lignin precursors, or secondary metabolites in order to indirectly scavenge H_2O_2 from the environment. Plant cells produce more of those enzymes when the concentration of Hg in the plant tends to rise, which is consistent with the current findings [58].

6.4 Non Enzymatic Defense Mechanism

Low molecular weight compounds such as glutathione, ascorbic acid etc. also plays a significant role in plant defense against mercury. Ascorbate increases with increase in mercury concentration. Ascorbate regulates various cellular functions such as cell division, cell differentiation, and senescence [74] and protects proteins and lipids [2]. Ascorbate content of 46 μmol g^{-1} FW was obtained in leaves of *Triticum aestivum* [58] under mercury stress. Glutathione is a water soluble thiol compound with low molecular weight which plays a major role in stress management as well as detoxification of xenobiotic compounds. It is directly involved in detoxification of peroxides formed by acting as a cofactor or conjugating with xenobiotic compounds [22, 52]. Glutathione content determines the tolerance of plants against any stress.

Phytochelatinsynthase facilitates in the formation of polymers with glutathione as precursor which will detoxify mercury since it entails thiol group to which the affinity of mercury is higher. Heavy metals enhance the production of phytochelatins and these phytochelatins will be absorbed easily by the plants resulting in phytoextraction of heavy metals such as Mercury [18]. These complexed metals will be transported to the plant cell and resulting in vacuoles ultimately leading to intracellular detoxification of heavy metals within the plant cell. Besides plants produces metal chelates, metallothionein, organic acids, polyphosphates etc. [3].

7 Rhizosphere Processes Facilitating Remediation

The efficiency of Phytoremediation depends on the plants with an effective microbial consortium associated with the root system which in turn determines the bioavailability of the pollutants. These factors not only decide the availability but also the benefit of the plant. Plants are responsible for the uptake of target pollutant via various uptake mechanisms which entails the transport across the root tissue plasma membrane, adsorption onto the root surfaces and root derived biomolecules which in turn reduces the bioavailability of these pollutants to the microorganisms. Root activities such as acidification, alkalinisation, modification of the redox potential, exudation of metal chelants and organic ligands are reported to the factors facilitating the metal availability [77].

About half of the photosynthates produced by plants are carried back to their roots during growth, while the remaining 12–40% are released as exudates in the rhizophere. These exudates contain proteins, carbohydrates, organic and amino acids, peptides, and carboxylic acids. These root secretions act as ligands, chelating heavy metal ions, and affecting the pH and Eh conditions in the rhizosphere in addition to providing energy to microorganisms. The primary factor influencing the mobilisation of metals in soils and their accumulation in plants is a change in pH and Eh conditions. Limiting metal ion uptake through roots is the first line of defence for a plant under stress. The bulk of metals enter roots through ion channels or energy-driven pumps. By altering the pH of the rhizosphere, plants restrict the uptake of metal ions, which causes heavy metal precipitation. Different root exudates that concentrate these metal ions in the apoplast and stop them from entering the cell are what cause the pH shifts [38]. Metal ion movement from symplastic to apoplastic space is actively aided by heavy metal transport and carrier protein. Root exudates especially low molecular weight organic acids influence the transport of mercury from soil to leaf tissues via xylem vessels [16] and can also influence the bioavailability by precipitation and complexation.

Commonly plant secreted organic acids are oxalic acid and citric acid which is involved in translocation and vacuolar sequestration of heavy metal. The organic acids secreted are positively correlated with mercury concentration in aerial and edible parts of four different winter wheat cultivars [36] whereas it can also use as ligands to adsorb the target metal on the soil surface. Oxalic acid, Malic acid, Citric acid and Fumaric acid were detected in *Poa annua*, *Medicago polymorpha* and *Malva sylvestris* which influences the bioavailability of the heavy metal [47]. *Echinochloa crusgalli* root exudates produce citric acid and oxalic acid, which are responsible for the bioaccumulation of Cd, Cu, and Pb [31].

In the Hg treated plant roots of *Medicago sativa*, granular deposition at the intercellular space, deformation of cortex cells, dense deposition in non-suberized areas of the endodermis, uneven thicker xylem vessels were exhibited and localization of Hg in the leaf veins can be observed [11]. These roots accumulate more Hg since its cell wall consists of structural proteins extension and expansions which is rich in cysteine group which serves as binding support for Hg. In addition to this, Hg tends to bind with plasma membrane protein such as aquaporin which blocks the movement of water inside the cell. In the Rhizosphere of Lavender, higher percentage of available Hg to total Hg was observed by [62] and antagonistic effect of Mn against mercury exhibited a strong negative correlation. They concluded that organic matter and rhizosphere Mn^{2+} blocks the Hg uptake by Lavender and also added that increased rhizosphere NO_3^{2-} concentration reduced the rhizosphere available concentration. Root architecture plays a critical role in Hg accumulation, transfer, partitioning etc. Six different Chinese sub-tropical tree species were studied to understand the impact of root branching structure, morphological and stoichiometric traits on mercury resulted in decreased mercury accumulation with increasing root order and found higher in cortex than stele due to availability of binding sites [38]. Higher order roots revealed less specific root area and weak absorptive potential. Root carbon and sulfur groups influence the Hg translocation to a greater extent. Hg uptake is limited by the complexation of Hg with cytosolic –SH groups belonging to phytochelatins and metallothioneins.

In addition to this, microorganisms are reported to increase the solubility and the oxidation state of the metal or metalloids through various processes such as production of organic ligands, microbial siderophores and the production of organic acids [77]. Apart from this, the nature of the pollutants and the soil conditions favors their accessibility to degradation or transformation. P-Actinobacteria was the frequent occurring organism in contaminated soil with 32.31% followed by P-proteobacteria with 31.17%, P-Actinobacteria with 26.16% and P-bacteroidetes with 27.20% as analyzed by Liu et al. [36] in the rhizosphere soils of *Oxalis corniculata.* Gram positive bacteria, Gram Negative bacteria and fungi are found to be positively correlated with mercury concentration in wheat glumes which signifies the translocation of mercury by secretion of organic acids and siderophores whereas protozoa was negatively correlated [37]. Application of humic acid along with VAM facilitated the plant growth and also inhibited Hg transfer in the edible crops such as Lettuce plant (*Lactuca sativa*) favoring Hg immobilization by stable complex formation [15].

8 Conclusion

Ecotoxicity of mercury has been getting worse due to the escalating anthropogenic activities causing build up in the aerial parts of the plant and in the rhizosphere due to the secretion of root exudates. Due to its contamination, even the growth, morphology and physiology were negatively impaired. Phytoremediation, which entails a spectrum of cost-effective plant-based methods that may be able to remediate Hg-contaminated soil as well as water systems, has been chosen as an ecologically acceptable solution for cleaning up heavy metal affected areas. Once mercury enters the plant system, it induces oxidative stress resulting in excess production of reactive oxygen species and causes lipid peroxidation which triggers the production of antioxidants who scavenge the free radical. Besides, Mercury prompts chromosomal aberration resulting in mutation and modifies the metabolism of phytochemicals.

It affects chlorophyll content thereby suppressing photosynthesis causing morphological distortion in plants. Advancement in biotechnology led to the development of genetically engineered plants with mercury resistant/tolerant gene for mercury sequestration. These transgenic plants expressed tolerance against mercury. Future investigation on different detoxification pathways of mercury within the plants cell will be an insight to explore the depth of their toxicity and pave way for better understanding in selecting hyperaccumulators for remediation.

References

1. Ahammad SJ, Sumithra S, Senthilkumar P (2018) Mercury uptake and translocation by indigenous plants. Rasayan J Chem 11(1):1–12
2. Akram NA, Shafiq F, Ashraf M (2017) Ascorbic acid-a potential oxidant scavenger and its role in plant development and abiotic stress tolerance. Front Plant Sci 613
3. Alamri SA, Siddiqui MH, Al-Khaishany MY, Nasir Khan M, Ali HM, Alaraidh IA, Alsahli AA, Al-Rabiah H, Mateen M (2018) Ascorbic acid improves the tolerance of wheat plants to lead toxicity. J Plant Interact 13(1):409–419
4. Ansari MKA, Ahmad A, Umar S, Iqbal M (2009) Mercury-induced changes in growth variables and antioxidative enzyme activities in Indian mustard. J Plant Interact 4(2):131–136
5. Arora A, Sairam RK, Srivastava GC (2002) Oxidative stress and antioxidative system in plants. Curr Sci 1227–1238
6. Asgher M, Khan MIR, Anjum NA, Khan NA (2015) Minimising toxicity of cadmium in plants—role of plant growth regulators. Protoplasma 252(2):399–413
7. Azad HN, Kafilzadeh F (2012) Physiological responses to mercury stress in the hydroponic cultures of safflower (Carthamus tinctorius L.) plants. J Biodivers Environ Sci (JBES) 2(9):12–20
8. Azevedo R, Rodriguez E (2012) Phytotoxicity of mercury in plants: a review. J Bot
9. Baya A, Heyst BV (2010) Assessing the trends and effects of environmental parameters on the behaviour of mercury in the lower atmosphere over cropped land over four seasons. Atmos Chem Phys 10(17):8617–8628
10. Boening DW (2000) Ecological effects, transport, and fate of mercury: a general review. Chemosphere 40(12):1335–1351
11. Carrasco-Gil S, Siebner H, LeDuc DL, Webb SM, Millán R, Andrews JC, Hernández LE (2013) Mercury localization and speciation in plants grown hydroponically or in a natural environment. Environ Sci Technol 47(7):3082–3090
12. Chattopadhyay S, Fimmen RL, Yates BJ, Lal V, Randall P (2012) Phytoremediation of mercury- and methyl mercury-contaminated sediments by water hyacinth (Eichhornia crassipes). Int J Phytorem 14(2):142–161
13. Chen J, Yang ZM (2011) Enhancement of tolerance of Indian mustard (Brassica juncea) to mercury by carbon monoxide. J Hazard Mater 186(2–3):1823–1829
14. Chen J, Shiyab S, Han FX, Monts DL, Waggoner CA, Yang Z, Su Y (2009) Bioaccumulation and physiological effects of mercury in Pteris vittata and Nephrolepis exaltata. Ecotoxicology 18(1):110–121
15. Cozzolino V, De Martino A, Nebbioso A, Di Meo V, Salluzzo A, Piccolo A (2016) Plant tolerance to mercury in a contaminated soil is enhanced by the combined effects of humic matter addition and inoculation with arbuscular mycorrhizal fungi. Environ Sci Pollut Res 23(11):11312–11322
16. DalCorso G, Manara A, Furini A (2013) An overview of heavy metal challenge in plants: from roots to shoots. Metallomics 5(9):1117–1132

17. Emamverdian A, Ding Y, Mokhberdoran F, Xie Y (2015) Heavy metal stress and some mechanisms of plant defense response. Sci World J
18. Fidalgo F, Azenha M, Silva AF, de Sousa A, Santiago A, Ferraz P, Teixeira J (2013) Copper-induced stress in S olanum nigrum L. and antioxidant defense system responses. Food Energy Secur 2(1):70–80
19. Gill SS, Khan NA, Tuteja N (2012) Cadmium at high dose perturbs growth, photosynthesis and nitrogen metabolism while at low dose it up regulates sulfur assimilation and antioxidant machinery in garden cress (*Lepidium sativum* L.). Plant Sci 182:112–120
20. Gimenez I, Waldbusser GG, Hales B, Keister JE (2018) Ocean acidification stress index for shellfish (OASIS): Linking Pacific oyster larval survival and exposure to variable carbonate chemistry regimes. Elem: Sci Anthropocene 6
21. Gomes MP, Le Manac'h SG, Maccario S, Labrecque M, Lucotte M, Juneau P (2016) Differential effects of glyphosate and aminomethylphosphonic acid (AMPA) on photosynthesis and chlorophyll metabolism in willow plants. Pestic Biochem Physiol 130:65–70
22. Hasanuzzaman M, Nahar K, Anee TI, Fujita M (2017) Glutathione in plants: biosynthesis and physiological role environmental stress tolerance. Physiol Mol Biol Plants 23:249–268. https://doi.org/10.1007/s12298-017-0422-2
23. Hasanuzzaman M, Bhuyan M, Zulfiqar F, Raza A, Mohsin SM, Mahmud JA, Fujita M, Fotopoulos V (2020) Reactive oxygen species and antioxidant defense in plants under abiotic stress: revisiting the crucial role of a universal defense regulator. Antioxidants 9(8):681
24. Hashemi SA, Tabibian S (2018) Application of Mulberry nigra to absorb heavy metal, mercury, from the environment of green space city. Toxicol Rep 5:644–646
25. Hu Y, Wang Y, Liang Y, Guo J, Gong H, Xu Z (2020) Silicon alleviates mercury toxicity in garlic plants. J Plant Nutr 43(16):2508–2517
26. Hussain K, Nabeesa S (2012) Bioaccumulation pattern of mercury in Bacopa monnieri (L.) Pennell. J Stress Physiol Biochem 8(2):10–21
27. Isah T (2019) Stress and defense responses in plant secondary metabolites production. Biol Res 52
28. Jansukaitiene I (2010) Impact of low concentration of cadmium on photosynthesis and growth of pea and barley. Environ Res Eng Manag 53(3):24–29
29. Jiang S-Y, Ma Z, Ramachandran S (2010) Evolutionary history and stress regulation of the lectin superfamily in higher plants. BMC Evol Biol 10(1):1–24
30. Kapoor D, Kaur S, Bhardwaj R (2014) Physiological and biochemical changes in Brassica juncea plants under Cd-induced stress. BioMed Res Int
31. Kim S, Lim H, Lee I (2010) Enhanced heavy metal phytoextraction by *Echinochloa crus-galli* using root exudates. J Biosci Bioeng 109:47–50
32. Kumar V, Singh J, Saini A, Kumar P (2019) Phytoremediation of copper, iron and mercury from aqueous solution by water lettuce (Pistia stratiotes L.). Environ Sustain 2:55–65
33. Li Y, Zhao J, Zhang B, Liu Y, Xu X, Li Y-F, Li B, Gao Y, Chai Z (2015) The influence of iron plaque on the absorption, translocation and transformation of mercury in rice (Oryza sativa L.) seedlings exposed to different mercury species. Plant Soil 398(1–2):87–97
34. Lima F, Martins G, Silva A, Vasques I, Engelhardt M, Cândido G, Pereira P, Reis R, Carvalho G, Windmöller C (2019) Critical mercury concentration in tropical soils: impact on plants and soil biological attributes. Sci Total Environ 666:472–479
35. Lin CH, Lerch RN, Kremer RJ, Garrett HE (2011) Stimulated rhizodegradation of atrazine by selected plant species. J Environ Qual 40(4):1113–1121
36. Liu N, Miao Y, Zhou X, Gan Y, Liu S, Wang W, Dai J (2018) Roles of rhizospheric organic acids and microorganisms in mercury accumulation and translocation to different winter wheat cultivars. Agr Ecosyst Environ 258:104–112
37. Liu Z, Wang L, Ding S, Xiao H (2018) Enhancer assisted-phytoremediation of mercury-contaminated soils by Oxalis corniculata L., and rhizosphere microorganism distribution of Oxalis corniculata L. Ecotoxicol Environ Saf 160:171–177
38. Luo Q, Wang S, Sun LN, Wang H, Bao T, Adeel M (2017) Identification of root exudates from the Pb-accumulator Sedum alfredii under Pb stresses and assessment of their roles. J Plant Interact 12(1):272–278

39. Lv S, Yang B, Kou Y, Zeng J, Wang R, Xiao Y, Li F, Lu Y, Mu Y, Zhao C (2018) Assessing the difference of tolerance and phytoremediation potential in mercury contaminated soil of a non-food energy crop, Helianthus tuberosus L.(Jerusalem artichoke). Peer J 6:e4325
40. Malar S, Sahi SV, Favas PJ, Venkatachalam P (2015) Mercury heavy-metal-induced physio-chemical changes and genotoxic alterations in water hyacinths [*Eichhornia crassipes (Mart.)*]. Environ Sci Pollut Res 22:4597–4608
41. Manikandan R, Sahi S, Venkatachalam P (2015) Impact assessment of mercury accumulation and biochemical and molecular response of Mentha arvensis: a potential hyperaccumulator plant. Sci. World J
42. Mao Q, Tang L, Ji W, Rennenberg H, Hu B, Ma M (2021) Elevated CO_2 and soil mercury stress affect photosynthetic characteristics and mercury accumulation of rice. Ecotoxicol Environ Saf 208:111605
43. Marrugo-Negrete J, Durango-Hernández J, Pinedo-Hernández J, Enamorado-Montes G, Díez S (2016) Mercury uptake and effects on growth in Jatropha curcas. J Environ Sci 48:120–125
44. Mei L, Zhu Y, Zhang X, Zhou X, Zhong Z, Li H, Li Y, Li X, Daud MK, Chen J, Zhu S (2021) Mercury-induced phytotoxicity and responses in upland cotton (Gossypium hirsutum L.) seedlings. Plants 10(8):1494
45. Mobin M, Khan NA (2007) Photosynthetic activity, pigment composition and antioxidative response of two mustard (Brassica juncea) cultivars differing in photosynthetic capacity subjected to cadmium stress. J Plant Physiol 164(5):601–610
46. Mohanty P, Matysik J (2001) Effect of proline on the production of singlet oxygen. Amino Acids 21(2):195–200
47. Montiel-Rozas M, Madejón E, Madejón P (2016) Effect of heavy metals and organic matter on root exudates (low molecular weight organic acids) of herbaceous species: An assessment in sand and soil conditions under different levels of contamination. Environ Pollut 216:273–281
48. Moreno FN, Anderson CW, Stewart RB, Robinson BH (2008) Phytofiltration of mercury-contaminated water: volatilisation and plant-accumulation aspects. Environ Exp Bot 62(1):78–85
49. Muddarisna N, Krisnayanti B, Utami S, Handayanto E (2013) Phytoremediation of mercury-contaminated soil using three wild plant species and its effect on maize growth. Appl Ecol Environ Sci 1(3):27–32
50. Nagajyoti PC, Lee KD, Sreekanth T (2010) Heavy metals, occurrence and toxicity for plants: a review. Environ Chem Lett 8(3):199–216
51. Naharro R, Esbrí JM, Amorós JA, Higueras PL (2020) Experimental assessment of the daily exchange of atmospheric mercury in Epipremnum aureum. Environ Geochem Health 42(10):3185–3198
52. Noctor G, Mhamdi A, Chaouch S, Han YI, Neukermans J, Marquez-Garcia BELEN, Queval G, Foyer CH (2012) Glutathione in plants: an integrated overview. Plant, Cell Environ 35(2):454–484
53. Parmar P, Patel M, Dave B, Subramanian R (2012) Identification of colocassia esculentum a novel plant Spp for the application of phytoremediation. Afr J Basic Appl Sci 4(3):67–72
54. Parmar P, Kumari N, Sharma V (2013) Structural and functional alterations in photosynthetic apparatus of plants under cadmium stress. Bot Stud 54(1):1–6
55. Pogrzeba M, Rusinowski S, Krzyżak J, Szada-Borzyszkowska A, McCalmont JP, Zieleźnik-Rusinowska P, Słaboń N, Sas-Nowosielska A (2019) Dactylis glomerata L. cultivation on mercury contaminated soil and its physiological response to granular sulphur aided phytostabilization. Environ Pollut 255:113271
56. Raj D, Kumar A, Maiti SK (2020) Brassica juncea (L.) Czern. (Indian mustard): a putative plant species to facilitate the phytoremediation of mercury contaminated soils. Int J Phytorem 22(7):733–744
57. Risch MR, DeWild JF, Krabbenhoft DP, Kolka RK, Zhang L (2012) Litterfall mercury dry deposition in the eastern USA. Environ Pollut 161:284–290
58. Sahu GK, Upadhyay S, Sahoo BB (2012) Mercury induced phytotoxicity and oxidative stress in wheat (Triticum aestivum L.) plants. Physiol Mol Biol Plants 18:21–31

59. Shaheen SM, Rinklebe J (2015) Phytoextraction of potentially toxic elements by Indian mustard, rapeseed, and sunflower from a contaminated riparian soil. Environ Geochem Health 37:953–967
60. Sharma P, Jha AB, Dubey RS, Pessarakli M (2012) Reactive oxygen species, oxidative damage, and antioxidative defense mechanism in plants under stressful conditions. J Bot. Article ID 217037
61. Sheetal K, Singh S, Anand A, Prasad S (2016) Heavy metal accumulation and effects on growth, biomass and physiological processes in mustard. Indian J Plant Physiol 21(2):219–223
62. Sierra M, Rodríguez-Alonso J, Millán R (2012) Impact of the lavender rhizosphere on the mercury uptake in field conditions. Chemosphere 89(11):1457–1466
63. Sinduja M, Avudainayagam S, Davamani V, Suganthi R (2018) Uptake of mercury by marigold and amaranthus on spiked soil. Madras Agric J 105(7–9):346–351
64. Singh H, Kumar D, Soni V (2020) Copper and mercury induced oxidative stresses and antioxidant responses of Spirodela polyrhiza (L.) Schleid. Biochem Biophys Rep 23:100781
65. Sitarska M, Traczewska T, Filarowska W, Hołtra A, Zamorska-Wojdyła D, Hanus-Lorenz B (2023) Phytoremediation of mercury from water by monocultures and mixed cultures pleustophytes. J Water Process Eng 52:103529
66. Smolinska B, Leszczynska J (2017) Photosynthetic pigments and peroxidase activity of Lepidium sativum L. during assisted Hg phytoextraction. Environ Sci Pollut Res 24:13384–13393
67. Sofo A, Scopa A, Nuzzaci M, Vitti A (2015) Ascorbate peroxidase and catalase activities and their genetic regulation in plants subjected to drought and salinity stresses. Int J Mol Sci 16(6):13561–13578
68. Suganthi R, Avudainayagam S, Davamani V, Banu KS, Chandrasekhar CN, Sivakumar U (2021) Indian mustard and Boston fern exhibits growth tolerance to increased dose of soil spiked mercury. Environ Ecol 39(3):525–532
69. Sytar O, Kumar A, Latowski D, Kuczynska P, Strzałka K, Prasad MNV (2013) Heavy metal-induced oxidative damage, defense reactions, and detoxification mechanisms in plants. Acta Physiol Plant 35(4):985–999
70. Tangahu BV, Sheikh Abdullah SR, Basri H, Idris M, Anuar N, Mukhlisin M (2011) A review on heavy metals (As, Pb, and Hg) uptake by plants through phytoremediation. Int J Chem Eng
71. Thekkeveedu R, Hegde S (2021) In vitro spore germination and phytoremediation of Hg and Pb using gametophytes of Pityrogramma calomelanos. Int J Phytorem 23(3):307–315
72. Tran TAT, Zhou F, Yang W, Wang M, Dinh QT, Wang D, Liang D (2018) Detoxification of mercury in soil by selenite and related mechanisms. Ecotoxicol Environ Saf 159:77–84
73. Chakraborty D, Choudhury B (2023) Toxic effects of mercury on crop plants and its phhysiological and biochemical responses. A review. Int J Adv Res 11(02):168–174. https://doi.org/10.21474/IJAR01/16236
74. Venkatesh J, Park SW (2014) Role of L-ascorbate in alleviating abiotic stresses in crop plants. Bot Stud 55:1–19
75. Wang J, Feng X, Anderson CW, Wang H, Zheng L, Hu T (2012) Implications of mercury speciation in thiosulfate treated plants. Environ Sci Technol 46(10):5361–5368
76. Wang J, Feng X, Anderson CW, Wang H, Wang L (2014) Thiosulphate-induced mercury accumulation by plants: metal uptake and transformation of mercury fractionation in soil-results from a field study. Plant Soil 375(1):21–33
77. Wenzel WW, Lombi E, Adriano DC (2004) Root and Rhizosphere Processes in Metal Hyperaccumulation and Phytoremediation. Heavy Met Stress Plants: Biomol Ecosyst 365
78. Zhang H, Feng X, Larssen T, Shang L, Li P (2010) Bioaccumulation of methylmercury versus inorganic mercury in rice (*Oryza sativa L.*) grain. Environ Sci Technol 44(12):4499–4504
79. Zhang L, Wong MH (2007) Environmental mercury contamination in China: sources and impacts. Environ Int 33(1):108–121

The Role of Plant Growth Promoting Bacteria on Mercury Remediation

M. Chandra Naik, Valeeta Marina Dsouza, Y. Ramesh, Mukul Kumar, and Nitish Kumar

Abstract Phytobial remediation is a cutting-edge technique that reduces mercury (Hg) poisoning of the environment by utilising plants and bacteria. Plant growth-promoting bacteria (PGPB) that aids phytoremediation have recently received a lot of attention for their ability to increase plant growth and metal tolerance while removing Hg on a massive scale. This chapter focuses on the PGPB traits that affect plants and the ways in which they respond to challenges brought on by Hg. We will cover a number of current instances of systems that increase the availability of Hg to plants and help them deal with Hg stressors. Despite the fact that phytoremediation has been shown to have limitations, encouraging outcomes have been achieved with consistent monitoring. An evaluation of the environmental consequences of PGPB is necessary before introducing PGPB-assisted phytoremediation of Hg in a field, especially with regard to the implications on indigenous bacteria.

Keywords Remediation mercury · Plant growth-promoting bacteria

M. C. Naik
Department of Crop Physiology, University of Agricultural Sciences, Raichur, Karnataka, India

V. M. Dsouza
Department of Plant Biotechnology, University of Agricultural Sciences, Raichur, Karnataka, India

Y. Ramesh
Department of Agronomy, University of Agricultural Sciences, Raichur, Karnataka, India

M. Kumar
Department of Botany and Plant Physiology, MBAC, Saharsa, Bihar, India
e-mail: drmukul.bau@gmail.com

N. Kumar (✉)
Department of Biotechnology, Central University of South Bihar, Gaya, Bijhar, India
e-mail: nitish@cub.ac.in; nitishbt@gmail.com

N. Kumar (ed.), *Mercury Toxicity Mitigation: Sustainable Nexus Approach*, Earth and Environmental Sciences Library, https://doi.org/10.1007/978-3-031-48817-7_11

1 Introduction

Mercury (Hg) is a naturally occurring toxic element in the Earth's crust, where it is perpetually ejected into space via volcanic eruptions and rock erosion. Hg mobilisation has, however, expanded dramatically since the industrial revolution due to the expansion of industrial and technical processes. According to a recent estimate by the United Nations Environment Programme, annual human Hg emissions could reach 2000–2500 tonnes. Due to its negative impacts on both environmental and human health, mercury leakage into the environment has drawn a lot of attention from researchers and governmental organisations. With the Minamata catastrophe in the second half of the twentieth century, publics became more aware of its toxicity. In addition, mercury has been placed third on the list of substance priority updated by the Agency for Toxic Substances and Disease Registry based on its occurrence, poisonousness, and potential for human exposure [22].

Mercury is a non-essential metal in natural systems, hence there are typically no specific mechanisms for its excretion and metabolism. Therefore, through the food chain, Hg eventually bio-accumulates and biomagnifies [19]. In the aquatic environment, where some naturally occurring microbes change the speciation of Hg to methyl-mercury (MeHg), which is quickly taken up by phytoplankton and then passed on to zooplankton, these processes are even more prominent. It eventually builds up in humans after accumulating in fish and other longer-living species [5, 17].

Vascular plants have developed sophisticated and numerous strategies to deal with hazardous heavy metals levels, including mercury. These comprise evading heavy metals, which limits their uptake, stabilising the metal inside the root by binding and sequestration, decreasing translocation to above the ground tissues, and sequestering or probably decreasing the Hg inside the leaves [28]. When it comes to non-essential elements like mercury, the majority of terrestrial plants act as excluders. By exuding organic acids like malate, citrate, and oxalate that chelate the metal ions in the soil solution and reduce the bioavailability, plants can reduction uptake of metal [25].

Cinnabar and metacinnabar, which are mineral forms of mercury, are absorbed by plants growing in Hg-contaminated soils, although the methods by which this occurs are not well understood. In field plants of *Marrubiumvulgare* taken from a heavily Hg contaminated site in Spain, Hg trafficking via the root tissues has been mapped. Hg binding to thiol-rich areas of the root apoplast revealed a depletion of the thiol pools that were available, initially in the apical portions of and secondary and primary roots (the main sites of precipitation), moving towards the outer layers and epidermis of cortical cells, and finally reaching the xylem vessels [6].

Plants serve as sinks for Hg species during phyto-extraction, so it's important to choose species that have robust cellular Hg-buffering capacities, extended root systems, and high biomass output. Hg^{2+} would be absorbed by roots and transported to shoots where oxidative stress-relieving enzymes, vacuolar sequestration, and chelation with thiol compounds are activated. Senescence indicates the moment when the Hg-loaded plant material is suitable for harvest and the soil has been decontaminated. The plant biomass should be recognised as dangerous, and any unintentional ingestion must be prevented throughout its life cycle in order to prevent the bio-magnification of the food chain. In order to implement this plan, the site would need to be properly managed and monitored. It could also be necessary to acquire the necessary tools for recovering metal from contaminated biomass [16].

In scenarios of global climate change, plants are more frequently and severely exposed to climatic stress episodes, such as high temperatures, droughts, and salinity, which restricts their development and performance. Additionally, the alteration in soil physicochemical-biological properties brought on by the direct (e.g., ion competition) or indirect (e.g., modification of soil biological and physicochemical properties) effects of climate change on bioavailability of metals in soils may prevent plants from adapting, making them more vulnerable to stress and warning the extensive use of phytoremediation.

Mercury phytoremediation relies on the ability of plant species to tolerate and remove the poisonous chemical from substrates over extended times in order to reduce the concentration of the metal to a safe level. Hg soil phytoremediation is also made more difficult by the fact that no plant species has yet been discovered as a Hg hyper-accumulator. Applying chemical compounds to soils to either increase or decrease Hg mobility is expensive, unsuited for large regions and volumes of polluted soils, and potentially detrimental to the ecosystem. A potential solution to these problems is to make use of the complex interactions among plants and the related rhizosphere and/or entophytic microbes (fungi or bacteria). With these views we can suggests that there are several possible ways to mitigate the problems associated with the mercury effected soils by effective use of different groups of microorganisms and one such way is that to identify the plant growth promoting bacteria, their improvement and their effective utilization for the reclamation of the mercury affected soils.

2 Plant Growth-Promoting Bacteria (PGPB)

Plant beneficial microorganisms (PBM), mainly plant growth-promoting bacteria (PGPB) make mutual associations with plants, removing the heavy metals toxicity, improving multimodal resistance of plants to metals and climatic stresses, and affecting the metals bioavailability in soils [20]. Plant growth-promoting bacteria are a diverse group of prokaryotes that live in a broad range of ecological niches. These species may be free-living in the rhizosphere (rhizobacteria), occupying root nodules (rhizobia), or residing inside the tissues of plants (endophytes). PGPB may serve as bio-fertilizers, phytostimulants, bio-alleviators, bio-pesticides, and bio-modifiers to promote plant growth. [20]. Under toxic metal stress, PGPB and allied plant species aid in the survival and growth of both partners. Bacterial systems that stimulate plant growth and defend against phytopathogens are many but often poorly understood. In general, PGPB supply vital nutrients (e.g., fixed P, N, and Fe), signals for systemic resistance induction, through the secretion of organicvolatile compounds, hormones such ethylene, asabscisic acid, jasmonate, gibberellins, indole-3-acetic acid cytokinins, enzymes such as, *chitinases, 1-aminocyclopropane-1-carboxylate, proteases, deaminase, lipases, cellulases, and* compounds such as siderophores or antibiotics or [10].

According to [8], a cluster of metal-resistant PGPB, including *Arthrobacter, Burkholderia, Pseudomonas, Bacillus, Azoarcus, Azospirillum, Klebsiella Azotobacter, Serratia, Alcaligenes, Enterobacter, Agrobacterium, Rhizobium* species, has the capacity for enhancing plant growth in metal-contaminated environments. The metal-resistant PGPB were discovered to improve plant metal resistance by increasing plant detoxification rates, enzymes produced by plant roots, and through soil pH adjustment [12]. Furthermore, by means of releasing chelating chemicals, and redox changes and acidification certain metal resistant PGPB can affect metal bioavailability and mobility and hence plant utilisation rate [33]. As a result, these metal-tolerant PGPB strains can be exploited as a potential candidate for metal phytoremediation to reduce the negative impact of metals and increase plant metal storage capability. A number of metal-resistant PGPB have been shown to boost plant phytoextraction/bioaccumulation capacity by secreting organic acids and siderophores, which enhance bioavailability of metals by lowering soil pH [21]. In contrast, some metal-resistant PGPB can release polymeric constituents (such as polysaccharides and glomalin) thatspeed up metal phyto-stabilization by limiting their mobility [20].

To eliminate pollutants and lessen their adverse impacts on the environment, bioremediation, a biotechnological technique that makes use of living organisms (algae, plants, fungi, or bacteria) or components of them [4]. Alternative systems that utilise microbes and plants are used to reduce the stress caused by environmental metal contamination and to limit the risks associated with it [1, 11, 27]. By bioaccumulating, adsorbing, and/or volatilizing mercury that has been introduced to a culture medium, mercury-resistant bacteria can lower the concentration and bioavailability of the metal [7, 26]. Mercury volatilization mechanisms are well-known and frequently linked to bacterial resistance to this metal. Due to their function in volatilization of mercury (Hg^0), the mer operon produces enzymes such NAD(P)H-dependent flavoproteinmercuric reductase(MerA) and organomercuriallyase (MerB), which are crucial for bacterial tolarance to inorganic and organic mercury forms [3, 30]. Themer operon is often linked with chromosome, plasmids, and transposon [14, 30]. Due to the persistence of multidrug-resistant bacteria, contamination of the environment by metal compounds exerts selection pressure on the preservation and spread of resistance of antibiotics components determined in mobile elements [2, 32]. Mercury resistance is generally connected with tolarance to a variety of metals (i.e. multi-metal tolerance) as well as to diverse grouops of antibiotics [2, 32].

3 Strategies of PGPB in Metal Contaminated Soil Remediation Under Distinct Climate Stresses

Producing phytohormones like auxin, cytokinin, gibberellin, abscisic acid, and ethylene as well as promoting the uptake of nutrients like nitrogen, phosphate, and potassium are two ways that PGPB promotes direct plant growth. The production of antibacterial chemicals and the coexistence of pathogens increase ISR, which indirectly stimulates plant development. This is so that pathogenic bacterial development is constrained by the relationship between PGPB's antibacterial activity and nutritional competition. Specific PGPB can exploit one or more of these strategies to enhance plant resilience to environmental stresses. In addition, PGPB can chelate, coordinate, and ion exchange metals. The mobilisation, stabilisation, and transformation of metals under climatic stress by PGPB can also enhance bioremediation effectiveness and lessen the impact of climatic stress by controlling plant ion balance and antioxidant enzyme activity.

Mercury volatilization, or the conversion of Hg^{2+} to Hg^0, is a key process of bacterial resistance that is carried out by proteins expressed at the mer operon [30]. Other mechanisms also play a role in bacterial resistance. In the bacterial strains with MIC more than 63 $\mu g\ mL^{-1}\ Hg^{2+}$, the occurrence of the merA and merB genes was linked with the presence of plasmids. In instances of horizontal gene transfer, antibiotic and metal resistance factors are typically linked to genes coded by plasmid [29]. According to [29], the ambient buildup of metals like cadmium (Cd), mercury, zinc (Zn), and copper encourages the co-selection of genes that are implicated in tolerance to both antibiotics and metals. Endophytic bacteria and environmental bacteria that have been isolated from contaminated environments both co-occur with multiresistance to metals and antibiotics [2, 15].

Weak understanding exists on the effectiveness, measurement, and potential mechanisms of endophytic bacterial phytoremediation of mercury. Eight of the 34 previously discovered endophytic bacterial strains with a wide range of metal and eleven antibiotic resistance: *Acinetobacter baumannii* BacI43, *Enterobacter* sp. BacI14, *Bacillus* sp. BacI34, *Pantoea*sp. BacI23, *Klebsiella pneumoniae* BacI20, *Pseudomonas* sp. BacI38, *Serratiamarcescens* BacI56, and *Pseudomonas* sp. BacI7. The other seven bacterial strains supported growth of maize on a substrate which is contaminated by mercury, with the exception of *Klebsiella pneumoniae* BacI20. But whereas *Bacillus* sp. BacI34 and *Acinetobacterbaumannii* BacI43 enhance entire dry biomass by about 47% Phytoremediation supported by endophytic bacteria helped to lessen substrate toxicity as measured in various model species. The endophytic bacterial strains chosen here are likely candidates for aided phytoremediation, which would help lessen the environmental toxicity of soils that are contaminated with mercury.

According to a study by Balan et al., the commercial mining of coal may have affected the mercury-tolerant bacteria *Pseudomonas frederiksbergensis* strain SS18 and *Pseudarthro bacteroxydans* strain MM20 that were isolated from the tundra habitat of Ny-lesund, Svalbard. *P. frederiksbergensis* strain SS18 appeared to be more tolerant of mercury than *P. oxydans* strain MM20, according to the minimum inhibitory concentration (MIC), mercury elimination, mercury biosorption, and resistance to antibiotics of these strains. Mercury removal and bio-sorption studies revealed that 90% of the mercury was removed from the culture media by the selected strains. Whereas the mercury biosorption assay revealed that mercury was accumulated in cell pellets of P. oxydans strain MM20 (22%), and *P. frederiksbergensis* strain SS18 (25%), respectivelyAlkyl halides, alkynes, alcoholic, aliphatic, and aromatic amines, alkanes, and nitrocompounds, primary amines, carboxylic acid, alkenes, and amide groups play a significant role in the development of mercury tolerance, according to the Fourier transform infrared study. This work indicates that bacterial isolates go through intricate structural and functional modifications in experimental settings in order to survive and immobilise hazardous substances like mercury.

Inoculation of four plant growth-promoting bacteria and their mixtures was done, among them strains such as *Pseudomonas baetica* and *Pseudomonas moraviensis* significantly improved the biological behaviour of *Lupinusalbus* var. orden Dorado by reducing the oxidative stress in mercury-polluted soils.

A indigenous to Hg– and other heavy metal-contaminated sites Hg-reducing bacterium *Photobacterium* sp. strain MELD1, showed promising phyto-extraction potential. *Vignaunguiculata* ssp. *sesquipedalis* inoculated with *Photobacterium* sp. and cultivated in contaminated soil (on normal 27 mg Hg/kg of soil) exhibited increased leaf number, seed number, and root length. Uptake of Hg in roots, and reduced Hg content in upper part in comparison to the un-inoculated control [23].

In a mixture of sand containing100 mg Pb kg^{-1} and 1 mg Hg kg^{-1}, two heavy metal-resistant rhizobacteria, *Brevundimonas diminuta* SF-S1-5 and *Alcaligenes faecalis* SF-S1-60, helped *Scirpusmu cronatus*to survive for 42 days in comparison to un-inoculated plants, the presence of bacteria increased phyto-accumulation up to 650 and 320 mg Pb kg^{-1} and up to 7.5 and 2.4 mg Hg kg^{-1} in the roots and shoots, respectively [13]. The presence of three strains of epiphytic bacteria enhanced the growth and Hg^{2+} absorption capabilities of *Salvinianatans* and *Lemna minor* cultivated in aqueous solution containing 0.3 mg $Hg(NO_3)2\ L^{-1}$ [31].

Hg-resistant *Enterobacter ludwigii*and *Klebsiella pneumoniae*, isolated from *Alternanth erasessilis* and *Cyperus esculentus*, growing on a Hg-contaminated site individually stimulated growth, lowered malondialdehyde content proline content, and electron leakagein commercial *Triticum aestivum*s eedlings under 75 μM $HgCl_2$ in hydroponics [24]. A group of five Hg/As-resistant bacteria were tested alongside with thiosulfate, a fertiliser that acts as a metal mobilising agent, to see if it could enhance phyto-extraction. The bacteria were chosen based on their capacity to produce IAA, ammonia, exo-polysaccharide, biofilm, or fix N2 [9]. Instead of implementing exotic, potentially invasive species, the bacterial possibilities for in situ Hg assisted-phytoremediation frequently come from decontaminated areas.

Brassica juncea seedlings cultivated in Cd, Cu, Pb, Zn, As, and Mo-contaminated soil can benefit from PGPB inoculation, according to a recent study. While it has been hypothesised that the non-indigenous inoculum may integrate into these intricate communities and intensify their interaction to produce a more effective community that synergistically resists heavy metals, it has also been suggested that over time, the non-indigenous inoculum does not significantly change the makeup of the resident bacterial communities. The use of non-indigenous species should be carefully reviewed and evaluated before organising large-scale field studies, though, because plant and bacterial interactions are complex [18].

4 Conclusion

The method by which PGPB enhances plant growth and resistance to abiotic and biotic stress, as well as the function in response to mercury toxicity and bioavailability, have all been covered in this chapter. This has important implications for understanding the interactions between external pressure variables and biological processes as well as the application of PGPB in phytoremediation of mercury-contaminated soils. However, there are restrictions in their utilization procedure because of the nature of PGPB itself. It is difficult to completely purge PGPB of contaminants due to their weak genetic stability and susceptibility to change. Second, the indigenous microorganisms and PGPB compete for survival, and eventually certain PGPB strains may be eradicated as a result of competitive failure. Finally, PGPB are quickly influenced by other variables such as soil pH, temperature outside, and so forth, which reduces the effectiveness of bioremediation. To further understand how the primary factors of metal, soil, microorganisms, and plants interact, more research is required.

References

1. Abhilash PC, Powell JR, Singh HB, Singh BK (2012) Plant–microbe interactions:novel applications for exploitation in multipurpose remediation technologies. Trends Biotechnol 30:416–420
2. Baker-Austin C, Wright MS, Stepanauskas R, McArthur JV (2006) Co-selection ofantibiotic and metal resistance. Trends Microbiol 14:176–182
3. Barkay T, Miller SM, Summers AO (2003) Bacterial mercury resistance from atoms to ecosystems. FEMS Microbiol Rev 27:355–384
4. Bharagava RN, Chowdhary P, Saxena G (2017) Bioremediation: an eco-sustainable green technology: its applications and limitations. In: Bharagava RN (ed) Environmental pollutants and their bioremediation approaches, 1st edn. CRC Press, pp 1–22
5. Cao L, Liu J, Dou S, Huang W (2020) Bio-magnification of methyl-mercury in a marine food web in Laizhou Bay (North China) and associated potential risks to public health. Mar Pollut Bull 150:110762
6. Carrasco-Gil S, Alvarez-Fernandez A, Sobrino-Plata J, Millan R, Carpena-Ruiz RO, Leduc DL, Andrews JC, Abadia J, Hernández LE (2011) Complexation of Hg with phytochelatins is important for plant Hg tolerance. Plant Cell Environ 34:778–791
7. Dash HR, Das S (2015) Bioremediation of inorganic mercury through volatilization and biosorption by transgenic Bacillus cereus BW-03 (pPW-05). Int Biodeterior Biodegrad 103:179–185
8. Enebe MC, Babalola OO (2018) The influence of plant growth promoting rhizo-bacteria in plant tolerance to abiotic stress: a survival strategy. Appl Microbiol Biotechnol 102:7821–7835

9. Franchi E, Rolli E, Marasco R, Agazzi G, Borin S, Cosmina P, Pedron F, Rosellini I, Barbafieri M, Petruzzelli G (2017) Phytoremediation of a multi contaminated soil: Mercury and arsenic phyto-extraction assisted by mobilizing agent and plant growth promoting bacteria. J Soils Sediments 17:1224–1236
10. Glick BR (2012) Plant growth-promoting bacteria: mechanisms and applications. Scientifica 963401
11. GREIPSSON, S., (2011) Phytoremediation. Nat. Educ. Knowl. 3:7
12. Guo JK, Muhammad H, Lv X, Wei T, Ren XH, Jia HL, Atif S, Hua L (2020) Prospects and applications of plant growth promoting rhizobacteria to mitigate soil metal contamination: a review. Chemosphere 246:125823
13. Hamzah A, Sarmani SB, Yatim NI (2015) Phytoremediation of Pb and Hg by using *Scirpusmucronatus*with addition of bacterial inoculums. J Radioanal Nucl Chem 304:151–155
14. Huang CC, Narita M, Yamagata T, Itoh Y, Endo G (1999) Structure analysis of a class II transposon encoding the mercury resistance of the Gram-positive bacterium *Bacillus megaterium*MB1, a strain isolated from Minamata Bay, Japan. Gene 234:361–369
15. Karmakar R, Bindiya S, Hariprasad P (2018) Convergent evolution in bacteria frommultiple origins under antibiotic and heavy metal stress, and endophytic conditionsof host plant. Sci Total Environ 650:858–867
16. Khalid S, Shahid M, Niazi NK, Murtaza B, Bibi I, Dumat CA (2017) Comparison of technologies for remediation of heavy metal contaminated soils. J Geochem Explor 182:247–268
17. Kocman D, Wilson SJ, Amos HM, Telmer KH, Steenhuisen F, Sunderland EM, Mason RP, Outridge P, Horvat M (2017) Toward an assessment of the global inventory of present-day mercury releases to freshwater environments. Int J Environ Res Public Health 14:138
18. Kong Z, Wu Z, Glick BR, He S, Huang C, Wu L (2019) Co-occurrence patterns of microbial communities affected by inoculants of plant growth-promoting bacteria during phytoremediation of heavy metal-contaminated soils. Ecotoxicol Environ Saf 183:109504
19. Kumari S, Jamwal R, Mishra N, Singh DK (2022) Recent developments in environmental mercury bioremediation and its toxicity: a review. Environ Nanotechnol Monit Manag 13
20. Ma Y, Oliveira RS, Freitas H, Zhang C (2016) Biochemical and molecular mechanisms of plant-microbe-metal interactions: relevance for phytoremediation. Front Plant Sci 7
21. Manoj SR, Karthik C, Kadirvelu K (2020) Understanding the molecular mechanisms for the enhanced phytoremediation of heavy metals through plant growth promoting rhizobacteria: a review. J Environ Manag 254:109779
22. Mason RP, Reinfelder JR, Morel FMM (1994) Bioaccumulation of mercury and methylmercury. in mercury as a global pollutant. In: Proceedings of the third international conference held in whistler. British Columbia, pp 10–14
23. Mathew DC, Ho YN, Gicana RG, Mathew GM, Chien MC, Huang CC (2015) A rhizosphere-associated symbiont, *Photobacterium* spp. strain MELD1, and its targeted synergistic activity for phytoprotection against mercury. PloS One 10:0121178
24. Mishra GI, Sapre S, Sharma A, Tiwari S (2016) Alle*via*tion of mercury toxicity in wheat by the interaction of mercury-tolerant plant growth-promoting rhizobacteria. J Plant Growth Regul 35:1000–1012
25. Montiel-Rozas MM, Madejon E, Madejon P (2016) Effect of heavy metals and organic matter on root exudates (low molecular weight organic acids) of herbaceous species: an assessment in sand and soil conditions under different levels of contamination. Environ Pollut 216:273–281
26. Ndu U, Barkay T, Mason RP, Schartup AT, Al-Farawati R, Liu J, Reinfelder JR (2015) The use of a mercury biosensor to evaluate the bioavailability of mercury-thiolcomplexes and mechanisms of mercury uptake in bacteria. PLoS ONE 10:0138333
27. Pattanayak B, Dhal NK (2014) Biotechnological potential of plant microbes interactionwith enhanced phytoremediation. J Environ Dev 8:676–682
28. Pinto AP, Simes I, Mota AM (2008) Cadmium impact on root exudates of sorghum and maize plants: a speciation study. J Plant Nutr 31:1746–1755
29. Seiler C, Berendonk TU (2012) Heavy metal driven co-selection of antibiotic resistancein soil and water bodies impacted by agriculture and aquaculture. Front Microbiol 3:399

30. Silver S, Phung LE (2005) A bacterial view of the periodic table: genes and proteins for toxic inorganic ions. J Ind Microbiol Biotechnol 32:587–605
31. Sitarska M, Traczewska T, Filyarovskaya V (2016) Removal of mercury (II) from the aquatic environment by phytoremediation. Desalination Water Treat 57:1515–1524
32. Summers AO, Wireman J, Vimy MJ, Lorscheider FL, Marshall B, Levy SB, Billard L (1993) Mercury released from dental" silver" fillings provokes an increase in mercury-and antibiotic-resistant bacteria in oral and intestinal floras of primates. Antimicrob AgentsChemother 37:825–834
33. Verma S, Kuila A (2019) Bioremediation of heavy metals by microbial process. Environ Technol Innov 14:100369

Treatment Methods for Mercury Removal From Soil and Wastewater

Iftikhar Ahmad, Noor Fatima, Effat Naz, Zia Ur Rahman Farooqi, and Laura Bulgariu

Abstract Mercury (Hg) contamination in soil and wastewater has frequently been reported, however, variable concentration was present in soil (0.003–4.6 mg/kg), and in sewage sludge (0.3–7.7 mg/L). To protect the environment and general welfare, it is imperative to address the treatment of Hg in wastewater and soil. In this analysis, we have examined a range of physical, chemical, and biological methods that have proven effective in the removal of Hg from soil and wastewater. These methods include adsorption (≤99%), membrane separation (≤99%) nano-remediation (≤99%), phytoremediation (≤96%) and bioremediation (≤96%) in water or wastewater, whereas soil washing (≤80%), chemical stabilization (≤90%), chemical extraction (≤80%), nano-remediation (≤99%), phytoremediation (≤96%), and bioremediation (≤100%) are the prominent methods for the treatment of Hg in soil. Based on these results, we conclude that biological methods are slow but the most efficient for treatment of Hg contaminated soil and wastewater. Biological methods (phytoremediation and bioremediation) remove Hg from soil and wastewater through reduction, accumulation, and volatilization mechanisms. The combination of one or two methods can be more effective compared to one method. However, further research is required in this direction to affirm practical mass application of these techniques for treatment of Hg contaminated soil and wastewater.

Keywords Soil · Bioremediation · Mercury · Treatment methods · Wastewater

I. Ahmad (✉) · N. Fatima
Department of Environmental Science, COMSATS University Islamabad, Vehari Campus, Vehari 61100, Pakistan
e-mail: iftikharahmad@cuivehari.edu.pk

E. Naz
Department of Botany, Faculty of Sciences, University of Agriculture, Faisalabad 37610, Pakistan

Z. U. R. Farooqi
Department of Soil and Environmental Sciences, MNS University of Agriculture, Multan, Pakistan

L. Bulgariu
Faculty of Chemical Engineering and Environmental Protection "Cristofor Simionescu", Department of Environmental Engineering and Management, Technical University Gheorghe Asachi of Iasi, 700050 Iaşi, Romania

N. Kumar (ed.), *Mercury Toxicity Mitigation: Sustainable Nexus Approach*, Earth and Environmental Sciences Library, https://doi.org/10.1007/978-3-031-48817-7_12

1 Introduction

Mercury (Hg) is a prevalent environmental pollutant with a global distribution, which exists in three distinct oxidation states namely elemental (Hg^0), mercurous (HgI), and mercuric (HgII) forms [85]. The determination of chemical speciation is crucial in assessing the degree of mobility and toxicity of Hg, which arises from both natural and human-induced emissions in various terrestrial, atmospheric, biotic, and aquatic settings [32, 65, 74, 102]. According to "Mercury and Air Toxicity Standards" (MATS), which were established by the US Environmental Protection Agency in 2011, coal-fired power stations must minimize the emission of hazardous pollutants, including mercury [7]. In 2017, the World Health Organization (WHO) designated Hg as one of the ten "chemicals of concern," acknowledging its significant global impact on human health. According to [95], the annual worldwide anthropogenic emissions of mercury are projected to be about 2000 Mg. Both organic and inorganic substances include Hg [111]. Mercury exhibits high mobility and undergoes deposition on various environmental compartments, including surface soils, aquatic bodies, and bottom sediments. This deposition occurs after its release into the atmosphere, which can be attributed to sources such as ores, fossil fuels, and mineral deposits [112]. The primary anthropogenic contributor of Hg stems from the emissions produced by coal-burning thermal power plants (TPPs). Conversely, natural sources of Hg comprise the evaporation from water bodies, such as oceans, and the degassing of the earth's crust [79]. Worldwide Hg emissions are significantly impacted by wildfires as well [50]. The natural emission of mercury in this region is further influenced by the geogenic process, which is attributed to the unique climatic conditions present [9]. The primary factors contributing to mercury pollution are atmospheric deposition and gold mining activities (Fig. 1). Southeast Asia is widely regarded as the primary contributor to mercury emissions on a global scale [75]. Pakistan is characterized by various geographical features, including the Himalayas mountains, the Hindukush and Karakoram long ranges, the flat-lying Indus Plain, the highland Baluchistan plateau, and the coastal Arabian Sea. Eqani et al. [31] selected a total of 22 locations for their study, the total concentrations of Hg in dust particles were 3 mg/kg.

Methylmercury (MeHg) is widely recognized as the most hazardous variant of mercury present in natural surroundings [47]. Methylmercury (Me–Hg) is widely recognized as the most concerning variant of mercury owing to its capacity to undergo bioaccumulation within the food web [11]. According to [38], the element Hg has detrimental impacts on various organs of the human body, including the kidneys, liver, lungs, and central nervous system, along with other aspects of human health (Fig. 2). Natural biomethylation processes are carried out by microorganisms in marine settings to generate Me–Hg. via the eating of fish, Me–Hg finally makes its way into humans via the food chain of aquatic creatures [70]. During the mid-1950s, an incident of fish contamination took place in Minamata Bay, Japan, leading to the exposure of individuals to organic Hg. The illness was subsequently designated as Minamata disease. The United Nations Minamata Convention for reducing the

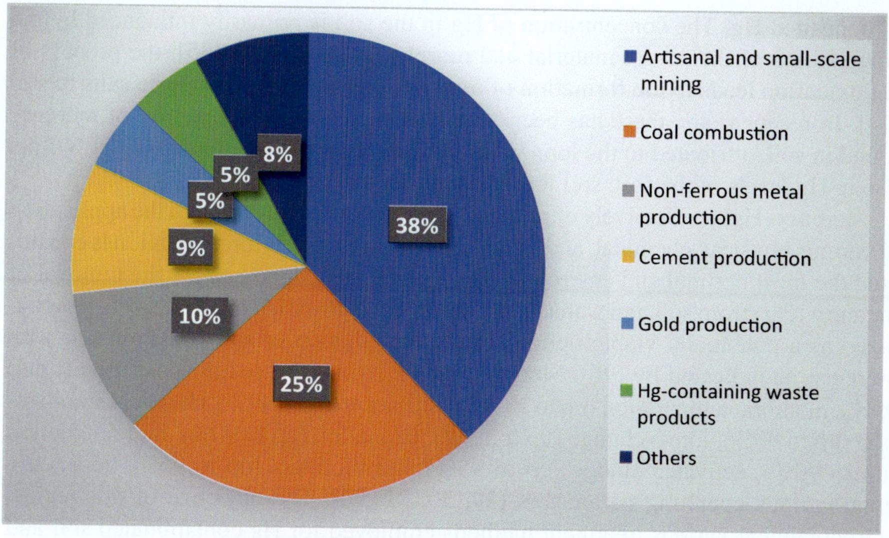

Fig. 1 Mercury metal can be accumulated in the environment through these sources

emission and use of Hg has been ratified by over 130 nations to effectively tackle the environmental health risk associated with mercury [67]. Individuals who were subjected to significant levels of Me–Hg experienced neurological defects because of prolonged mercury toxicity. These defects manifested as ataxia, muscular weakness, numbness in the limbs, impairments in speech, chewing, and swallowing, as well as heightened and accelerated tendon reactions [33, 61]. Females and individuals in the elderly population exhibit a higher propensity for experiencing renal function impairment induced by Hg [54, 55].

Since excessive mercury exposure via soil may result in major health issues, soil contamination has drawn a lot of attention in recent years [24]. As per the findings of [105], the presence of organic matter in soil serves as a mechanism for the retention

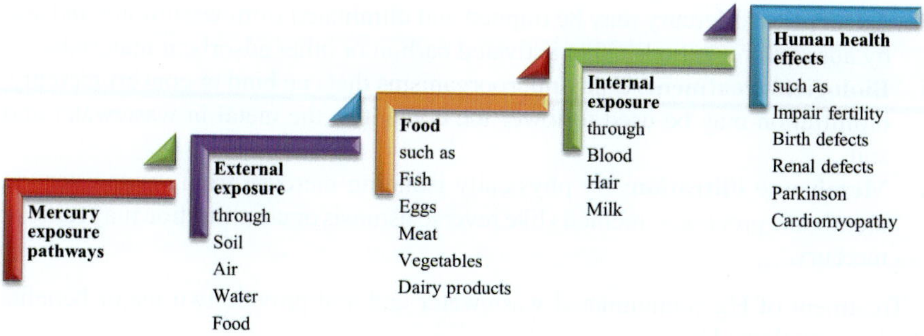

Fig. 2 Mercury exposure pathways and their impact on human health

of deposited Hg. The concentration of Hg in the soil is primarily influenced by the composition of the parent material and morphogenetic factors [35], the process of Hg oxidation leads to the formation of minerals and inorganic mercuric salts in soil [74]. In recent years, there has been a notable increase in the quantity of mercury found in soil, attributed to the long-distance migration of this element [112]. With a rise in Hg levels in the soil, soil fertility declines [48].

Mercury (Hg) is effectively eliminated from soil and water through the application of various physical, chemical, and biological techniques. Phytoremediation is considered the most optimal and environmentally sustainable approach for the removal of mercury (Hg) from soil and water. The plant species *Brassica juncea* has been recognized as a potentially viable option for the remediation of soil contaminated with mercury, as indicated by [96]. Mercury can be removed through adsorption (up to 99%), membrane separation (up to 98%), distillation (up to 99%), chemical stabilization (up to 90%), ion exchange (up to 95%), ozonation (up to 90%), bioremediation (up to 90%), activated sludge (up to 95%) and biofilters (up to 95%) depending upon various operating parameters [39, 50, 72, 90]. The objective of this review is to document various treatment methods employed for Hg contaminated soil and wastewater.

2 Treatment of Mercury Contamination

To safeguard human health and preserve the environment, it is imperative to undertake the treatment of wastewater that contains mercury, with the aim of reducing its concentration to levels deemed acceptable. Due to mercury's toxicity and possible effects on ecosystems and human health, wastewater pollution with mercury is a serious environmental problem. Several treatment techniques may be used to remediate mercury pollution in wastewater, including:

i. **Precipitation and Coagulation**: By forming insoluble compounds, chemical additives may be utilized to precipitate and remove mercury from wastewater and soil.
ii. **Adsorption**: Mercury may be trapped and eliminated from wastewater and soil by adsorption methods using activated carbon or other adsorbent materials.
iii. **Biological treatment**: Using microorganisms that can bind or convert mercury, biofiltration may be used to lower the content of the metal in wastewater and soil.
iv. **Membrane Filtration**: By physically isolating mercury from wastewater via membrane processes, methods like reverse osmosis or ultrafiltration may remove mercury.

Treatment of Hg contaminated wastewater and soil provide two major benefits which are explained here.

i. **Resource recovery**: A precious metal that can be recovered and used again is mercury. To recover Hg as a valuable resource and avoid the need for further mining and processing of virgin materials, lead in wastewater may be treated.
ii. **Economic benefits**: For businesses that specialize in water treatment technology as well as for industries that utilize recovered Hg as a raw material, treating Hg in wastewater may open new business prospects.

3 Treatment of Mercury Contaminated Water/Wastewater

Mercury contamination of water and wastewater is increasing due to discharge of untreated industrial effluent into water bodies. The mechanistic efficacy of several physical, chemical, and biological treatment techniques for removing Hg from water and wastewater under various operating circumstances is summarized in Table 1 and Fig. 3.

3.1 Adsorption

The utilization of adsorption as a technique for the elimination of heavy metals from wastewater is prevalent due to its advantageous attributes, including its user-friendly nature, high efficacy, and cost-effectiveness. One of the wastewater treatment methods commonly employed in commercial settings for the removal of Hg is adsorption, which offers a moderate level of effectiveness. An adsorbent material that can attract and hold onto the target metal ions is used in the process. Activated carbon, zeolites, silica gel, chitosan, and other organic and inorganic chemicals are some of the substances that may absorb mercury [18, 72]. Humic acid application to Hg contaminated water caused adsorption at the rate of 123 mg/g to remove Hg from water [108]. Mercury can bind on the surfaces of adsorbent material which have ability to remove 60–99% Hg from wastewater [18, 72]. The researchers employed a mesoporous zsm-5 material, to which they introduced 2-hydroxy acetophenone-4 N-pyrrolidine thiosemicarbazone (HAPT), resulting in the development of a novel adsorbent to eliminate mercury (II) ions from a laboratory-synthesized aqueous solution [1].

The ability to selectively adsorb Hg ions, as well as their price, usability, and accessibility, led to the choice of these materials. During the adsorption process, the adsorbent material is typically introduced to the wastewater and mixed for a certain period. During this period, the Hg ions in the wastewater are attracted to the surface of the adsorbent material and are maintained there by several processes, including chemical bonding and electrostatic attraction. Once the adsorption process is complete and the material has been removed from the wastewater, the Hg ions are kept in the adsorbent material. After that, the Hg ions may be recovered or eliminated from the adsorbent substance. The variety of adsorbents that may be

Table 1 Treatment methods for efficient removal of mercury from water and wastewater

Technique	Operating conditions	Mechanisms of removal	Removal efficiency	Reference
Adsorption	Contact time: 1–24 h, pH: 2–8	Attachment of mercury to adsorbent surfaces	60–99%	Chen et al. [18], Nadeem et al. [72], Wang et al. [108]
Microfiltration	Pressure: 1–10 bar, Temperature: 5–50 °C, pH: 4–10	Physical straining and size exclusion	50–99%	Islam et al. [46]
Membrane separation	Pressure, membrane type, molecular weight cut-off	Separation of mercury through size-based filtration	98%	Huang et al. [43]
Membrane Distillation	Temperature, pressure, volatilization of mercury	Evaporation and condensation of mercury vapors	89–96%	Hasan [40], Alkhudhiri et al. [6]
Ozonation	Temperature, Conc. of O_3, pH	Oxidation, Absorption	≤93%	Huang et al. [44]
Nano-remediation	pH, Conc, of NPs, Size, Surface area	adsorption, ion exchange, chelation, oxidation	75–99%	Wang et al. [110], Gil-Díaz et al. [36]
Phytoremediation	Plant species Growth medium	Phytoextraction, Phytostabilization	≤96%	Tiodar et al. [101], Sitarska et al. [93]
Bioremediation	Dependent on microbial species and site conditions	Transformation of Hg into less toxic forms by microorganisms	≤96%	Sinha and Khare [92], Sanjaya et al. [86]

employed in the treatment process is one of the benefits of using an adsorption unit. A variety of adsorbents are available and used depending on how well they bind to the pollutants that need to be removed. Additionally, improvement might be attained by synthesizing additional substances onto the adsorbent's surface. The following materials might be used as adsorbents:

3.1.1 Carbon Material

The adsorption method might use a variety of carbon material types. Carbon materials have been widely utilized in the field of wastewater treatment due to their exceptional properties that allow for the achievement of high recovery values during the removal of various contaminants. Numerous carbonaceous materials have been extensively investigated for their efficacy in adsorbing mercury from aqueous solutions. These materials encompass a wide range of carbon-based structures, such as activated carbons, carbon fibers, porous carbons, carbon nanotubes, graphene, and alkynyl carbons [20, 110]. The process of adsorption commonly occurs through chemical or mechanical means. The physical process is governed by the presence of weak Van Der Waals interactions between the adsorbent and adsorbate. The chemical reaction is driven by the link established between the functional groups commonly attached to the surface of the adsorbent and the adsorbate. The functional groups present in the attached compounds, particularly those containing oxygen, have a substantial impact on the efficient elimination of mercury from wastewater. This is the reason why chemisorption is frequently favored over physisorption, as highlighted by [97]. The polyacrylate modified carbon was efficiently removed Hg from wastewater with adsorption efficiency of 76 mg/g biomass [3]. Corn straw biochar absorbed 42% of Hg from aqueous solution, which was increased to 77% after addition of sodium sulfide to biochar [98]. This removal was attributed to the large surface area of adsorbent material. Similarly, chemically modified sugarcane bagasse applied in wastewater to treat Hg, which was found to remove >90% Hg under different operating conditions [37].

3.1.2 Zeolites

Zeolites have been widely utilized in the field of water treatment, specifically for the purpose of eliminating heavy metal ions from artificially produced wastewater. [99] conducted a study wherein they demonstrated the utilization of zeolites derived from coal fly ash, which were subsequently doped with silver nanoparticles, for the purpose of Hg(II) removal from water, which was removed 99% of Hg by the synthetic adsorbent. However, additional research is necessary to gain a comprehensive understanding of the mechanisms by which mercury ions adsorb and to evaluate the economic considerations associated with the utilization of silver nanoparticles in zeolites derived from coal fly ash. In another study, fly-ash derived synthetic zeolite can remove up to 99% Hg from wastewater [20], which was increased during wet desulfurization process. Natural zeolite efficiently removes Hg from water when combined with silver nanocomposite through ion exchange and oxidation–reduction mechanisms [45]. The removal efficiency of zeolite-silver nanocomposite was comparable with available commercial adsorbent materials employed for Hg removal.

3.2 *Membrane Separation*

A physical treatment method that effectively removes Hg from wastewater is membrane separation. A permeable membrane is used in a variety of physical treatment techniques known as membrane separation to remove various components from liquids. In the membrane separation procedure, a liquid stream is forced through a porous membrane by applying pressure on it. A membrane's pores have a specified size, allowing some particles to flow through while blocking others. Depending on the size of the holes, several membrane technologies, including size exclusion, are used in membrane separation. The membrane separation technique has advanced over the last 20 years from small-scale lab research to its use in significant industrial applications [94]. This material finds application in various industrial sectors of importance, such as lithium-ion batteries, fuel cells utilizing membranes, reverse osmosis (RO) membrane separation for desalination purposes, membrane reactors for wastewater treatment, as well as in endeavors related to water and energy sustainability [52]. Water purification is one of the crucial functions of membrane technology. Desalination provides all home and industrial water needs in the Middle East, particularly in Kuwait and Qatar [34]. In recent years, the possibility of removing mercury ions from wastewaters via membrane filtration has also been investigated. Conventional methods of wastewater treatment may prove ineffective due to the presence of mercury in various wastewater streams, either in particulate or colloidal states. [103] conducted a study in which they evaluated the efficacy of microfiltration (MF), ultrafiltration (UF), nanofiltration (NF), and reverse osmosis (RO) in achieving the required discharge limit (1.3 ng Hg/L) from oil refinery effluent. They found MF and UF are effective membrane separation techniques which lower Hg level up to desired limit of 1.3 ng Hg/L. Traditional membrane separation techniques like RO and NF often need a lot of energy and money while UF may operate under reduced pressures, its ability to effectively remove small ionic chemicals is compromised due to the larger size of the membrane pores. In order to facilitate the aggregation of small-sized pollutants with macromolecules and the subsequent formation of complexes, water-soluble polymers are introduced into the feed solution. These complexes effectively sequester the contaminants while still permitting the passage of water [43]. They used UF and polymer-enhanced ultrafiltration (PEUF) techniques to remove Hg from wastewater, later was found to achieve 99% Hg removal rate. In a separate study conducted by [8] focused on the efficacy of batch and continuous contact UF systems in the removal of Hg(II) from water. The study also explored the potential benefits of incorporating FeS as an assisting agent in the filtration process.

The treatment of wastewater with MF is a proven membrane separation technique. It entails utilizing a MF membrane to rid wastewater of particles, germs, and other contaminants. The Hg in wastewater may be successfully treated using MF, which effectively removes particles and pollutants of various sizes, including Hg particles, with holes that vary in size from 0.1 to 10 microns. Microfiltration may be employed as a pretreatment step before other methods that target dissolved mercury species in the removal of mercury. For instance, MF may assist in removing suspended Hg-containing particles or colloidal Hg from the water, lowering the total Hg level in the process. Thiol-functionalized membrane can remove Hg from the spiked wastewater below the detection limit [46]. The process is initially started from pre-treatment of wastewater through UF membrane and then Hg removal through adsorption on MF, the modified cysteamine membrane was capable to remove 97% of Hg from wastewater. Similarly, [57] reported real time application of MF to remove Hg from wastewater. They achieved 90% removal rate using MF without any addition of other techniques or chemicals. Similar kind of results obtained for Hg removal during desulfurization process. In general, membrane filtration techniques have exhibited significant potential in the remediation of wastewater containing Hg. These methods are favorable options for the treatment of industrial and municipal wastewater due to their effectiveness, affordability, and ecological friendliness. Membrane distillation is a promising approach for eliminating mercury from water. How distillation may be used to remove mercury from water is as follows:

- Distillation setup: A boiling flask, a condenser, and a collecting vessel make up a distillation apparatus. The mercury is removed from the tainted water by heating it in the boiling flask, which causes the water to evaporate.
- Evaporation: Water vaporizes as it heats, rising into the condenser. The vapor cools down in the condenser, returning to liquid form as a result.
- Collection: The purified water that has condensed from the vapor drops into the collecting container. The mercury stays in the boiling flask or creates a residue that may be further processed since it is a heavy metal with a higher boiling point.
- Mercury separation: To get rid of any last-remaining impurities or solid particles, mercury may be further processed using methods like centrifugation or filtering.

It's crucial to remember that distillation mostly eliminates mercury from water, making it ineffective for treating soil directly. Hasan [40] successfully applied membrane distillation technique to remove Hg (up to 89%) from dairy wastewater. Similarly, [6] reported Hg removal (up to 96%) by the application of air gap membrane distillation from water. Due to energy needs, probable mercury vapor emissions, and the presence of other pollutants, distillation may not be appropriate in all situations. When choosing the best treatment technique to remove mercury, it is essential to consider the unique features of mercury-contaminated water, consult an expert, and adhere to environmental standards.

3.3 Ion Exchange

Ion exchange is a technique for removing mercury from wastewater. Ion exchange is a method that is often used to remove different pollutants from water, including heavy metals. The exchange of target ions in this example, mercury ions in water for ions with a comparable charge linked to a solid ion exchange resin is the basic idea behind ion exchange [110]. The ion exchange technique may be used to remove mercury from wastewater in the following ways:

- Ion exchange resin selection: A suitable ion exchange resin that can specifically bind mercury ions is selected. While being stable under wastewater circumstances, the resin should exhibit a high affinity and selectivity for mercury ions.
- Column setup: As part of the treatment system, a column filled with the chosen ion exchange resin is inserted. Mercury-containing wastewater is pumped through the column encountering the resin.

- Ion exchange process: As wastewater passes over the resin bed, mercury ions stick to the resin, while other ions in the water are removed or exchanged with the bound ions on the resin.
- Regeneration: As time passes, the mercury ions in the ion exchange resin cause it to lose efficiency. To get rid of the built-up mercury and restore the resin's ability to get rid of more mercury, regeneration is necessary. To release the bonded mercury ions and restore the resin's functionality, this is commonly accomplished by washing the resin with a regenerant solution, such as acids or other chemicals.
- Disposal or Recovery: To avoid contaminating the environment, the mercury-rich regenerant solution produced from the regeneration process must be properly treated and disposed of. Alternatively, depending on the application, the mercury may be extracted from the regenerant solution using additional procedures such as precipitation or electrochemical techniques.

Industrial wastewater can be efficiently treated with the help of ion exchange resins, this technique effectively removed Hg ions from the wastewater [100]. A few minerals such as vermiculite and montmorillonite remove Hg through ion exchange mechanism [110]. Zeolite act as ion exchanger because it contains Si and Al in its structure, when Si is replaced with Al, a negative charge is created that attract Hg [12]. In another study, a synthetic ion exchange resins showed an adsorption capacity of 527 mg Hg/g [110]. However, the choice of resin, contact period, wastewater makeup, and beginning mercury content are just a few examples of the variables that affect how well mercury may be removed by ion exchange. For effective mercury removal, process parameters are often optimized by laboratory testing or pilot-scale research.

3.4 *Ozonation*

A potent oxidizing agent, ozone (O_3) may react with a variety of contaminants, including mercury. Ozone treatment of carbon material improved its efficiency by 134 times to capture Hg [59], this is because of C–O functional group that present on ozone treated.

carbon material. When ozone is added to wastewater, it performs chemical interactions with mercury that cause the metal to be oxidized and transformed into less hazardous forms, or they help the metal be removed from the water. It is crucial to remember that the effectiveness of ozonation might change based on the composition and mercury content of the wastewater. To assess the appropriateness and efficacy of ozonation for a specific application, it is advised to undertake pilot-scale experiments or confer with specialists in wastewater treatment [2]. Mercury may be removed by ozonation by several techniques, including:

- Direct oxidation: Ozone may oxidize divalent mercury (Hg^{2+}) to generate oxidized mercury species such as HgO or Hg(OH) or directly react with elemental mercury (Hg^0).

- Indirect oxidation: Through interactions with water molecules, ozone may produce reactive hydroxyl radicals (·OH). The mercury species that are present in the wastewater may then be oxidized by these hydroxyl radicals.

In the context of wastewater treatment systems, oxidized mercury species have the potential to undergo either precipitation or sorption processes, leading to their attachment to suspended particles or activated carbon. The process of extracting mercury from water is facilitated by this method, for instance, application of ozone improved the oxidation of elemental Hg into mercuric species and achieved the Hg removal efficiency (up to 93%), which was due to absorption of Hg by the wet photocatalytic membrane reactor coated with Fe–TiO_2 polyvinylidene fluoride [44].

4 Treatment of Mercury Contaminated Soil

Soil is a major sink of Hg after release into air and discharge into water bodies. Mercury cannot be degraded however, it can be adsorbed, leach down, transformed (oxi-red reactions) and precipitated, thus it can be either immobilized or removed from the medium. The mechanistic efficacy of several physical, chemical, and biological treatment techniques for removing Hg from soil under various operating circumstances is summarized in Table 2.

4.1 Soil Washing

Water is used in soil washing, a physical method of removing mercury from the soil, to reduce the amount of mercury present [112]. The idea behind the soil washing method is that the majority of hazardous compounds bond to fine soil particles (such as clay and silt), and that washing with water would remove both the pollutants and the particles. Chemical extraction is the technique of removing Hg from soil using chemicals; it may be combined with physical separation. The decreased amount of dirt that must be further disposed of or treated is this technology's principal benefit. However, when Hg and soil particles have a strong connection, this approach is impractical [112]. Soil washing technique to removed Hg depends on various soil properties, such as particle size distribution, clay concentration, humic content, heterogeneity, and others, play a significant role in influencing the efficacy and cost-effectiveness of soil washing techniques [21, 89]. Mercury removal from soil improved by the addition of different chemical solutions acid such as HCl, and HNO_3 [110], they also found 90% Hg recovery by the application of sodium thiosulfate in Hg-contaminated soil. It's crucial to remember that the efficiency of soil washing relies on a variety of elements, including the properties of the soil, the quantity and kind of mercury present, and the washing solution and additives used. To improve the soil washing procedure

Table 2 Treatment methods for efficient removal of Hg from soil

Technique	Operating conditions	Mechanisms of removal	Removal efficiency	Reference
Soil washing	Water-based solution and mechanical agitation, Optimal pH and temperature conditions	Solubilization and/or dispersion of mercury from soil particles	50–80%	Effendi et al. [27]
Chemical Stabilization	pH control (typically alkaline conditions), Addition of sulfur or other stabilizing agents	Formation of stable, less toxic mercury compounds	70–90%	Mercury [64], Wang et al. [110]
Chemical Extraction	Addition of chelating agents or surfactants, Control of pH and temperature	Formation of soluble complexes for extraction	40–80%	Sahle-Demessie et al. [84], Mulligan et al. [71]
Nano-remediation	pH, Conc, of NPs, Size, Surface area	Adsorption, ion exchange, electrostatic attraction, chelation, oxidation	$\leq$99%	Wang et al. [107]
Phytoremediation	Dependent on plant species and site conditions	Phytoextraction, Phytovolatilization	Variable	Petro-Souza et al. [78], Sanjaya et al. [86]
Bioremediation	Dependent on microbial species and site conditions	Reduction, Volatilization	Variable	Ekyastuti and Setyawati [28], Sanjaya et al. [86]

and guarantee successful cleanup of mercury-contaminated soil, pilot studies and site-specific assessments are often carried out.

4.2 *Chemical Stabilization*

For the removal of mercury from the soil, one approach that is often utilized is chemical stabilization. With this technique, stabilizing agents—like sulfur—are added to the polluted soil. Mercury and the stabilizing agents combine to generate stable and less hazardous mercury compounds, including mercuric sulfide (HgS). This transformation makes mercury less mobile and bioavailable, essentially immobilizing it in the soil and reducing its likelihood of seeping into groundwater or being absorbed by plants. According to [64], chemical stabilization normally removes mercury from soil with an efficacy of 70–90%. The Mercury Study Report to Congress (Volume VI: Remediation) by the United States Environmental Protection Agency (EPA) offers comprehensive documentation on the application of chemical stabilization as a means of remediating mercury contamination. This approach is widely recognized as a viable strategy for mitigating mercury pollution in soil. In order to reduce mercury exposure to the environment, stabilization methods use chemical complexation to immobilize mercury in polluted locations and reduce solubility [109]. According to [42], the presence of Hg(0) in soil that is contaminated can undergo a chemical stabilization process where it reacts with sulfur-containing reagents like elemental sulfur, pyrite (FeS_2), or thiosulfate. This reaction leads to the formation of highly insoluble HgS. The major drawback of stabilization method is stabilization of Hg within soil, which requires regular monitoring [110]. Humic acid application to soil caused 96% stabilization of Hg due to formation of stable complexes with sulfate [108].

4.3 Chemical Extraction

Chemical extraction uses extracting chemicals to separate pollutants from soils, sludges, and sediments, lowering the number of hazardous wastes that must be treated. This technique has been employed for the remediation of soils contaminated with heavy metals [76]. In order to mitigate the environmental impact of extracting fluid discharge and facilitate the recovery of the extracting agent, the removal of heavy metals from the extracting solution is undertaken subsequent to the extraction process. The extraction solution can be effectively purified by employing precipitation, followed by a subsequent separation process, to eliminate the solubilized heavy metals. Chemical extraction is a frequently employed method for the remediation of mercury-contaminated soil. This method involves the utilization of chelating agents or surfactants to induce the formation of soluble complexes of mercury through a selective chemical reaction. These compounds increase mercury's mobility, enabling its extraction from the soil matrix. The recovered mercury may subsequently be removed from the solution using further processing steps. Chemical extraction may remove mercury from soil with an effectiveness that varies from 40 to 80% [84]. In order to improve extraction efficiency, recent research has concentrated on choosing chelating agents and surfactants as well as managing temperature and pH conditions. Studies have investigated how well different extraction techniques, such as the use of DTPA (diethylenetriaminepentaacetic acid) and EDTA (ethylenediaminetetraacetic acid), remove mercury from polluted soil [51]. Sequential chemical extraction (SCE) techniques are commonly employed for the chemical speciation of heavy metals in sludge. The experimental procedure necessitates the utilization of various chemical extractants in a series of reagents that exhibit escalating potency. It is expected that a distinct chemical variant of the metal will undergo dissolution in each stage of the extraction process [60]. Depending on the individual metal and its content in the soil, there are several procedures involved in extracting metal from soil. Here is a broad description of what happens:

- Soil sampling and analysis: Laboratory testing is used to determine the metal content of soil samples taken from the place of interest. The presence and concentration of target metals are determined with the use of this analysis.
- Soil pretreatment: To improve the extraction process, the soil may sometimes go through pretreatment. Depending on the kind of soil and metal pollutants, pretreatment techniques may include soil washing, size reduction, or drying.
- Leaching: Leaching is a typical process for removing metals from soil. In order to selectively dissolve the target metals while minimizing the extraction of unwanted components, it includes the use of suitable leaching agents or solvents. Depending on the metal being targeted, leaching agents may be acids, bases, or complexing agents.
- Separation: After the metal has been dissolved and leached into the solution, the remaining soil solids are removed. The solid particles in the leachate may be removed using methods like filtering or sedimentation.
- Metal recovery and purification: After being recovered from the leachate, the metal is next purified using procedures like electrowinning, solvent extraction, or precipitation. These techniques assist in separating the metal from other substances found in the leachate, producing a pure version of the metal.

After metal recovery, the residual solution and solid wastes may need to be treated to get rid of any leftover pollutants before disposal. To adhere to environmental requirements and guarantee correct disposal, a variety of treatment techniques, such as precipitation, neutralization, or filtration, may be used.

5 Treatment of Mercury Contaminated Soil and Water

5.1 Nano-Remediation

Nanoparticles (NPs), nanosheet or nanocomposite are getting attention to treat Hg contaminated water, wastewater, and soil [110]. Nanomaterials are the most widely used nanotechnology for treatment of Hg contaminated water, carbon nanotubes, magnetite NPs, mesoporous silica, and O, S and N-based functional groups are being used for Hg removal [5, 66, 110, 114]. The mechanisms behind Hg removal could be adsorption, ion exchange, electrostatic attraction, chelation, and oxidation by these NPs. It is because high surface area of these NPs, presence of various functional groups, ease of metal recovery, easy synthesis, and modification [110]. Application of nanotechnology increased removal efficiency of Hg from water [5, 36, 81, 110, 114] such as MoS_2 nanosheets achieved (up to 99% with adsorption capacity of 2563 mg/g), carbon nanotubes (up to 92% with adsorption capacity of 59 mg/g), magnetite NPs (up to 98% with adsorption capacity of 52 mg/g), mesoporous silica (up to 95% with adsorption capacity of 43 mg/g), nanocellulose (up to 98% with adsorption capacity of 75 mg/g). Application of nano zerovalent Fe NPs immobilized Hg (up to 98%) in water, adsorption was the dominant mechanism of Hg removal from water revealed by X-ray photoelectron spectroscopy [36]. Magnetite-graphene oxide NPs composite removed Hg at the rate of 17 mg/g from water, the dominant mechanisms responsible for Hg removal was adsorption on OH and –COO functional groups revealed by FTIR spectra [22]. In addition to Fe, Cu-oxide NPs developed to treat Hg contaminated water, results revealed 75% Hg removed by CuO-NPs from water contaminated with 200 μg/L Hg [82].

Application on nanoparticles also immobilized Hg in soil for instance, [30] studied sorption and release of Hg by the application of water treatment residual nanoparticles, results revealed that this treatment stabilized Hg in soil due to inner sphere sorption through OH group. Similarly, water treatment residual nanoparticles applied in calcareous soils and results conferred 93% Hg immobilization due to stable complex formation as $Hg(OH)_2$ amor, OH group was responsible of Hg immobilization by chemisorption process on water treatment residual nanoparticles revealed by FTIR spectroscopy [66]. Dissolve organic matter (DOM) increased the stability of selenium NPs that caused Hg immobilization in soil [107]. Intermediate molecular weight DOM had an imperative role for binding of Se-NPs which improved Hg bioremediation in soil. The molecular weight of DOM had central role in remediation of HG in soil, for instance, molecular weight of DOM (<1000 Da) stabilized 97% Hg, ≤3500 Da stabilized 98% Hg, ≤10,000 Da stabilized 99% Hg, and >10,000 Da stabilized 95% Hg [109].

5.2 *Phytoremediation*

Phytoremediation refers to the utilization of plants for the purpose of collecting, detoxifying, and assimilating heavy metals present in contaminated soil and water (Fig. 3). Depending on the features of the wastewater and the toxins present, phytoremediation may be administered in a variety of methods. In a built-in wetland, for instance, the wastewater is channeled through several plant beds where the plants absorb the mercury via their roots. The plants in a Phyto reactor are grown in a container filled with polluted water, and the mercury is eliminated naturally by the plants. Through a process known as phytoextraction, these plants may bind mercury in their roots, stems, and leaves. Certain plant species, including *Panicum coloratum* (TF ≤ 10.94), *Datura stramonium* (TF ≤ 8.30), *Cyperus eragrostis* (TF ≤ 3.60), *Persicaria lapathifolia* (TF ≤ 3.10), and *Phragmites australis* (TF ≤ 1.99) have been proven to be efficient in removing Hg from water [101, 62]. In a hydroponic solution, Hg contamination was developed in the range of 50–60 mg/L, thereafter *Sesbania grandiflora* was grown on this medium. Results revealed maximum stabilization of Hg within plant roots with TF < 1 [58, 101]. Some of aquatic plant species are known as good scavenger of Hg in water such *Eichornia crassipes, Myriophyllum aquaticum, Pistia stratiotes, Ludwigina palustris, Scirpus tabernaemontani, Mentha aquatica* and *Colocasia esculenta,* these species can remove more than 95% of Hg from water/wastewater [101]. Some other pleustophytes were studied for their role in phytoextraction of Hg from water, the researcher reported that *Lemna minor* had shown bioconcentration factor (BCF) in the range (275–561) with removal efficiency (RE) (68–94%) whereas *Salvinia natans had shown* BCF in the range (280–567) with RE (61–91%).

in monoculture [93]. During mixed culture of both plants, BCF of Hg was in the range (216–856) with RE (82–96%). They concluded that both plant species were found to be efficient at extracting Hg from water, however, the mixed culture of both plants was the most effective [93]. The constructed wetland was filled with biochar and gravel to improve phytoremediation of Hg from wastewater. Results revealed that biochar filled constructed wetland showed higher removal of Hg through enhanced reduction, accumulation in the *Lythrum salicaria* plant and volatilization processes [16].

Selected plants are used in phytoremediation, which involves phytostabilization, phytoextraction, and phytovolatilization, to remove Hg pollution from the soil. In addition to being employed as a final decontamination step, it is also utilized in conjunction with other treatment methods [87, 110]. To mitigate the migration of Hg through soil erosion and deflation, the process of phytostabilization is employed. This technique involves the immobilization of Hg within the soil by means of Hg absorption and subsequent storage in the roots of plants, or through the precipitation of Hg within the root zone. For example, previous studies have demonstrated that certain willow species can accumulate bioavailable mercury in their root system. This accumulation leads to a decrease in the concentration of bioavailable mercury in the rhizosphere, while the overall level of total mercury in the soil remains relatively stable [102]. Recent studies have indicated that the formation of a Hg–Se insoluble complex in the rhizospheres and/or roots is the probable mechanism responsible for the restriction of bioaccessibility, absorption, and translocation/bioaccumulation of Hg by selenium (Se) [116]. According to [77] phytoextraction refers to the mechanism by which plants absorb and transport mercury (Hg) through their roots into the above-ground parts, such as shoots, which can then be harvested and incinerated. By enhancing Hg bioavailability before plant accumulation, chemicals like potassium iodide (KI) and ammonium thiosulfate $[(NH_4)2S_2O_3]$ may be employed to aid in phytoextraction [106].

Phytovolatilization is a remediation technique that is specific to Hg due to its excessive volatility, for instance, Hg is absorbed by plant roots, moved via the xylem, and then discharged from cellular tissues into the atmosphere [105]. The fundamental benefit of phytovolatilization is that Hg is removed from soils without having to be harvested and disposed of. While previous safety assessment studies have demonstrated that the widespread dispersion and subsequent dilution of Hg through phytovolatilization surpass potential hazards, concerns regarding the safety of this strategy may persist [69]. Inoculation of *Lecythophora* sp., fungus to LB medium supplemented with Hg caused > 86 removal of Hg by volatilization [15], moreover, its combined application with biochar decreased Hg uptake in lettuce from soil. The selection of plant species utilized in phytoremediation plays a pivotal role in determining the overall effectiveness of Hg remediation. The most popular and visually pleasing method of treating soil is phytoremediation since it is both ecologically and aesthetically acceptable. It doesn't need expensive heat treatment, harmful chemicals for the environment, extensive transportation, or expensive machinery [77]. However, if phytoremediation is chosen as the remedial strategy, the following considerations must be made:

- In the context of phytostabilization, it is imperative to consider certain criteria when selecting candidate plants. One crucial factor is the presence of deep root systems, which aid in the stabilization of the soil. However, the toxicity of Hg to plant roots and the overall survival rate of the plants should be carefully evaluated.
- Plant species that possess a substantial amount of biomass and exhibit physiological mechanisms that facilitate the absorption of Hg play a crucial role in the process of phytoextraction. It is worth noting that, thus far, no plant species have been identified as hyper-accumulators of mercury [105], however, recently many plant species such as *Axonopus compressus, Lepidium sativum, Brassica juncea, Oxalis corniculate, Arabidopsis thaliana, E. crassipes, M. aquaticum, P. stratiotes, L. palustris,* L. minor, S. natans [56, 101, 93]. In addition, when plants are consumed, biological exposure to Hg may occur; herbivorous animals may introduce Hg pollutants into the food chain. Concerns exist about the management of Hg-containing biomass, which may need more work and raise costs.
- Phytoremediation of Hg from soil depends upon environmental factors, available fraction of Hg in soil, parent material, plant species, and soil properties [56].
- There exists a specific set of plant species that possess the capacity to effectively absorb Hg, and the extent of their remediation benefits is contingent upon the depth of their root zones [87].
- Secondary contamination might result from polluting the atmosphere owing to excessive volatility of elemental Hg. There is a scarcity of studies that have documented the utilization of plants for the process of phytovolatilization of mercury [105].
- The phytoremediation process is characterized by a significant time investment and necessitates multiple growing seasons, in contrast to excavation/disposal or incineration, which can be completed within a matter of weeks to months.

5.3 Bioremediation

Microbial cells are used in the bioremediation process to remove heavy metals from water through sorption, accumulation, sequestration, volatilization, and oxidation–reduction reactions (Fig. 3). Sinha and Khare [92] propose that the phenomenon of volatilization involves the utilization of bacteria that possess resistance to Hg and possess operons responsible for the binding, transportation, and detoxification of Hg(II) and organic Hg compounds, ultimately converting them into elemental Hg. This serves as a preventive measure against the accumulation of Hg within the food chain. However, it has been demonstrated that microbial volatilization can occur independently of the mer-operon encoded system. Some microorganisms can reduce Hg^{2+} to elemental Hg and decrease its toxicity in the medium, for instance, *Klebsiella pneumoniae* immobilized on Ca-alginate remove Hg (up to 89%) in a fluidized bed reactor through reduction of Hg^{2+} to elemental Hg and its subsequent volatilization to atmosphere [115]. Application of alginate immobilized *Enterobacter* sp. was effectively removed Hg (up to 100%) from industrial effluent [92]. These cells can be used in multiple cycles to treat Hg contaminated wastewater. In another study, [92] treated Hg contaminated water with the help of *Bacillus cereus* cells immobilized on Ca-alginate and found 104 mg Hg/g biomass. Microorganisms having Hg tolerance could be a good candidate for treatment of Hg contaminated.

water. Here two microbes were applied to treat Hg contaminated water, a bacterium *Mycolicibacterium peregrinum* and a fungus *Cladosporium halotolerans* that can grow up to 3000 mg Hg/L and can remove 77 and 91% Hg from water, respectively [86].

The bacteria such as *Azotobacter* spp., *B. subtilis, B. cereus, Burkholderia* spp., *Brevundimonas vesicularis, Enterobacter* spp., *Fusobacterium aquatile, K. pneumoniae, M. peregrinum, Pseudomonas aeruginosa,* and fungus such as *Aspergillus* sp., *Curvular*ia geniculate, *Cladosporium halotolerans, Cladosporium cladosporioides, Lindgomycetaceae,* and *Westerdykella* [86, 78, 28] had showed Hg resistance and had practical application to remove Hg from the medium, the reported removal efficiency was range (56–100%). Although there have been discoveries of several microorganisms capable of degrading organic mercury (Hg) pollutants in soil, several challenges have been identified. These obstacles include limited accessibility of Hg to microorganisms, the existence of other harmful substances that may impede the function of Hg-tolerant microbes, inadequate nutrient availability, and insufficient biochemical capacity for effective biodegradation [49]. Consequently, researchers have conducted investigations into the utilization of genetically modified microorganisms for the purpose of biologically treating soils contaminated with mercury [41]. As an illustration, the Hg regulator mer R, which exhibits a significant affinity, was engineered to specifically target the outer membrane of *Escherichia coli* cells. This targeted localization resulted in a six-fold increase in the absorption capacity for Hg compared to the corresponding wild-type cells [10].

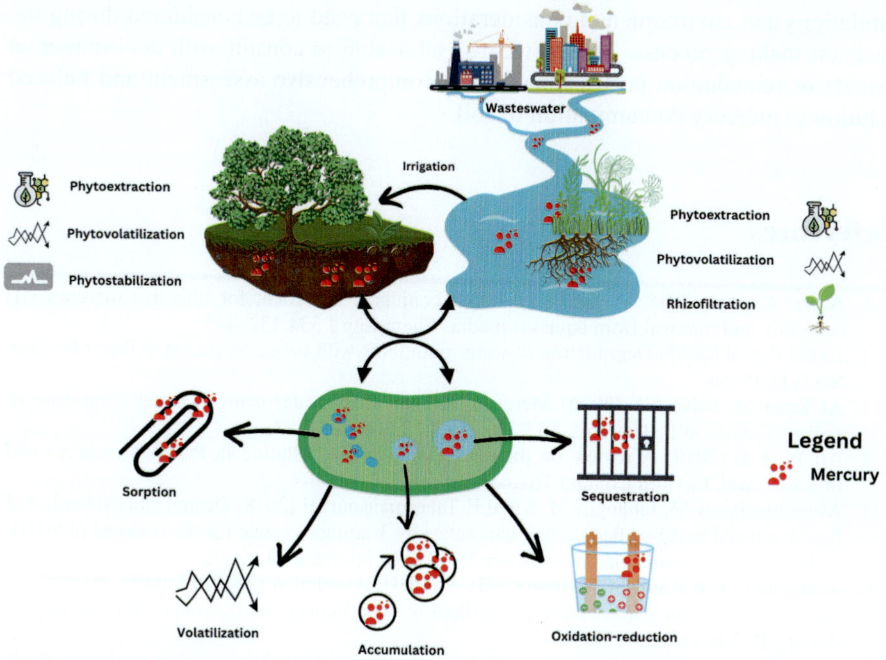

Fig. 3 Mechanistic pathways indicating phytoremediation (extraction, stabilization, rhizofiltration, and volatilization induced by plants) and bioremediation (sorption, accumulation, sequestration, volatilization, and oxidation–reduction reactions induced by microorganisms viz. bacteria, fungi, and algae) of Hg in soil and wastewater

6 Conclusion

The issue of mercury contamination in municipal and industrial wastewater is a matter of significant importance. Numerous studies have shed light on the issue of Hg pollution in the environment, which has been found to have significant implications for both human and plant health. To tackle the problem of Hg pollution, numerous treatment methods can be opted with variable efficiency for Hg removal from wastewater and soil. The treatment methods such as adsorption, membrane separation, ion-exchange and ozonation are normally employed for Hg treatment in water or wastewater, whereas soil washing, chemical stabilization, and chemical extraction are the prominent methods for the treatment of Hg in soil. However, nano-remediation (99%), phytoremediation ($\leq$96%), and bioremediation ($\leq$100%) could be employed for successful removal of Hg from both soil as well as water/wastewater. Here, we conclude that biological methods are the most efficient at removing Hg from soil and water by reduction, accumulation, and volatilization mechanisms. It is imperative to acknowledge that the choice of a particular methodology is contingent upon a multitude of factors, encompassing the magnitude and type of mercury pollution, soil properties, and site-specific circumstances. Additionally, some methods may have limitations and environmental considerations that need to be considered during the decision-making process. Therefore, it is advisable to consult with environmental experts or remediation professionals for a comprehensive assessment and tailored solution to mercury contamination in soil.

References

1. Abbas K et al (2018) A ligand anchored conjugate adsorbent for effective mercury (II) detection and removal from aqueous media. Chem Eng J 334:432–443
2. Akbal F et al (2015) Degradation of water pollutants with ozone oxidation. J Water Environ Sci 6(1):45–56
3. Al-Yaari M, Saleh TA (2022) Mercury removal from water using a novel composite of polyacrylate-modified carbon. ACS Omega 7(17):14820–14831
4. Ali W et al (2019) A review on the status of mercury pollution in Pakistan: sources and impacts. Arch Environ Contam Toxicol 76:519–527
5. Alimohammady M, Jahangiri M, Kiani F, Tahermansouri H (2018) Design and evaluation of functionalized multi-walled carbon nanotubes by 3-aminopyrazole for the removal of Hg(II) and As(III) ions from aqueous solution. Res Chem Intermed 44:69–92
6. Alkhudhiri A, Hakami M, Zacharof MP, Abu HH, Alsadun A (2020) Mercury, arsenic and lead removal by air gap membrane distillation: experimental study. Water 12:1574. https://doi.org/10.3390/w12061574
7. Ancora MP et al (2016) Meeting minamata: cost-effective compliance options for atmospheric mercury control in Chinese coal-fired power plants. Energy Policy 88:485–494
8. Azimi A et al (2017) Removal of heavy metals from industrial wastewaters: a review. ChemBioEng Rev 4(1):37–59
9. Azizullah A et al (2011) Water pollution in Pakistan and its impact on public health—a review. Environ Int 37(2):479–497
10. Bae W et al (2001) Genetic engineering of Escherichia coli for enhanced uptake and bioaccumulation of mercury. Appl Environ Microbiol 67(11):5335–5338
11. Bailon MX et al (2018) Total mercury, methyl mercury, and heavy metal concentrations in Hyeongsan river and its tributaries in Pohang city, South Korea. Environ Monit Assess 190:1–16
12. Barman MK, Bhattarai A, Saha B (2023) Applications of ion exchange resins in environmental remediation. https://doi.org/10.1002/vjch.202300027
13. Beckers F, Rinklebe J (2017) Cycling of mercury in the environment: Sources, fate, and human health implications: a review. Crit Rev Environ Sci Technol 47(9):693–794
14. Bridges CC, Zalups RK (2017) Mechanisms involved in the transport of mercuric ions in target tissues. Arch Toxicol 91(1):63–81
15. Chang J, Duan Y, Dong J, Shen S, Si G, He F, Yang Q, Chen J (2019) Bioremediation of Hg-contaminated soil by combining a novel Hg-volatilizing Lecythophora sp. fungus, DC-F1, with biochar: Performance and the response of soil fungal community. Sci Total Environ 671:676–684
16. Chang J, Peng D, Deng S, Chen J, Duan C (2022) Efficient treatment of mercury(II)-containing wastewater in aerated constructed wetland microcosms packed with biochar. Chemosphere 290:133302
17. Chen J et al (2020) Inhibition of tyrosinase by mercury chloride: spectroscopic and docking studies. Front Pharmacol 11:81
18. Chen PH, Hsu CF, Tsai DDW, Lu YM, Huang WJ (2014) Adsorption of mercury from water by modified multi-walled carbon nanotubes: adsorption behaviour and interference resistance by coexisting anions. Environ Technol 35(15):1935–1944. https://doi.org/10.1080/09593330.2014.886627

19. Clarkson TW, Magos L (2006) The toxicology of mercury and its chemical compounds. Crit Rev Toxicol 36(8):609–662
20. Czarna D, Baran P, Kunecki P, Panek R, Żmuda R, Wdowin M (2018) Synthetic zeolites as potential sorbents of mercury from wastewater occurring during wet FGD processes of flue gas. J Clean Prod 172:2636–2645
21. Dermont G et al (2008) Soil washing for metal removal: a review of physical/chemical technologies and field applications. J Hazard Mater 152(1):1–31
22. Diagboya PN, Olu-Owolabib BI, Adebowale KO (2015) Synthesis of covalently bonded graphene oxide–iron magnetic nanoparticles and the kinetics of mercury removal. RSC Adv 5:2536–2542
23. Ding W et al (2019) Responses of the grass Paspalum distichum L. to Hg stress: a proteomic study. Ecotoxicol Environ Saf 183:109549
24. Dong X et al (2010) A novel approach for soil contamination assessment from heavy metal pollution: a linkage between discharge and adsorption. J Hazard Mater 175(1–3):1022–1030
25. Duan Y et al (2016) Application of a reactive adsorbent-coated support system for removal of mercury (II). Colloids Surf, A 509:623–630
26. Duan Y et al (2016) Synthesis, characterization, and application of pyrite for removal of mercury. Colloids Surf, A 490:326–335
27. Effendi AJ et al (2020) Optimizing soil washing remediation of mercury contaminated soil using various washing solutions and solid/liquid ratios. In: E3S web of conferences, edp sciences
28. Ekyastuti W, Setyawati TR (2015) Identification and in vitro effectiveness test of four isolates of mercury-resistant bacteria as bioaccumulation agents of mercury. Procedia Environ Sci 28:258–264
29. Elhami S, Shafizadeh S (2016) Removal of Mercury (II) using modified Nanoclay. Mater Today: Proc 3(8):2623–2627
30. Elkhatib E, Moharem M, Mahdy A, Mesalem M (2017) Sorption, release and forms of mercury in contaminated soils stabilized with water treatment residual nanoparticles. Land Degrad Dev 28:752–761
31. Eqani SAMAS et al (2016) Mercury contamination in deposited dust and its bioaccumulation patterns throughout Pakistan. Sci Total Environ 569:585–593
32. Fitzgerald WF et al (2007) Marine biogeochemical cycling of mercury. Chem Rev 107(2):641–662
33. Gall JE et al (2015) Transfer of heavy metals through terrestrial food webs: a review. Environ Monit Assess 187:1–21
34. Ghaffour N (2009) The challenge of capacity-building strategies and perspectives for desalination for sustainable water use in MENA. Desalin Water Treat 5(1–3):48–53
35. Gil C et al (2010) Determination and assessment of mercury content in calcareous soils. Chemosphere 78(4):409–415
36. Gil-Díaz M, Rodríguez-Alonso J, Maffiotte CA, Baragaño D, Millán R, Lobo MC (2021) Iron nanoparticles are efficient at removing mercury from polluted waters. J Clean Prod 315:128272
37. Giraldo S, Robles I, Ramirez A et al (2020) Mercury removal from wastewater using agroindustrial waste adsorbents. SN Appl Sci 2:1029
38. Guallar E et al (2002) Mercury, fish oils, and the risk of myocardial infarction. N Engl J Med 347(22):1747–1754
39. Hargreaves AJ et al (2016) Mercury and antimony in wastewater: fate and treatment. Water Air Soil Pollut 227:1–17
40. Hasan AA (2022) Mercury removal from dairy wastewater by using membrane distillation. J Eng Sustain Dev 26. https://doi.org/10.31272/jeasd.26.2.1
41. He Z et al (2011) Removal of mercury from sediment by ultrasound combined with biomass (transgenic Chlamydomonas reinhardtii). Chemosphere 83(9):1249–1254
42. He F et al (2015) In situ remediation technologies for mercury-contaminated soil. Environ Sci Pollut Res 22:8124–8147

43. Huang Y et al (2015) Removal of mercury (II) from wastewater by polyvinylamine-enhanced ultrafiltration. Sep Purif Technol 154:1–10
44. Huang ZS, Wei ZS, He YM, Pei JL, Xiao XL, Tang MR, Yu S (2018) Combined membrane photocatalytic ozonation and wet absorption of elemental mercury. Atmos Pollut Res 9:230–237
45. Inglezakis VJ, Kudarova A, Guney A, Kinayat N, Tauanov Z (2023) Efficient mercury removal from water by using modified natural zeolites and comparison to commercial adsorbents. Sustain Chem Pharm 32:101017
46. Islam MS, Vogler RJ, Al Hasnine SMA, Hernández S, Malekzadeh N, Hoelen TP, Hatakeyama ES, Bhattacharyya D (2020) Mercury removal from wastewater using cysteamine functionalized membranes. ACS Omega 5(35): 22255–22267. https://doi.org/10.1021/acsomega.0c02526
47. Jagtap R, Maher W (2015) Measurement of mercury species in sediments and soils by HPLC–ICPMS. Microchem J 121:65–98
48. Kabata-Pendias A (2010) Trace elements in soils and plants. CRC Press
49. Krämer U, Chardonnens A (2001) The use of transgenic plants in the bioremediation of soils contaminated with trace elements. Appl Microbiol Biotechnol 55:661–672
50. Kumar A et al (2018) Mercury from wildfires: Global emission inventories and sensitivity to 2000–2050 global change. Atmos Environ 173:6–15
51. Kördel W et al (2013) Incorporating availability/bioavailability in risk assessment and decision making of polluted sites, using Germany as an example. J Hazard Mater 261:854–862
52. Le NL, Nunes SP (2016) Materials and membrane technologies for water and energy sustainability. Sustain Mater Technol 7:1–28
53. Lee M-R et al (2017) Blood mercury concentrations are associated with decline in liver function in an elderly population: a panel study. Environ Health 16(1):1–8
54. Li Y et al (2013) Blood mercury concentration among residents of a historic mercury mine and possible effects on renal function: a cross-sectional study in southwestern China. Environ Monit Assess 185:3049–3055
55. Li P et al (2015) Human inorganic mercury exposure, renal effects and possible pathways in Wanshan mercury mining area, China. Environ Res 140:198–204
56. Liu Z, Chen B, Wang L, Urbanovich O, Nagorskaya L, Li X, Tang L (2020) A review on phytoremediation of mercury contaminated soils. J Hazard Mater 400:123138
57. Liu C, Farooq K, Doll B, Venkatadri R (2013) Economical and reliable mercury reduction in refinery and power plant discharge wastewater with robust microfiltration membrane technology. Desalin Water Treat 51:4980–4986
58. Malar S, Sahi SV, Favas PJC, Venkatachalam P (2015) Assessment of mercury heavy metal toxicity-induced physiochemical and molecular changes in Sesbania grandiflora L. Int J Environ Sci Technol 12:3273–3282
59. Manchester S, Wang X, Kulaots I, Gao Y, Hurt RH (2008) High capacity mercury adsorption on freshly ozone-treated carbon surfaces. Carbon N Y 46(3):518–524. https://doi.org/10.1016/j.carbon.2007.12.019
60. Marchioretto M et al (2002) Heavy metals extraction from anaerobically digested sludge. Water Sci Technol 46(10):1–8
61. Matsuyama A et al (2018) Chemical characteristics of dissolved mercury in the pore water of Minamata Bay sediments. Mar Pollut Bull 129(2):503–511
62. Mbanga O, Ncube S, Tutu H, Chimuka L, Cukrowska E (2019) Mercury accumulation and biotransportation in wetland biota affected by gold mining. Environ Monit Assess 191:186
63. Meng M et al (2019) An integrated model for input and migration of mercury in Chinese coastal sediments. Environ Sci Technol 53(5):2460–2471
64. Mercury AIOA (1997) Mercury study report to congress volume II: an inventory of anthropogenic mercury emissions in the united states. Environ Prot
65. Mergler D et al (2007) Methylmercury exposure and health effects in humans: a worldwide concern. AMBIO: J Hum Environ 36(1):3–11

66. Moharem M, Elkhatib E, Mesalem M (2019) Remediation of chromium and mercury polluted calcareous soils using nanoparticles: Sorption–desorption kinetics, speciation and fractionation. Environ Res 366–373
67. Monitoring A, Symon C (2013) AMAP/UNEP technical background report for the global mercury assessment 2013: final technical report; output
68. Montoya AJ et al (2019) Adsorption of gaseous elemental mercury on soils: Influence of chemical and/or mineralogical characteristics. Ecotoxicol Environ Saf 170:98–106
69. Moreno FN et al (2005) Mercury volatilisation and phytoextraction from base-metal mine tailings. Environ Pollut 136(2):341–352
70. Moriarity RJ et al (2020) Subsistence fishing in the Eeyou Istchee (James Bay, Quebec, Canada): a regional investigation of fish consumption as a route of exposure to methylmercury. Chemosphere 258:127413
71. Mulligan C et al (2001) Remediation technologies for metal-contaminated soils and groundwater: an evaluation. Eng Geol 60(1–4):193–207
72. Nadeem FZ, Sajid Z, Jilani MI, Raza A, Abbas F (2016) New generation super adsorbents – a review. Int J Chem Biochem Sci 10:95–105
73. Neufeld RD, Hermann ER (1975) Heavy metal removal by acclimated activated sludge. Journal (Water Pollution Control Federation) 310–329
74. O'Connor D et al (2019) Mercury speciation, transformation, and transportation in soils, atmospheric flux, and implications for risk management: a critical review. Environ Int 126:747–761
75. Pacyna EG et al (2010) Global emission of mercury to the atmosphere from anthropogenic sources in 2005 and projections to 2020. Atmos Environ 44(20):2487–2499
76. Paff SW et al (1994) Acid extraction treatment system for treatment of metal contaminated soils, US Environmental Protection Agency, Superfund Innovative Technology Evaluation
77. Petruzzelli G et al (2013) Phytoremediation towards the future: focus on bioavailable contaminants. Plant-Based Remediat Process 273–289
78. Pietro-Souza W, de Campos PF, Mello IS, Stachack FFF, Terezo AJ, Cunha CND, White JF, Li H, Soares MA (2020) Mercury resistance and bioremediation mediated by endophytic fungi. Chemosphere 240:124874. https://doi.org/10.1016/j.chemosphere.2019.124874
79. Rahman Z, Singh VP (2019) The relative impact of toxic heavy metals (THMs)(arsenic (As), cadmium (Cd), chromium (Cr)(VI), mercury (Hg), and lead (Pb)) on the total environment: an overview. Environ Monit Assess 191:1–21
80. Raj D, Maiti SK (2019) Sources, toxicity, and remediation of mercury: an essence review. Environ Monit Assess 191:1–22
81. Ram B, Chauhan GS (2018) New spherical nanocellulose and thiol-based adsorbent for rapid and selective removal of mercuric ions. Chem Eng J 331:587–596
82. Raul PK, Das B, Umlong IM, Devi RR, Tiwari G, Kamboj DV (2022) Toward a feasible solution for removing toxic mercury and chromium from water using copper oxide nanoparticles. Front Nanotechnol 4:805698. https://doi.org/10.3389/fnano.2022.805698
83. Sahito ZA et al (2021) Arsenic and mercury uptake and accumulation in oilseed sunflower accessions selected to mitigate co-contaminated soil coupled with oil and bioenergy production. J Clean Prod 291:125226
84. Sahle-Demessie E et al (2000) Solvent extraction and soil washing treatment of contaminated soils from wood preserving sites: Bench-scale studies. Remediat J 10(3):85–109
85. Saiz-Lopez A, Travnikov O, Sonke JE et al (2020) Photochemistry of oxidized Hg(I) and Hg(II) species suggests missing mercury oxidation in the troposphere. PNAS 117:30949–30956. https://doi.org/10.1073/pnas.1922486117
86. Sanjaya WTA, Khoirunnisa NS, Ismiani S, Hazra F, Santosa DA (2021) Isolation and characterization of mercury-resistant microbes from gold mine area in Mount Pongkor, Bogor District, Indonesia. Biodiversitas 22:2656–2666
87. Sao K et al (2014) A review on heavy metals uptake by plants through biosorption. Int Proc Econ Dev Res 75:78

88. Shan Y et al (2019) Removal of elemental mercury from flue gas using microwave/ultrasound-activated Ce–Fe magnetic porous carbon derived from biomass straw. Energy Fuels 33(9):8394–8402
89. Sierra C et al (2011) Feasibility study on the use of soil washing to remediate the As–Hg contamination at an ancient mining and metallurgy area. J Hazard Mater 196:93–100
90. Singh RJ, Martin CE, Barr D, Rosengren RJ (2019) Immobilised apple peel bead biosorbent for the simultaneous removal of heavy metals from cocktail solution, Cogent Environ Sci 5:1. https://doi.org/10.1080/23311843.2019.1673116
91. Sinha A, Khare SK (2012) Mercury bioremediation by mercury accumulating Enterobacter sp. cells and its alginate immobilized application. Biodegradation 23:25–34
92. Sinha A, Pant KK, Khare SK (2012) Studies on mercury bioremediation by alginate immobilized mercury tolerant Bacillus cereus cells. Int Biod Biodegrad 71(1):1–8. https://doi.org/10.1016/j.ibiod.2011.12.014
93. Sitarska M, Traczewska T, Filarowska W, Hołtra A, Zamorska-Wojdyła D, Hanus-Lorenz B (2023) Phytoremediation of mercury from water by monocultures and mixed cultures pleustophytes. J Water Process Eng 52:103529
94. Strathmann H (2001) Membrane separation processes: current relevance and future opportunities. AIChE J 47(5):1077–1087
95. Streets DG et al (2011) All-time releases of mercury to the atmosphere from human activities. Environ Sci Technol 45(24):10485–10491
96. Su Y et al (2008) Phytoextraction and accumulation of mercury in three plant species: Indian mustard (Brassica juncea), beard grass (Polypogon monospeliensis), and Chinese brake fern (Pteris vittata). Int J Phytorem 10(6):547–560
97. Sun X et al (2011) Density functional study of elemental mercury adsorption on surfactants. Fuel 90(3):1061–1068
98. Tan G, Sun W, Xu Y, Wang H, Xu N (2016) Sorption of mercury (II) and atrazine by biochar, modified biochars and biochar based activated carbon in aqueous solution. Bioresour Technol 211:727–735
99. Tauanov Z et al (2018) Synthetic coal fly ash-derived zeolites doped with silver nanoparticles for mercury (II) removal from water. J Environ Manag 224:164–171
100. Tauanov Z, Shah D, Inglezakis V, Jamwal PK (2018) Hydrothermal synthesis of zeolite production from coal fly ash: a heuristic approach and its optimization for system identification of conversion. J Clean Prod 182:616–623
101. Tiodar ED, Văcar CL, Podar D (2021) Phytoremediation and microorganisms-assisted phytoremediation of mercury-contaminated soils: challenges and perspectives. Int J Environ Res Public Health 18(5):2435. https://doi.org/10.3390/ijerph18052435
102. Ullrich SM et al (2001) Mercury in the aquatic environment: a review of factors affecting methylation. Crit Rev Environ Sci Technol 31(3):241–293
103. Urgun-Demirtas M et al (2012) Achieving very low mercury levels in refinery wastewater by membrane filtration. J Hazard Mater 215:98–107
104. Wang Y et al (2005) Changes in Hg fractionation in soil induced by willow. Plant Soil 275:67–75
105. Wang J et al (2012) Remediation of mercury contaminated sites–a review. J Hazard Mater 221:1–18
106. Wang J et al (2014) Thiosulphate-induced mercury accumulation by plants: metal uptake and transformation of mercury fractionation in soil-results from a field study. Plant Soil 375:21–33
107. Wang J et al (2019) Sulfur-modified organoclay promotes plant uptake and affects geochemical fractionation of mercury in a polluted floodplain soil. J Hazard Mater 371:687–693
108. Wang Y, Huang Z, Sheng L, Ma Y (2023) Effect of modified humic acid residue on the adsorption and passivation of Hg^{2+}/Pb^{2+} in solution and soil. J Mol Liq 377:121581
109. Wang X, Pan X, Gadd GM (2019) Soil dissolved organic matter affects mercury immobilization by biogenic selenium nanoparticles. Sci Total Environ 658:8–15
110. Wang L, Houa D, Caoa Y, Ok YS, Tack FMG, Rinklebe J, O'Connora D (2020) Remediation of mercury contaminated soil, water, and air: a review of emerging materials and innovative technologies. Environ Int 134:105281

111. Xinmin Z et al (2006) Mercury in the topsoil and dust of Beijing City. Sci Total Environ 368(2–3):713–722
112. Xu J et al (2015) Sources and remediation techniques for mercury contaminated soil. Environ Int 74:42–53
113. Yadav M et al (2021) Physical and chemical methods for heavy metal removal. Pollut Water Manag: Resour, Strat Scarcity 377–397
114. Zarei S, Niad M, Raanaei H (2018) The removal of mercury ion pollution by using Fe_3O_4-nanocellulose: synthesis, characterizations and DFT studies. J Hazard Mater 344:258–273
115. Zeroual Y, Moutaouakkil A, Blaghen M (2001) Volatilization of mercury by immobilized bacteria (Klebsiella pneumoniae) in different support by using fluidized bed bioreactor. Cur Microbiol 43(1):322–327
116. Zhao J et al (2013) Selenium inhibits the phytotoxicity of mercury in garlic (Allium sativum). Environ Res 125:75–81

Printed by Printforce, the Netherlands